Studies on Plant Demography

A Festschrift for John L. Harper

Photograph by Douglas Gowan

John L. Harper

Studies on Plant Demography

A Festschrift for John L. Harper

EDITED BY

JAMES WHITE
Department of Botany
University College
Dublin, Ireland

1985

ACADEMIC PRESS
Harcourt Brace Jovanovich, Publishers
London Orlando San Diego
New York Austin Montreal Sydney
Tokyo Toronto

ACADEMIC PRESS INC. (LONDON) LTD.
24–28 Oval Road
LONDON NW1 7DX

United States Edition published by
ACADEMIC PRESS, INC.
Orlando, Florida 32887

British Library Cataloguing in Publication Data

Studies on plant demography: a festschrift for
John L. Harper.
1. Plant populations
I. White, James II. Harper, John L.
581.5'248 QK911

Library of Congress Cataloging in Publication Data
Main entry under title:

Studies on plant demography.

"Publications of John L. Harper": p.
Includes index.
1. Plant populations–Addresses, essays, lectures.
2. Vegetation dynamics–Addresses, essays, lectures.
3. Harper, John L.–Addresses, essays, lectures.
I. Harper, John L. II. White, James.
QK910.S78 1985 581.5'248 85-9833
ISBN 0–12–746630–4 (alk. paper)
ISBN 0–12–746631–2 (paperback)

PRINTED IN THE UNITED STATES OF AMERICA

85 86 87 88 9 8 7 6 5 4 3 2 1

Contents

3. Fires and Emus: The Population Ecology of Some Woody Plants in Arid Australia

James C. Noble

4. Differences in Life Histories between Two Ecotypes of *Plantago lanceolata* L.

J. M van Groenendael

5. Variation and Differentiation in Populations of *Trifolium repens* in Permanent Pastures

Roy Turkington

III The Demographic Interpretation of Plant Form: Application to Plant Competition and Production

13. On the Astogeny of Six-cornered Clones: An Aspect of Modular Construction

A. D. Bell

14. The Importance of Plant Form as a Determining Factor in Competition and Habitat Exploitation

Peter H. Lovell and Patricia J. Lovell

15. Modular Demography and Form in Silver Birch

Michelle Jones

16. Modular Demography and Growth Patterns of Two Annual Weeds (*Chenopodium album* L. and *Spergula arvensis* L.) in Relation to Flowering

Lucie Maillette

V The Influence of Pathogens and Predators on Plant Populations

VI Plant Reproductive Biology

Contributors

Numbers in parentheses indicate the pages on which the authors' contributions begin.

F. A. Bazzaz (373), Department of Organismic and Evolutionary Biology, The Biological Laboratories, Harvard University, Cambridge, Massachusetts 02138, USA

A. D. Bell (187), School of Plant Biology, University College of North Wales, Bangor, Gwynedd LL57 2UW, Wales

Marguerite A. Bough (143), Department of Plant Sciences, The University of Western Ontario, London, Ontario, Canada N6A 5B7

J. J. Burdon (313), Division of Plant Industry, CSIRO, Canberra City, A.C.T. 2601, Australia

Paul B. Cavers (143), Department of Plant Sciences, The University of Western Ontario, London, Ontario, Canada N6A 5B7

Paul Alan Cox (359), Department of Botany and Range Science, Brigham Young University, Provo, Utah 84602, USA

Rodolfo Dirzo (343), Departamento de Ecología, Instituto de Biología, Universidad Nacional Autónoma de México, 04510 Mexico City, Mexico

Miguel Franco[1] (257), School of Plant Biology, University College of North Wales, Bangor, Gwynedd LL57 2UW, Wales

J. M. van Groenendael[2] (51), Institute for Ecological Research, Department of Dune Research Weever's Duin, 2232 EG Oostvoorne, The Netherlands

H. J. Harvey (111), Department of Applied Biology, University of Cambridge, Cambridge CB2 3DX, England

Ad H. L. Huiskes (83), Delta Institute for Hydrobiological Research, 4401 EA Yerseke, The Netherlands

Michelle Jones (223), School of Plant Biology, University College of North Wales, Bangor, Gwynedd LL57 2UW, Wales

Patricia J. Lovell (209), Department of Botany, University of Auckland, Auckland, New Zealand

Peter H. Lovell (209), Department of Botany, University of Auckland, Auckland, New Zealand

[1]Present address: Departamento de Ecología, Instituto de Biología, Universidad Nacional Autónoma de México, 04510 Mexico City, Mexico.

[2]Present address: Department of Vegetation Science, Plant Ecology and Weed Science, Agricultural University, 6708 PD Wageningen, The Netherlands.

Jon Lovett Doust (327), Department of Biology, Amherst College, Amherst, Massachusetts 01002, USA

Lesley Lovett Doust (327), Department of Biological Sciences, Mount Holyoke College, South Hadley, Massachusetts 01075, USA

Richard N. Mack (127), Department of Botany, Washington State University, Pullman, Washington 99164, USA

Lucie Maillette[3] (239), Centre de recherches écologiques de Montréal, Montreal, Quebec, Canada H3C 3J7

Miguel Martínez-Ramos (17), Estación de Biología Tropical Los Tuxtlas, Instituto de Biología, Universidad Nacional Autónoma de México, San Andrès Tuxtla, Veracruz, Mexico

Robert E. L. Naylor (95), School of Agriculture, The University, Aberdeen AB9 1UD, Scotland

James C. Noble (33), Division of Wildlife and Rangelands Research, CSIRO, Deniliquin, N.S.W. 2710, Australia

John Ogden (3), Department of Botany, University of Auckland, Auckland, New Zealand

Daniel Piñero (17), Departamento de Ecología, Instituto de Biología, Universidad Nacional Autónoma de México, 04510 Mexico City, Mexico

E. G. Reekie (373), Department of Organismic and Evolutionary Biology, The Biological Laboratories, Harvard University, Cambridge, Massachusetts 02138, USA

José Sarukhán (17), Departamento de Ecología, Instituto de Biología, Universidad Nacional Autónoma de México, 04510 Mexico City, Mexico

Atie W. Stienstra[4] (83), Delta Institute for Hydrobiological Research, 4401 EA Yerseke, The Netherlands

B. R. Trenbath[5] (171), Centre for Environmental Technology, Imperial College, London SW7 1LU, England, and Elm Farm Research Centre, Hamstead Marshall, Near Newbury, Berkshire RG15 OHR, England

R. S. Tripathi (157), Department of Botany, School of Life Sciences, North-Eastern Hill University, Shillong 793014, India

Roy Turkington (69), Botany Department, University of British Columbia, Vancouver, British Columbia, Canada V6T 2B1

A. R. Watkinson (275), School of Biological Sciences, University of East Anglia, Norwich NR4 7TJ, England

James White (291), Department of Botany, University College, Dublin 4, Ireland

[3]Present address: Centre d'études nordiques, Université Laval, Sainte-Foy, Quebec, Canada G1K 7P4.

[4]Present address: Department of Plant Ecology, Lange Nieuwstraat 106, 3512 PN Utrecht, The Netherlands.

[5]Present address: 12 New Road, Reading, Berkshire RG1 5JD, England.

Foreword

The influence of John Lander Harper on plant biology, ecology and evolution has been profound. During the 24 years between his first significant publication on comparative plant ecology and the appearance in 1977 of his monumental volume ''Population Biology of Plants'', his was the guiding spirit that made botanists realize the primary importance of Darwinian natural selection for plant evolution. He is without doubt the father of a new discipline: comparative autecology of plants, which stands between physiological autecology and synecology or systems ecology. For attacking problems connected with the adaptive significance of those genetic differences that are of greatest importance to the evolutionist, this intermediate approach is of paramount importance. Moreover, it is not truly scientific until the more casual observations and anecdotal accounts of the traditional naturalist are replaced by precise, quantitative data and their careful analysis using the most modern methods. John Harper has led the way to this conversion of naturalists' observations to a precise scientific discipline. He has shown us that plant competition not only exists in nature, but also can be repeated and analysed by means of carefully controlled experiments, taking into account both genetic differences between species and the diversity of microhabitats.

Throughout his career, but particularly during his years of maturity, John Harper has stressed the importance of applying to agricultural practices the discipline that he has fostered. He has the almost unique distinction of being a farmer's son who has become a leading scientist by developing to new heights that branch of science that is most relevant to farming. The present volume contains essays that reflect his influence on plant population biology in a manner that befits well his third of a century of leadership.

I cannot close this foreword without adding a personal note. We are all aware of the great contribution that John Harper has made, not only to science itself, but also to communion and collaboration between scientists. I have been aware of John's accomplishments in this direction during the 25 years that I have known him, including visits to Bangor, where I not only have been stimulated by seeing him and his associates in action, but also with Mrs. Stebbins have thoroughly enjoyed the gracious hospitality of John and Borgny at their home. To end on a note of jollity, I repeat a delightful anecdote that he told me near the beginning of his visit to Davis, California, in 1959–1960.

Soon after he arrived, John went to the nearby city of Sacramento to rent a car. While he was riding in a taxicab between the bus depot and the car lot, the friendly cabby commented on an accent that was strange to him. ''You're not American, are you?'' ''No.'' ''Where're you from?'' ''Oxford, England.'' ''How long you been here?'' ''About a week.'' ''Well, for a furriner who's been here such a short time, you sure speak English good.''

John, this volume represents another milestone in your scientific career. I feel sure that there will be many more before it is over.

Your old friend,

G. Ledyard Stebbins

Preface

This volume of studies on plant demography has been compiled to mark the 60th birthday of Professor John L. Harper. The contributors are only some of the much larger number of his students and scientific collaborators, first at Oxford and since 1960 at Bangor. The exigencies of publishing imposed a restriction on the number of papers and only those on plant population biology could be included, to ensure coherency of content. These constraints have excluded from this volume several of his former students who would wish, however, to be associated with it. Of the many visitors to John Harper's research school at Bangor only those who had published at least one paper with him during the period 1970–1982 were invited to contribute, again because of the limits of available space.

The papers come from botanists of eight nationalities in nine countries, which hints at the wide diffusion of the characteristic Harperian approach to plant ecology. They represent work in progress and reflect the range of topics which has engaged John Harper's interest over the past 30 years. I have exercised only a light editorial hand, mostly on technical matters, and have not sought to impose conformity to a particular style of presentation. These studies comprise a rich if somewhat diverse feast from which, I hope, all students of plant demography, whatever their principles or prejudices, will derive some pleasure.

We present this Festschrift to John Harper with a toast for his health and happiness.

James White

Profile of John L. Harper

John Lander Harper was born on 27 May 1925 into a farming family and his early interests in natural history began to crystallize when he went to Lawrence Sheriff School, a boys' grammar school in Rugby. Over a relatively short period this small school, and its Natural History Society in particular, produced five career biologists: Gerald Thompson, who later founded Oxford Scientific Films; A. J. Cain, now Professor of Genetics at Liverpool; F. A. L. Clowes, presently Reader in Botany at Oxford; C. C. Mcready; and John L. Harper. It was not by coincidence that this abundance of talent moved into biology, for the society was led by Wilfred Kings, who imbued its members with enthusiasm for field and laboratory work as well as book learning. Public recognition of the influence of Kings is given in the dedication of *Population Biology of Plants:* 'to W. K., Teacher and Friend'. He kept up a correspondence with his pupils until his death at the age of 90.

John Harper and Lionel Clowes were often to be seen as schoolboys on bicycles on their way to explore ponds, ditches and other sites of biological interest. At unrecorded ages they observed and counted the buttercups of Lilbourne or Catthorpe Farms, and Mrs. Harper tolerated, 'for the sake of science', a wide variety of animals not usually encouraged as pets. This deep interest in natural history has never disappeared, but, more importantly, it provided the sources whence the later professional studies in population biology had their real foundations. John still assumes that all proper biologists match his enthusiasm for and knowledge of natural history. We are, of course, remembering the time when DNA was unknown and the study of biology was almost entirely the study of organisms; animals for dissection were bid for at the cattle market and almost everything needed for practical school biology was collected from the wild or from the farm.

From the sixth form at Sheriff School, John entered Magdalen College, Oxford, in 1943 as a demy. In the Botany School, many of the staff were away and there were few undergraduates. Staff–student ratios are not available but it is rumoured that John found himself as the only undergraduate present in some lectures. Apocryphally, this situation still occurs in some universities, but for different reasons. In the 1940s, and still, education at Oxford had its foundations in practical work and through the tutorial. One of his tutors recalls that 'his prose style was graphic and his quick wit led him to write interesting essays often with wild hypotheses', thus providing sound evidence for the hypothesis that phylogeny recapitulates ontogeny or more clumsily that students are taught in the ways their teachers learned. John's undergraduate career received a considerable boost at the beginning of his final year when the war had ended and there were gathered in the Botany School such outstanding men as Lionel Clowes and Christopher Mcready from Lawrence Sheriff School, and John Warren Wilson and John Burnett, who struck sparks off one another.

First Class Honours in Botany in 1946 led to postgraduate research under the supervision of Dr. J. L. Harley. This period was not very productive and, although John has retained more than a passing interest in root-surface populations of bacteria and fungi, he has rarely been tempted to return to it seriously. The M.A. and D.Phil. in 1950 were interdigitated with a short period in the West Indies which yielded the first two publications of his career—on Panama disease of banana. This short experience of tropical agriculture excited him greatly. We move now to the beginning of Harperian ecology!

Around this time, when some of John's students-to-be were still young and others were but twinkles in their father's eye, Professor G. E. Blackman, of the Department of Agriculture, University of Oxford, was searching for an able botanist who was ambitious for research. John Harper was formally interviewed and appointed as Demonstrator. His roots in agriculture and farming meant that he could identify with everyone on the staff and with the students, many of whom were from farming stock. Teaching and tutorials demanded much time and both were pursued with great vigour and success. The range of subjects taught was quite incredible by modern-day standards although the opportunity (forced or otherwise) to integrate subjects and to pursue diverse research interests was a real advantage. Early in the 1950s John had two relatively unrelated research interests, one of which was to be phased out during the decade. This was a study with Phyllis Landragin and J. W. ('Lud') Ludwig on the influence of the environment on seed and seedling mortality. In a series of eight papers and three shorter communications, the interactions between germination vigour, the environment and pathogenicity were explored successfully, using maize, at a time when the introduction of this crop to Britain was being seriously examined by G. E. Blackman's Unit of Experimental Agronomy. The work is still being quoted more than 25 years after it was published.

The other increasingly dominant part of the Unit of Experimental Agronomy was concerned with weed control and it was not an accident that John gave increasing attention and thought to the biology of weeds. In part this was because Ken Woodford, John Fryer, Jimmy Elliott and Keith Holly were vigorous, interactive and even demanding, but in part too because G. E. Blackman had a tiny slush fund; he was prepared to find subsistence allowances for the vacation employment of occasional students. Most of the work for *Some aspects of the ecology of buttercups in permanent grassland* was done in this way. One of us, having had exciting, highly argumentative ('challenge everything he says') and lengthy tutorials (sometimes three hours instead of the one hour the college paid for) in the smoke-ridden room that was a box carved out of *the* laboratory and having smoked half his packet of Balkan Sobranie ovals, was invited to spend two weeks of the coldest Easter vacation on record in the field chasing buttercups by running transects up ridges and down into furrows where the water table was measurably above ground. Subsequently, the acclaimed public emergence of John Harper as weed biologist and ecologist took place in Margate at the First British Weed Control Conference in 1953. The trio of Joan Thurston (wild oats), Alec Lazenby (rushes) and John Harper (buttercups) excited that conference in the heady days when there were fewer than 10 chemicals available for weed control.

The year 1954 saw a very eligible bachelor taken off the shelf by the charms of Borgny and they set up house not too far away from Mary and Robin Snow, with whom John had happily lodged *en famille* for some years in a very cultured and caring home. The Harper family later grew when Belinda, Claire and Jonathan were born. Brought up under the influence of Benjamin Spock, the family has been a model of interaction with closeness despite the pressures which are nowadays thrust on successful academic and public figures.

John's first research grant was from the Nature Conservancy and through it Ian MacNaughton began the study of the five British species of *Papaver*. This was a natural development from the earlier work on the three commonest species of *Ranunculus* and marked the real beginning of the phase of 'closely related species'; the origins of this are in the report of a Symposium of the British Ecological Society, in which there is more to read between the lines than on them. In 1956 and 1959 there appeared three papers which marked the beginning of John Harper's original thinking on weeds and their control: the evolution of resistance to herbicides, ecological aspects of weed control and ecological significance of dormancy. Two of these papers are still extensively quoted in the 1980s.

The comparative biology of closely related species living in the same area brought together Ian MacNaughton, whose studies were mainly genetical, apart from an accidentally classic competition experiment; John Clatworthy, a Rhodes Scholar who focussed on competition, struggling to make *Lemna, Salvinia* and *Trifolium* grow properly; and myself, who had been overwhelmed by David Lack's *The Natural Regulation of Animal Numbers,* read for the first time during the last months of National Service. Those were exciting days. We met regularly as a group in the evenings at the Harper home, sometimes presenting our own studies but more often than not trying to thrash out many of the more philosophical aspects of the Gaussian hypothesis. The summary of these deliberations was finally published in 1961 but not before two prestigious and subsequently reformed British journals had declined to do so, presumably because the treatment was so far removed from the current mainstream of ecological research in the plant sciences. But prophets are often rebuffed in their own country.

Before John left Oxford in 1960 he organized what was later recognized as the first Symposium of the British Ecological Society: *The Biology of Weeds*. He had local support but the Society itself did not then have the financial strength nor the innovative imagination which it has since acquired. From 1959 to 1960 he took a sabbatical leave in California and it was there that he chose *Bromus rigidus* and *B. madritensis* as the closely related species and dropped them into patterns on soil surfaces. It was at this time that he gratefully absorbed the direct influence of Ledyard Stebbins. Later he presented *Approaches to the study of plant competition,* a keynote paper, in Fred Milthorpe's Symposium for the Society of Experimental Biology. The Oxford period had been increasingly productive for him; by the time it came to a close many of the foundations of later work had been laid but most of the work was unfinished. In the 1950s in Oxford, John had enjoyed outstanding support from G. E. Blackman, who had a genuine interest in his work and who through a mixture of avuncular and paternal traits promoted it without ever interfering. Other colleagues in the Agriculture Department, the Unit, the Zoology Department, the Biochemistry Department, the Edward Grey Institute and the Bureau of Animal Population collaborated openly and genuinely to encourage a young lecturer. John left behind Harper's Field at the University Farm, Wytham—hence the later need for Henfaes.

Thus was the scene set for John Harper to take up the Chair of Agricultural Botany at University College of North Wales, Bangor. Responsibility for a department, for administration or for management had not been part of his education, for the Professor (and Miss Talbot) ran the Oxford Department. In October 1960, Agricultural Botany in Bangor had one professor, four full-time members of academic staff and one demonstrator. There were four final-year students but they read for a degree in Botany with Agricultural Botany. The principal teaching glasshouses were relics of bygone days sited where the Tower Block now sits, and field facilities were at Victoria Park, where the tennis courts were later made. Almost everything was about to change, and rapidly! Agricultural Botany almost immediately became a degree subject in its own right; after a decent interval, this divorce from Botany led to re-marriage with a different contract. The search for field station facilities found Treborth and plans were immediately laid to develop there. Courses were redesigned and experimental plant science, in particular, was introduced in lectures and projects. Between 1961 and 1962 the number of full-time research students rose from 6 to 12. Many of them were supervised by the new professor. In diverse ways several of these new graduate students played their parts in the 1960s in the development of the study of the population biology of plants. The department in those days was very cohesive: it was possible to cajole everybody to clear away stones from an experimental plot, and the whole department turned up for research seminars, open-ended sessions on Friday afternoons. John Harper cultivated this aura but quite unobtrusively. Small and growing was beautiful and very, very successful. Between 1962 and 1964 there were other major developments. Penyffridd was released by NAAS and the UGC purchased it. Agricultural Botany, with others, shared the new site, and Treborth was given over to the Botany Department. The M.Sc. in Crop Protection was introduced, to be followed later by the M.Sc. in Ecology; Chris Marshall was added to the staff and Bob Whitbread replaced Margaret Mence as plant pathologist. By October 1964 David Machin had arrived for the first time as statistician to the faculty but was living in the school and Peter Lovell had been appointed; there were 22 full-time research students registered for the session, many of them from overseas. Growth continued apace and the painstaking process of collecting the data for the development of population biology went on, punctuated (or synthesized) by invited reviews from the conductor: *The individual in the population* (1964) and *A Darwinian approach to plant ecology* (1967), the latter being the Presidential Address to the British Ecological Society.

If parts of this commentary read like the history of a department or group, so be it; the head of department was the director and truly the leader. New ideas for teaching, research and science were thrown out continuously; some were rejected but many were taken up, some inadvertently!

The theme, or mixture, of original scientific papers which were records of research and published jointly by student and supervisor, and the papers written by John Harper himself, many of these gems of scholarship rather than conventional reviews, summarises the strategy of John's approach to the publication of his contributions to science, at least between 1952 and 1985. Throughout his scientific career he has been an internationalist, recognizing the importance of other groups like those led by Kira in Japan and de Wit in the Netherlands. He has openly acknowledged two influences from nearer home: the Welsh Plant Breeding Station in the 1940s and John Maynard Smith in later years. But, of course, the overwhelming influence has been that of Charles Darwin.

When they set up home at Dwygyfylchi in 1960, John and Borgny began to make a garden and after early battles with sheep, weeds and moss, a garden did emerge, not only a beautiful garden but a plantsman's delight, for the aspect, the climate and the soil are not dissimilar from those of the famous garden at Bodnant, over the mountain. A measure of the success is given by the fact that on his 55th birthday John received, from his children, a chain-saw. Many plantsmen grow trees at too high a density but few of them, I suspect, see research programmes when, in their own gardens, branches and buds begin to die when individuals grow into close proximity. Fewer still begin to challenge the sacred concepts of apical dominance—and that from a man who knew Robin Snow better than most.

By the mid-1960s it had become very apparent that the departments of Botany and Agricultural Botany were not only overlappping in their teaching and research but that united together they would be the largest and most effective group of plant biologists in the United Kingdom. That Bangor was the first place to achieve a fusion of departments, thus reversing the tradition of departments arriving by fission, was due to the skills of John Harper. It was a fresh and different sort of challenge although the techniques used to achieve it were akin to those of his science. The School of Plant Biology came into existence in October 1967 and naturally John became the undisputed head. At a stroke, the numbers of members of staff, academic, technical and secretarial, doubled.

Making the new school effective took much time and energy but science was never neglected. Research students from home and abroad, postdoctorals from abroad and many shorter-stay visitors, from abroad in particular, continued to arrive to work with the head of school. Reproductive strategy, competition, demography and population dynamics were main themes but time was also found for the preparation of reviews, with Peter Lovell and Keith Moore on shapes and sizes of seeds, and on plant demography. As with life, there were periods of special excitement and times of routine. Lady Luck sometimes smiled and sometimes frowned. Nevertheless, the head, the school and population biology continued to flourish and develop. *Population Biology of Plants* was published in 1977, much of the writing being done during an earlier, very productive sabbatical leave in Montpellier. It was a review and synthesis of the state of the art but not a swan-song: even in its first chapter new concepts are outlined and one of these in particular (modular construction) has been subsequently developed.

John Harper enjoys discussions about science immensely and, being an exciting speaker, has always been in demand for symposia and meetings. The times in the West Indies, California and Montpellier have been mentioned already but there were, in addition, other periods, many of them spent in the United States and Europe where he went to enthuse, stimulate and be stimulated. His energy levels have always been best recharged by being mentally or physically energetic. The combination of discussion, gardening and music has been the recipe. He was generally recognized as a brilliant scientist in the United States long before he even began to be recognized generally in Europe, including Great Britain. Nevertheless, since the old order inevitably changes, proper recognition began to come, even at home. The Presidency of the British Ecological Society in 1966, appointment to the Natural Environment Research Council in 1971 (a position he held for 10 years) and then the accolade of Fellowship of the Royal Society in 1978 were followed by appointment to the Agricultural Research Council (1980) and an honorary D.Sc. from the University of Sussex in 1984, 2 years after John took early retirement (in part to escape the traumas of Britain's attempt to make its universities second class but more to allow full-time science and public service). In 1984, too, he was twice honoured in the United States: by the coveted honour of Foreign Membership of the National Academy of Science and by the nomination 'Eminent Ecologist' of the Ecological Society of America.

Anyone who expects to have scientific papers published must be prepared to do a stint in an editorial role. John Harper's first was the Editorship of *Biology of Weeds* but subsequently he served on the Editorial Board of the *Journal of Ecology,* was Editor-in-Chief of *Agroecosystems* from 1974 to 1981 and became an assistant editor of the *Proceedings of the Royal Society B* in 1980. Currently, he co-edits *Oecologia.* As he is clearly a glutton for punishment, his extraordinarily quick brain and pen have allowed him to carry the loads without detriment to his own science, although sometimes the handwriting has needed experienced experts to decipher it.

I have tried to sketch a profile which touches on some of the aspects of John Harper's life, career and science. Others must attempt the appraisal; I am too close. The formal public record of the science follows as a bibliography but it does not adequately represent the even greater contribution to science made through the continual stimulation and encouragement of all those with whom he came into contact as teacher, supervisor, collaborator or colleague.

G. R. Sagar
School of Plant Biology
University College of North Wales
Bangor, Gwynedd, Wales

Scientific Publications of John L. Harper

1950

Harper, J. L. (1950). Studies in the resistance of certain varieties of banana to Panama disease. I. Internal factors for resistance and antibiotics. *Plant and Soil,* **2,** 374–382.

Harper, J. L. (1950). Studies in the resistance of certain varieties of banana to Panama disease. II. The rhizosphere. *Plant and Soil,* **2,** 383–394.

1951

Osborn, E. M. & Harper, J. L. (1951). Antibiotic production by growing plants of *Leptosyne maritima. Nature,* **167,** 685.

1953

Harper, J. L. & Sagar, G. R. (1953). Some aspects of the ecology of buttercups in permanent grassland. *Proceedings of the British Weed Control Conference.* **1,** 256–263.

1954

Harper, J. L. (1954). Influence of temperature and soil water content on the seedling blight of maize. *Nature,* **173,** 391.

Harper, J. L. (1954). Some relationships between the soil water table and the comparative ecology of *Ranunculus bulbosus, R. repens* and *R. acris. International Botanical Congress Report,* **2,** 207–208.

1955

Harper, J. L. (1955). The influence of the environment on seed and seedling mortality. VI. The effects of the interaction of soil moisture content and temperature on the mortality of maize grains. *Annals of Applied Biology.* **43,** 696–708.

Harper, J. L. (1955). Problems involved in the extension of maize cultivation into northern temperate regions. *World Crops,* **7,** 93–96.

Harper, J. L., Landragin, P. A. & Ludwig, J. W. (1955). The influence of the environment on seed and seedling mortality. I. The influence of time of planting on the germination of maize. *New Phytologist,* **54,** 107–131.

Harper, J. L., Landragin, P. A. & Ludwig, J. W. (1955). The influence of the environment on seed and seedling mortality. II. The pathogenic potential of the soil. *New Phytologist,* **54,** 119–131.

Harper, J. L. & Landragin, P. A. (1955). The influence of the environment on seed and seedling mortality. IV. Soil temperature and maize grain mortality with special reference to cold test procedure. *Plant and Soil,* **6,** 360–372.

1956

Harper, J. L. (1956). Studies in seed and seedling mortality. V. Direct and indirect influences of low temperatures on the mortality of maize. *New Phytologist,* **55,** 35–44.

Harper, J. L. (1956). The evolution of weeds in relation to resistance to herbicides. *Proceedings of the 3rd British Weed Control Conference,* **1,** 179–188.

Acheson, R. M., Harper, J. L. & McNaughton, I. H. (1956). Distribution of anthocyanin pigments in poppies. *Nature,* **178,** 1283–1284.

1957

Harper, J. L. (1957). Biological flora of the British Isles. *Ranunculus acris* L., *R. repens* L. and *R. bulbosus* L. *Journal of Ecology,* **45,** 289–342.

Harper, J. L. (1957). Ecological aspects of weed control. *Outlook on Agriculture,* **1,** 197–205.

Harper, J. L. & Wood, W. A. (1957). Biological flora of the British Isles. *Senecio jacobaea* L. *Journal of Ecology,* **45,** 617–637.

Ludwig, J. W., Bunting, E. S. & Harper, J. L. (1957). The influence of the environment on seed and seedling mortality. III. The influence of aspect on maize germination. *Journal of Ecology,* **45,** 205–224.

1958

Harper, J. L. (1958). The ecology of ragwort (*Senecio jacobaea*) with special reference to control. *Herbage Abstracts,* **28,** 151–157.

Harper, J. L. (1958). Famous Plants—8—The Buttercup. *New Biology,* **26,** 30–46.

Ludwig, J. W. & Harper, J. L. (1958). The influence of the environment on seed and seedling mortality. VII. Depth of sowing of maize. *Plant and Soil,* **10,** 37–48.

Ludwig, J. W. & Harper, J. L. (1958). The influence of the environment on seed and seedling mortality. VIII. The influence of soil colour. *Journal of Ecology,* **46,** 381–389.

1959

Harper, J. L. (1959). The ecological significance of dormancy and its importance in weed control. *IVth International Congress of Crop Protection, Hamburg (1957),* **1,** 415–420.

Harper, J. L. & Chancellor, A. P. (1959). The comparative biology of closely related species living in the same area. IV. *Rumex:* Interference between individuals in populations of one and two species. *Journal of Ecology,* **47,** 679–695.

1960

Harper, J. L. (1960). Factors controlling plant numbers. *The Biology of Weeds. British Ecological Society Symposium,* **1,** 119–132.

Harper, J. L. & McNaughton, I. H. (1960). The inheritance of dormancy in inter- and intra-specific hybrids of *Papaver. Heredity,* **15,** 315–320.

McNaughton, I. H. & Harper, J. L. (1960). The comparative biology of closely related species living in the same area. I. External breeding barriers between *Papaver* species. *New Phytologist,* **58,** 15–26.

McNaughton, I. H. & Harper, J. L. (1960). The comparative biology of closely related species living in the same area. II. Aberrant morphology and a virus-like syndrome in hybrids between *Papaver rhoeas* L. and *P. dubium* L. *New Phytologist,* **59,** 27–41.

McNaughton, I. H. & Harper, J. L. (1960). The comparative biology of closely related species living in the same area. III. The nature of barriers isolating sympatric populations of *Papaver dubium* and *P. lecoquii. New Phytologist,* **59,** 129–137.

Sagar, G. R. & Harper, J. L. (1960). Factors affecting the germination and early establishment of plantains (*Plantago lanceolata, P. media* and *P. major*). *The Biology of Weeds. British Ecological Society Symposium,* **1,** 236–245.

1961

Harper, J. L. (1961). Approaches to the study of plant competition. *Mechanisms in Biological Competition. Symposium of the Society for Experimental Biology,* **15,** 1–39.

Harper, J. L., Clatworthy, J. N., McNaughton, I. H. & Sagar, G. R. (1961). The evolution and ecology of closely related species living in the same area. *Evolution,* **15,** 209–227.

Harper, J. L. & Gajic, D. (1961). Experimental studies on the mortality and plasticity of a weed. *Weed Research,* **1,** 91–104.

Sagar, G. R. & Harper, J. L. (1961). Controlled interference with natural populations of *Plantago lanceolata, P. major* and *P. media. Weed Research,* **1,** 163–176.

1962

Acheson, R. M., Jenkins, C. L., Harper, J. L. & McNaughton, I. H. (1962). Floral pigments in *Papaver* and their significance in the systematics of the genus. *New Phytologist,* **61,** 256–260.

Clatworthy, J. N. & Harper, J. L. (1962). The comparative biology of closely related species living in the same area. V. Inter- and intra-specific interference with cultures of *Lemna* spp. and *Salvinia natans. Journal of Experimental Botany,* **13,** 307–324.

Harper, J. L. & McNaughton, I. H. (1962). The comparative biology of closely related species living in the same area. VII. Interference between individuals in pure and mixed populations of *Papaver* species. *New Phytologist,* **61,** 175–188.

McNaughton, I. H. & Harper, J. L. (1962). Interference in mixed poppy populations. *Annals of Applied Biology,* **50,** 352.

1963

Harper, J. L. (1963). The biological role of soil water—introduction. *Welsh Soils Discussion Group Report No. 4, Soil Moisture,* 52–54.

Harper, J. L. & Clatworthy, J. N. (1963). The comparative biology of closely related species. VI. Analysis of the growth of *Trifolium repens* and *T. fragiferum* in pure and mixed populations. *Journal of Experimental Botany,* **14,** 172–190.

1964

Harper, J. L. (1964). The individual in the population. *Journal of Ecology,* **52** (Suppl.), 149–158.

Cavers, P. B. & Harper, J. L. (1964). Biological flora of the British Isles. No. 98. *Rumex obtusifolius* L. and *R. crispus* L. *Journal of Ecology,* **52,** 737–766.

McNaughton, I. H. & Harper, J. L. (1964). Biological flora of the British Isles. No. 99. *Papaver* L. *Journal of Ecology,* **52,** 767–793.

Sagar, G. R. & Harper, J. L. (1964). Biological flora of the British Isles. No. 95. *Plantago major* L., *P. media* L. and *P. lanceolata* L. *Journal of Ecology,* **52,** 189–221.

1965

Harper, J. L. (1965). The nature and consequence of interference among plants. *Genetics Today, 11th International Congress of Genetics,* **2,** 465–482.

Harper, J. L. (1965). Establishment, aggression and cohabitation in weedy species. *Genetics of Colonising Species* (Ed. by H. G. Baker & G. L. Stebbins), pp. 243–268. Academic Press, New York.

Govier, R. N. & Harper, J. L. (1965). Angiospermous hemiparasites. *Nature,* **205,** 722–723.

Harper, J. L., Williams, J. T. & Sagar, G. R. (1965). The behaviour of seed in soil. I. The heterogeneity of soil surfaces and its role in determining the establishment of plants from seed. *Journal of Ecology,* **53,** 273–286.

Williams, J. T. & Harper, J. L. (1965). Seed polymorphism and germination. I. The influence of nitrates and low temperatures on the germination of *Chenopodium album. Weed Research,* **5,** 141–150.

1966

Harper, J. L. (1966). The reproductive biology of the British poppies. *Reproductive Biology and Taxonomy of Vascular Plants* (Ed. by J. G. Hawkes), pp. 26–39. Botanical Society of the British Isles, London.

Cavers, P. B. & Harper, J. L. (1966). Germination polymorphism in *Rumex crispus* and *Rumex obtusifolius. Journal of Ecology,* **54,** 367–382.

Harper, J. L. & Benton, R. (1966). The behaviour of seeds in soil. II. The germination of seeds on the surface of a water supplying substrate. *Journal of Ecology,* **54,** 151–166.

1967

Harper, J. L. (1967). A Darwinian approach to plant ecology. *The Teaching of Ecology. Journal of Ecology,* **55,** 247–270.

Harper, J. L. (1967). The teaching of experimental ecology. *Symposium of the British Ecological Society,* **7,** 135–146.

Cavers, P. B. & Harper, J. L. (1967). Studies in the dynamics of plant populations. I. The fate of seed and transplants introduced into various habitats. *Journal of Ecology,* **55,** 59–71.

Cavers, P. B. & Harper, J. L. (1967). The comparative biology of closely related species living in the same area. IX. *Rumex:* The nature of adaptation to a sea-shore habitat. *Journal of Ecology,* **55,** 73–82.

Harper, J. L. & Obeid, M. (1967). The influence of seed size and depth of sowing on the establishment and growth of varieties of fiber and oil seed flax. *Crop Science,* **7,** 527–532.

Litav, M. & Harper, J. L. (1967). A method for studying the spatial relationships between the root systems of two neighbouring plants. *Plant and Soil,* **26,** 389–392.

Obeid, M., Machin, D. & Harper, J. L. (1967). Influence of density on plant to plant variation in fiber flax, *Linum usitatissimum* L. *Crop Science,* **7,** 471–473.

1968

Harper, J. L. (1968). The regulation of numbers and mass in plant populations. Population Biology and Evolution (Ed. by R. C. Lewontin), pp. 139–158. Syracuse University Press, New York.

Hatto, J. & Harper, J. L. (1968). The control of slugs and snails in British cropping systems, especially grassland. *INCRA Project No. 115A,* 25 pp.

Putwain, P. D., Machin, D. & Harper, J. L. (1968). Studies in the dynamics of plant populations. II.

Components and regulation of a natural population of *Rumex acetosella* L. *Journal of Ecology,* **57,** 327–356.

1969

Harper, J. L. (1969). The role of predation in vegetational diversity. *Diversity and Stability in Ecological Systems. Brookhaven Symposia in Biology,* **22,** 48–62.

1970

Harper, J. L. (1970). The population biology of plants. *Population Control* (Ed. by A. Allison), pp. 32–44. Penguin Books, Harmondsworth.

Harper, J. L. (1970). Grazing, fertilizers and pesticides in the management of grasslands. *The Scientific Management of Animal and Plant Communities for Conservation.* 11th Symposium of the British Ecological Society. (Ed. by E. Duffey and A. S. Watt), pp. 15–31. Blackwell Scientific Publications, Oxford.

Harper, J. L., Lovell, P. H. & Moore, K. G. (1970). The shapes and sizes of seeds. *Annual Review of Ecology and Systematics,* **1,** 327–356.

Harper, J. L. & Ogden, J. (1970). The reproductive strategy of higher plants. I. The concept of strategy with special reference to *Senecio vulgaris* L. *Journal of Ecology,* **58,** 681–698.

Ellern, S. J., Harper, J. L. & Sagar, G. R. (1970). A comparative study of the distribution of the roots of *Avena fatua* and *A. strigosa* in mixed stands using a ^{14}C-labelling technique. *Journal of Ecology,* **58,** 865–868.

Putwain, P. D. & Harper, J. L. (1970). Studies in the dynamics of plant populations. III. The influence of associated species on populations of *Rumex acetosa* L. and *R. acetosella* L. in grassland. *Journal of Ecology,* **58,** 251–264.

White, J. & Harper, J. L. (1970). Correlated changes in plant size and number in plant populations. *Journal of Ecology,* **58,** 467–485.

1971

Harper, J. L. & White, J. (1971). The dynamics of plant populations. *Proceedings of the Advanced Study Institute (Oosterbeek)* pp. 41–63. Center for Agricultural Publication and Documentation, Wageningen.

1972

Harper, J. L. (1972). Projects and their assessment in degree examinations. *Journal of Biological Education,* **6,** 318–321.

Putwain, P. D. & Harper, J. L. (1972). Studies in the dynamics of plant populations. V. Mechanisms governing the sex ratio in *Rumex acetosa* and *R. acetosella. Journal of Ecology,* **60,** 113–129.

Ross, M. A. & Harper, J. L. (1972). Occupation of biological space during seedling establishment. *Journal of Ecology,* **60,** 77–88.

1973

Haizel, K. A. & Harper, J. L. (1973). The effects of density and the timing of removal on interference between barley, white mustard and wild oats. *Journal of Applied Ecology,* **10,** 23–31.

Sarukhán, J. & Harper, J. L. (1973). Studies on plant demography: *Ranunculus repens* L., *R. bulbosus* L. and *R. acris* L. I. Population flux and survivorship. *Journal of Ecology,* **61,** 675–716.

Tripathi, R. S. & Harper, J. L. (1973). The comparative biology of *Agropyron repens* (L.) Beauv. and *A. caninum* (L.) Beauv. I. The growth of mixed populations established from tillers and from seeds. *Journal of Ecology,* **61,** 353–368.

Trenbath, B. R. & Harper, J. L. (1973). Neighbour effects in the genus *Avena.* I. Comparison of crop species. *Journal of Applied Ecology,* **10,** 379–400.

1974

Harper, J. L. (1974). Agricultural ecosystems. Editorial. *Agro-Ecosystems,* **1,** 1–6.

Harper, J. L. (1974). Soil sterilisation and disinfection: The perturbation of ecosystems. *Agro-Ecosystems,* **1,** 105–106.

Harper, J. L. (1974). A centenary in population biology. *Nature,* **252,** 526–527.

Harper, J. L. & White, J. (1974). The demography of plants. *Annual Review of Ecology and Systematics,* **5,** 419–463.

Kays, S. & Harper, J. L. (1974). The regulation of plant and tiller density in a grass sward. *Journal of Ecology,* **62,** 97–105.

1976

Harper, J. L. (1976). Enseignement superieur agronomique, recherche et development. *Cahiers des Ingenieures Agronomes,* **310,** 13–14.

Bazzaz, F. A. & Harper, J. L. (1976). Relationship between plant weight and numbers in mixed populations of *Sinapis alba* (L.) Rabenh. and *Lepidium sativum* L. *Journal of Applied Ecology,* **13,** 211–216.

Cahn, M. & Harper, J. L. (1976). The biology of the leaf-mark polymorphism in *Trifolium repens* L. I. Distribution of phenotypes at a local scale. *Heredity,* **37,** 309–325.

Cahn, M. & Harper, J. L. (1976). The biology of the leaf mark polymorphism in *Trifolium repens* L. II. Evidence for the selection of leaf marks by rumen fistulated sheep. *Heredity,* **37,** 327–333.

1977

Harper, J. L. (1977). *Population Biology of Plants.* Academic Press, London.

Harper, J. L. (1977). The contribution of terrestrial plant studies to the development of the theory of ecology. *The Changing Scenes in Natural Sciences, 1776–1976. Academy of Natural Sciences, Special Publication,* **12,** 139–157.

Harper, J. L. (1977). Plant relations in pastures. *Plant Relations in Pastures.* (Ed. by J. R. Wilson), pp. 3–14. Commonwealth Scientific and Industrial Organization, Melbourne.

Bazzaz, F. A. & Harper, J. L. (1977). Demographic analysis of the growth of *Linum usitatissimum. New Phytologist,* **78,** 193–208.

Mack, R. N. & Harper, J. L. (1977). Interference in dune annuals: spatial pattern and neighbourhood effects. *Journal of Ecology,* **65,** 345–363.

1978

Harper, J. L. (1978). The demography of plants with clonal growth. *Structure and Functioning of Plant Populations.* (Ed. by A. H. J. Freysen & J. W. Woldendorp), pp. 27–48. North-Holland Publ., Amsterdam.

Watkinson, A. R. & Harper, J. L. (1978). The demography of a sand dune annual: *Vulpia fasciculata.* I. The natural regulation of populations. *Journal of Ecology,* **66,** 15–33.

1979

Harper, J. L. & Bell, A. D. (1979). The population dynamics of growth form in organisms with modular construction. *Population Dynamics, Symposium of the British Ecological Society,* **20,** 29–52.

Bazzaz, F. A., Carlson, R. W. & Harper, J. L. (1979). Contribution to reproductive effort by photosynthesis of flowers and fruit. *Nature,* **279,** 554–555.

Huiskes, H. L. & Harper, J. L. (1979). The demography of leaves and tillers of *Ammophila arenaria* in a dune sere. *Oecologia Plantarum,* **14,** 435–446.

Noble, J. C., Bell, A. D. & Harper, J. L. (1979). The population biology of plants with clonal growth. I. The morphology and structural demography of *Carex arenaria. Journal of Ecology,* **67,** 983–1008.

Onyekwelu, S. S. C. & Harper, J. L. (1979). Sex ratio and niche differentiation in spinach (*Spinacia oleracea*). *Nature,* **282,** 609–611.

Turkington, R. & Harper, J. L. (1979). The growth, distribution and neighbour relationships of *Trifolium repens* in a permanent pasture. I. Ordination, pattern and contact. II. Inter- and intra-specific contact. III. The establishment and growth of *Trifolium repens* in natural and perturbed sites. IV. Fine-scale biotic differentiation. *Journal of Ecology,* **67,** 201–254.

1980

Harper, J. L. (1980). Plant demography and ecological theory. *Oikos,* **35,** 244–253.

Burdon, J. J. & Harper, J. L. (1980). Relative growth rates of individual members of a plant population. *Journal of Ecology,* **68,** 953–957.

Dirzo, R. & Harper, J. L. (1980). Experimental studies on slug–plant interactions. II. The effect of grazing by slugs on high density monocultures of *Capsella bursa-pastoris* and *Poa annua. Journal of Ecology,* **68,** 999–1011.

Lovett Doust, J. & Harper, J. L. (1980). The resource costs of gender and maternal support in an andromonoecious umbellifer, *Smyrnium olusatrum* L. *New Phytologist,* **85,** 251–264.

Vernet, P. & Harper, J. L. (1980). The cost of sex in seaweeds. *Biological Journal of the Linnean Society,* **13,** 129–138.

1981

Harper, J. L. (1981). The meaning of rarity. *The Biological Aspects of Rare Plant Conservation* (Ed. by Hugh Synge), pp. 189–203. Wiley, New York.

Harper, J. L. (1981). The concept of population in modular organisms. *Theoretical Ecology. Principles and Applications* (Ed. by R. M. May), pp. 53–77, 2nd Edition. Blackwell Scientific Publications, Oxford.

1982

Harper, J. L. (1982). After description. *The Plant Community as a Working Mechanism* (Ed. by E. I. Newman). B.E.S. Special Publication Series, No. **1,** pp. 11–25. Blackwell Scientific Publications, Oxford.

Harper, J. L. (1982). The biology of white clover (*Trifolium repens*) in a permanent pasture. *2nd Nat. Symp. on Nitrogen Fixation, Helsinki, Finland,* 255–260.

Dirzo, R. & Harper, J. L. (1982). Experimental studies on slug–plant interactions. III. Differences in the acceptability of individual plants of *Trifolium repens* to slugs and snails. *Journal of Ecology,* **70,** 101–118.

Dirzo, R. & Harper, J. L. (1982). Experimental studies on slug–plant interactions. IV. The performance of cyanogenic and acyanogenic morphs in the field. *Journal of Ecology,* **70,** 119–138.

1983

Harper, J. L. (1983). A Darwinian plant ecology. *Evolution from Molecules to Men* (Ed. by D. S. Bendall), pp. 323–345. Cambridge University Press, London.

Fowler, N., Zasada, J. & Harper, J. L. (1983). Genetic components of morphological variation in *Salix repens*. *New Phytologist,* **95,** 121–131.

1984

Mithen, R., Harper, J. L. & Weiner, J. (1984). Growth and mortality of individual plants as a function of 'available area'. *Oecologia (Berl.),* **62,** 57–60.

1985

Harper, J. L. (1985). Modules, branches and the capture of resources. *Population Biology of Clonal Organisms* (Ed. by J. B. C. Jackson, L. W. Buss & R. E. Cook). Yale University Press, New Haven.

Turkington, R., Harper, J. L., de Jong, P. & Aarssen, L. W. (1985). A reanalysis of interspecific association in an old pasture. *Journal of Ecology,* **73,** 123–131.

I

Dynamics and Evolution of Plant Populations in Natural or Seminatural Environments

1

Past, Present and Future: Studies on the Population Dynamics of Some Long-lived Trees

John Ogden

Department of Botany
University of Auckland
Auckland, New Zealand

I. INTRODUCTION

> One of the consequences of the development of the theory of vegetational climax has been to guide the observer's mind forwards. Vegetation is interpreted as a stage on the way to something. It might be more healthy and scientifically more sound to look more often backwards and search for the explanation of the present in the past, to explain systems in relation to their history rather than their 'goal' (Harper 1977 p. 628).

This statement is made in the context of forest dynamics and implies that study of the demography of long-lived tree populations should involve a search for the past influences which have contributed to their present structure. In the same context, on the other hand, ''Ecology, if it is to gain maturity as a science, must become predictive . . .'' and matrix methods are suggested as one of the most promising approaches (Harper 1977 p. 625).

The pattern of distribution of individuals, age–frequency distribution and presence of decomposing remnants of former generations may contribute clues about the past history of a forest stand (Henry & Swan 1974). Tree-ring sequences in individual trees may variously record past competitive relationships, climate or traumatic events such as fires or wind-throw, which have also affected population mortality and recruitment (Fritts 1976). Predictions of the future of tree populations and accounts of their history have generally relied upon the interpretation of existing size–frequency distributions (Daubenmire 1968), sometimes aided by small samples of increment cores to establish the 'significance' of the relationship between size and age (Veblen, Schlegel & Escobar

STUDIES ON PLANT DEMOGRAPHY:
A FESTSCHRIFT FOR JOHN L. HARPER

1980). Less commonly, predictive approaches have incorporated size-specific growth, fecundity and mortality rates, which require observations on marked individuals over many years if they are to be reliably estimated (Enright & Ogden 1979).

Despite the different techniques required, reconstruction of past performance and prediction of future states are complementary approaches. Both must set our understanding of the behaviour of tree populations on realistic time scales, normally centuries and sometimes millenia. Moreover, this time scale itself rules out much experimentation, and dictates that most aspects of the study be performed in a natural setting. Frequently this setting will be a mixed forest community, so that the influence of neighbours of other species must be considered, and the approach becomes comparative. Attention almost inevitably shifts from the numerical structure of a species population to its role in relation to other members of the community.

In this account I present descriptions of the natural history of some long-lived trees in the forests of Tasmania and New Zealand. I assume that the rules governing the behaviour of herbaceous monocultures can also be applied to trees (White 1980) and that different 'safe sites' (Harper, Williams & Sagar 1965) or 'regeneration niches' (Grubb 1977) are important in the dynamics of mixtures. My approach relies on tree-rings for historical reconstruction and transition matrices for approximate predictions from current structures and processes. I emphasise the interpretation of population behaviour and life-cycle strategies on a Holocene time scale.

II. SOME METHODOLOGICAL CONSIDERATIONS

"It is wholly unrealistic and very dangerous to assume any relation between the size of trees and their age, other than the vague principle that the largest trees in a canopy are likely to be old" (Harper 1977 p. 634). Similar skewed size–frequency distributions can represent stable self-replicating 'climax' populations (size $\approx$ age), even-aged thinning hierarchies (size $\not\approx$ age) or some intermediate condition. Accurately ageing a substantial number of trees should separate the extreme possibilities, but cannot eliminate the potential confusion between a permanent self-replicating member of a stand and one which is gradually infiltrating an area. Moreover species populations may sometimes comprise copses of even age scattered in a multi-aged matrix determined by safe-site frequency and dispersal characteristics (Horn 1981).

Trees are normally aged by taking increment cores. Allowances are made for any 'missing radius' when the core does not bisect the pith, and for the variable time taken to reach coring height. Consequently the age obtained must be regarded as simply the best estimate. Moreover, diameter measurement is also subject to error, at least in the case of trees with large irregular boles.

A consequence of these errors is that most age–diameter scatter diagrams contain more 'variance' than is apparent; each point floats in a drop of uncertainty which increases with tree diameter. A small sample of inaccurately aged trees from widely separated size classes may provide a 'highly significant' age–diameter relationship but obscure a multiple cohort structure derived from several distinct and more or less even-aged waves of regeneration. If each cohort develops the log-normal size–frequency characteristic of a monoculture and thins through time, then a composite size–frequency distribution will be heavily weighted by the most recent cohort, and older modes may be immersed.

The concept of 'relatively even-aged' deserves comment because the size and density relationships of even-aged populations of both herbs and trees have been shown to follow well-defined rules (White 1980). In natural situations 'even-aged' has meaning only in the context of the 'normally attainable age', best regarded as that at which those few survivors that gain dominance in their habitat tend to die (Harper & White 1974). Thus the normally attainable age for mountain beech (*Nothofagus solandri*) in the montane forests of New Zealand is probably about 250 years, although a majority die earlier, and maximum longevity is in excess of 300 years (Wardle 1970b). In the case of kauri (*Agathis australis*) normal age is probably six or seven centuries, although some trees live for twice this span (Ecroyd 1982; Ahmed 1984). 'Relatively even-aged' in such populations refers to the situation in which a site is 'fully stocked' with seedlings over a time period which is considerably less than the normally attainable age. In the case of *Nothofagus,* for example, a stand may develop from suppressed seedlings following wind-throw of the canopy trees, and the age range of the population will reflect the age range of the 'advance growth' seedlings, commonly 20 or 30 years (Kirkland 1961; Wardle 1970a). Seedlings of kauri become established in pioneer *Leptospermum* communities arising after fire, but the optimum establishment period may be quite restricted (<60 years) (Mirams 1957; Lloyd 1960). Later recuits commence life at a competitive disadvantage and are likely to be 'weeded out' before the original cohort reaches maturity. These two examples and others suggest that for many purposes a stand can be regarded as 'even-aged', or a cohort defined, if most of the individuals in it fall within an age range of not much more than 10% of the 'normally attainable age'. Although strict definition of these terms is not feasible, reliable data on age distribution and longevity are required for predictive models.

III. SOME EXAMPLES

A. Pencil Pine (*Athrotaxis cupressoides*)

At the upper limit for tree growth in western Tasmania, often above the regional *Eucalyptus* timberline, are scattered stands of the endemic pencil pine.

The stands occur beside tarns and on the damp shady backwalls of cirques, where the ice lingered in the waning phases of the last glacial period, and snow accumulates each winter. The regional hazard, fire, rarely penetrates these cool rainy heights and some of the trees are over a thousand years old (Ogden 1978a). The old trees carry poor cone crops, seedlings are almost absent, and the distribution and structure of the populations suggest that regeneration is rare and periodic. I have studied the dendrochronology and ecology of these populations throughout Tasmania, and those on Mount Field in particular.

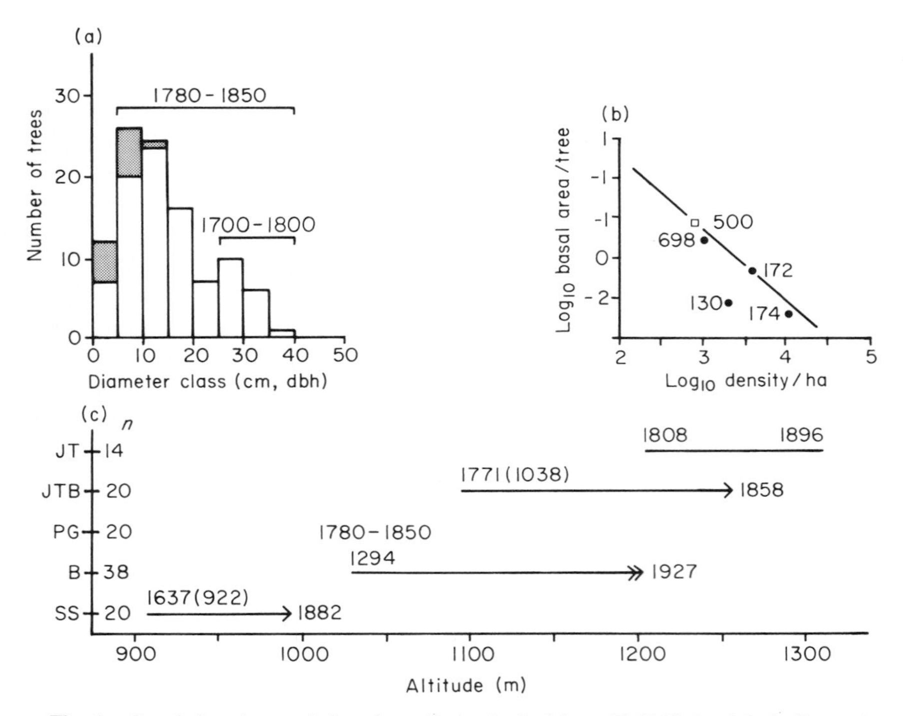

Fig. 1. Population characteristics of pencil pine in the Mount Field National Park, Tasmania. a, Structure of a young stand, Pandani Grove (121 stems measured, 21 cored). Estimated germination dates for individuals in different classes are given. Stippling indicates standing dead trees. b, 'Thinning line' for pencil pine stands on Mount Field and one stand (square) at Pine Lake (grid ref. 750790): $y = 1.5 - 0.9x$. Median ages of trees estimated from core samples give an approximate age ranking for the stands. c, Establishment dates in five populations. Bars indicate the altitudinal ranges of the samples aged. The age range is given with extreme outliers in brackets. Single arrows imply a negative relationship between age and altitude, significant at $P < 0.05$; double arrows at $P < .01$. The population identifiers and number (n) of trees aged are given on the vertical axis. The sites are as follows (grid references in brackets): SS, above Lake Seal (649750-651752); B, Belcher valley (c. 628737); PG, Pandani Grove (664734); JTB, Johnstone tarn burn (640754); JT, above Johnstone tarn (641753). (Original data.)

At Pandai Grove most of the largest trees germinated after 1780, and the recruitment phase lasted about 70 years (Fig. 1a). During this period the trees grew slowly, reaching an average height of only 1 m after 55 years, when they presumably began to break through the shrubby heath characteristic of open sites at this altitude. The abundance of frost-damaged rings suggests that there was no protective over-storey, although subalpine forest of *Eucalyptus* and *Nothofagus* spp. occurs close by today. Subsequently growth rates increased, individuals came into contact, and thinning commenced. No recruitment has taken place for a century and dead trees comprise about 12% of the basal area.

At higher altitudes the younger stands also show monomodal size frequencies and no relationship between age and diameter. Cores taken from trees at different altitudes within these populations suggest upwards migration from lake margins onto the adjacent slopes (Fig. 1c). Most of this migration may have culminated late last century, but the highest and youngest stand (Johnstone Tarn) has added a few recruits since. The lack of dead stems and the low density and basal area in this stand suggest it is still under-stocked for its age, although it carries a similar number of 'genets' to a stand of veterans nearby. With the exception of this stand all the populations have basal areas ($m^2 \cdot ha^{-1}$) far higher than those of the timberline Eucalypts (Ogden & Powell 1979). Despite widely different ages they all appear to be close to carrying capacity; mortality must prevail in future as the survivors get bigger (Fig. 1b).

I envisage pencil pine as a species expanding its range during the cold variable climate of Pleistocene Tasmania and currently 'sitting out' the Holocene in scattered stands above timberline. Such stands have been present throughout their current range on Mount Field for the last thousand years, but the altitudinal locus of regeneration appears to have shifted. Gradual migrations have occurred and local 'even-aged' stands developed. Great longevity allows the species to track environmental variability on a scale of centuries, regenerating mainly in those periods when favourable combinations of circumstances permit. Dendrochonological study of the oldest stand has provided evidence of long-term climatic fluctuations but cannot yet precisely define their nature (LaMarche *et al.* 1979).

B. King Billy Pine (*Athrotaxis selaginoides*)

Generally at slightly lower altitudes than pencil pine, and only where rainfall is high and fire frequency low, occurs its close congener the king billy pine. Trees of this species occur as scattered veteran emergents above a canopy of *Nothofagus cunninghamii,* as isolated clumps of younger trees, or, in the absence of pencil pine, as a krumholtz zone at timberline (Ogden, 1978a).

As with pencil pine, population age structures suggest periods of recruitment followed by prolonged self-thinning. For example, on Mount Anne most of the king billy trees in a few hectares between 750 and 840 m altitude appear to have germinated during the sixteenth century, and many have since been eliminated (Fig. 2a). Scattered remnants of a previous generation occur below the stand, and there are gnarled krumholtz trees on the exposed ridges above.

At Swift Creek the size–frequency distribution also suggested two generations (Fig. 2b). The stag-headed veterans were scattered over several hectares. Most had rotten centres, and were untouched when the younger trees were logged. Extrapolation from increment cores suggested they germinated during or before

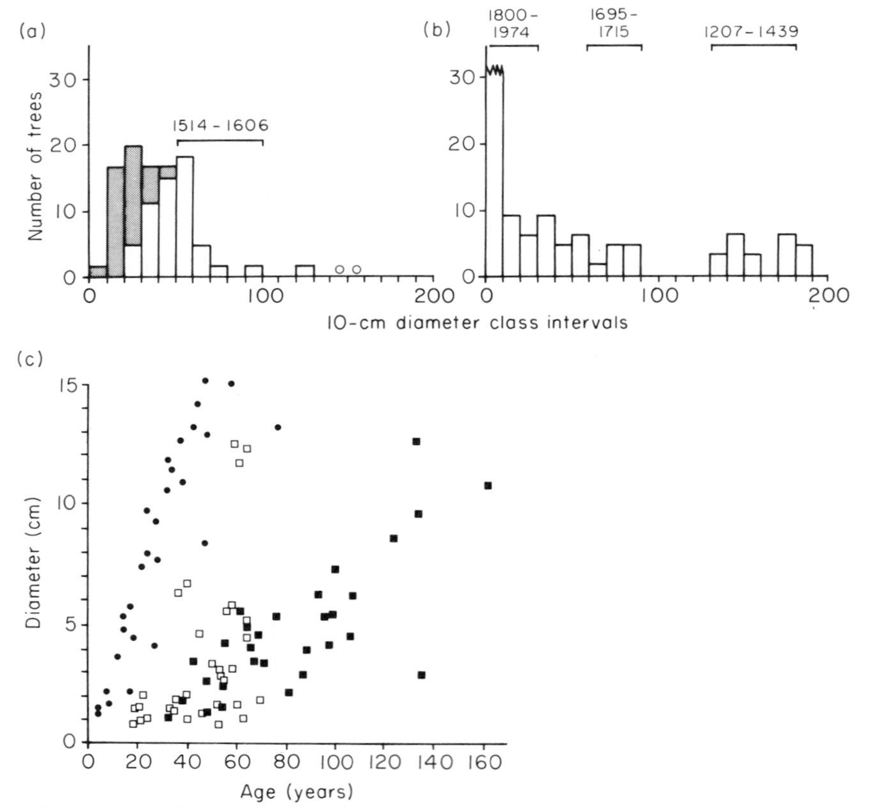

Fig. 2. Population structure in king billy pine in Tasmania. a, Size structure on Mount Anne (grid ref. 537473). Stippling indicates standing dead trees. Circles: two larger trees at slightly lower altitude. b, Size structure at Swift Creek (grid ref. 186912). Estimated germination dates for individuals in different classes are given. c, Growth rates of young trees at Swift Creek and Mount Read (grid ref. 774675). Black circles: Swift Creek, currently dominant trees, growth until 15 cm dbh reached (1718–1800). Black squares: Swift Creek, currently suppressed saplings <15 cm dbh. Open squares: current saplings growing in post-fire regeneration on Mount Read. Note that the fastest-growing saplings on Mount Read have growth rates approaching those which characterised the early development of the stand at Swift Creek. (Original data.)

the fifteenth century. The logged trees formed a small clump (<1 ha). Cross sections reliably demonstrated that all had originated between 1695 and 1715. These trees had been fast-growing when young, with rates comparable to those in a currently young post-fire stand on Mount Read (Fig. 2c). Their growth rates contrasted markedly with those of younger trees (mostly post-1800) growing beneath them or in the surrounding *Nothofagus* forest. The presence of soil charcoal and fire scars on some of the veterans strongly implied that the logged stand had originated from rapid regeneration following a local fire, spreading down from the drier *Eucalyptus*-covered ridges above (Jackson 1968).

C. Red Beech (*Nothofagus fusca*)

William Colenso, an infamous clergyman and one of New Zealand's pioneer plant collectors, is commemorated in the name of a mountain in the central Ruahine ranges of the North Island. On the western slopes of this mountain between 820 and 1112 m the forest canopy is composed almost solely of red beech (Ogden 1971a). Similar extensive monospecific forests occur at mid-altitudes elsewhere in the country.

Red beech can reach diameters of up to 3 m and heights of 30 m and has impressive buttresses. The 'normally attainable age' is probably between 350 and 450 years. Small core samples from trees in different size classes give significant correlations between age and diameter (Ogden 1978b). Mast seeding occurs at intervals of from 3 to 7 years. The seedlings are 'light-demanding', and canopy recruitment appears to be restricted to tree-fall gaps or larger disturbed areas.

On Mount Colenso size–frequency distributions differed between samples, but were often multimodal (Fig. 3). As with *Athrotaxis* in Tasmania, recent mortality, represented by standing dead trees, was concentrated in the size classes immediately below the modes. In the upper stand the mossy remains of former giants provided mute testimony to previous wind-throw, although the structure and mortality pattern of the new generation implied a competitive hieararchy.

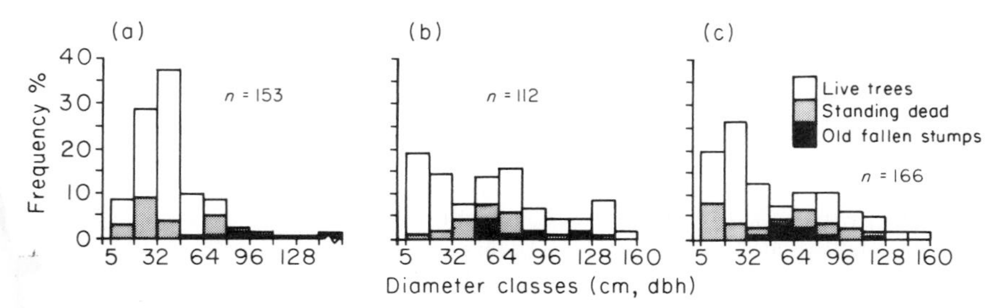

Fig. 3. Size–frequency distribution of red beech in Mount Colenso. a, 7112 m. b, 990 m. c, 838 m. Stippling indicates standing dead trees; black, fallen dead. (Redrawn from Ogden 1971b.)

Following a mast year (1971) the germinating density was estimated at 479×10^3 ha^{-1}. The majority of these seedlings was beneath the dense, shoulder-high cover of the fern *Dicksonia lanata,* which forms extensive clones covering much of the forest floor. In this situation light levels are usually below compensation point, at 1–4% relative light intensity (June 1974). The half-lives of seedling populations in different micro-sites were estimated from marked individuals of the 1971 cohort, and older seedlings, recounted at monthly intervals (Fig. 4). Mast seeding, occurring at intervals of less than 10 years, coupled with the measured survival rates, appears quite capable of maintaining the favourable log micro-sites permanently stocked with seedlings (June & Ogden 1975). The seedling populations under ferns, on the other hand, are ephemeral; seedlings on bare or moss-covered mineral soil have an intermediate life-expectancy. Observations of older seedling populations on logs and soil confirmed the differential survivorship. Established seedlings on logs had a half-life of 10.5 years, and ring counts showed that they sometimes reached 25 years while still less than 50 cm tall.

The presence of large canopy gaps, abundant fallen logs and few saplings contributed to an impression of a population in decline. Using established seedling numbers as the input, however, and the measured rates of growth and survival in different size classes, population growth rates (λ) close to 1 were obtained for all stands (June & Ogden 1978; Enright & Ogden 1979).

Logs may be available for colonisation by beech seedlings for 200 years (June 1974). Only those in canopy gaps receive relative light intensities high enough (35%) for rapid growth by red beech. Thus recruitment from safe sites on the forest floor to maturity in the canopy requires the coincidence of two mosaics: safe sites and light gaps. If these co-occur long enough (50 years?), thinning

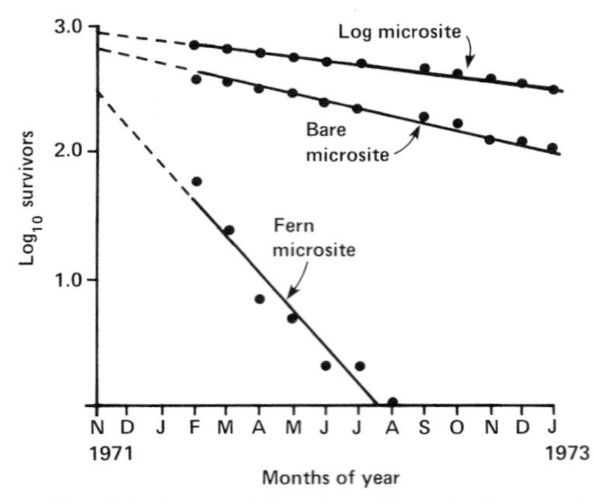

Fig. 4. Survivorship of first-year red beech seedlings in plots on three microsites. (Redrawn from June & Ogden 1975.)

occurs in the densely stocked seedling patch and a new canopy tree rises rooted in the remains of a previous generation. The whole process apparently occurs in roughly synchronous 'waves' following disturbance, but the time scale is such that many canopy openings are not fully healed before the next disturbance.

D. Mixed Beech Forests (*Nothofagus fusca, N. solandri, N. menziesii*)

The Rangataua lava flow is a forest wilderness area in the Tongariro National Park in the centre of the North Island of New Zealand. The flow is Pleistocene in age, and extends from near timberline (c. 1250 m), here formed by almost pure stands of stunted mountain beech (*N. solandri* var. *cliffortioides*) to c. 750 m, where it is clothed by mixed beech–podocarp forest reaching 40 m in height.

Six 1-ha stands were chosen at different altitudes. In each, 200 trees >10 cm diameter at breast height (dbh) were sampled by using modified point-centred-quarter (pcq) methods (Ogden & Powell 1979). These data provided basic stand descriptions—composition, density and basal area. Combinations of the three beech species formed the bulk of the canopy at all altitudes. As on Mount Colenso standing dead trees comprised an impressive proportion (13–33%) of the total basal area.

For each tree in the pcq sample we recorded a potential replacement, defined rather subjectively as the closest or most vigorous sapling within 5 m of its base. Also, whenever possible, we recorded the sapling composition of canopy gaps. These data allowed the construction of species replacement matrices for each stand (Horn 1975; Enright & Ogden 1979). When the predicted proportions of the three beech species are compared to those currently present the aggressive potential of silver beech (*N. menziesii*) is apparent (Fig. 5). In contrast, red beech (*N. fusca*) seems destined to decline in all stands. Moreover, the rate of red beech

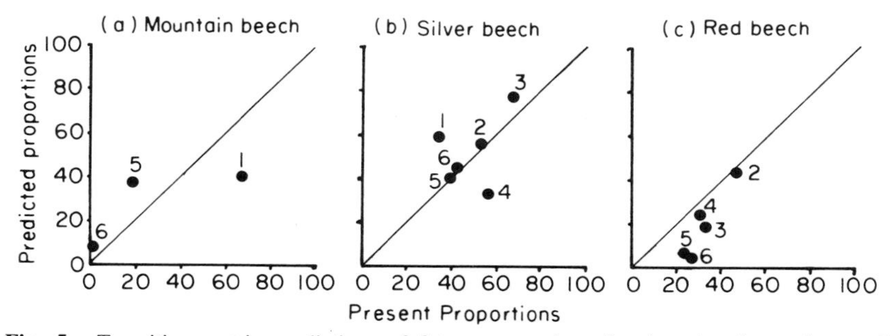

Fig. 5. Transition matrix predictions of future proportions for three beech species on the Rangataua lava flow. a, Mountain beech. b, Silver beech. c, Red beech. Points above the 45° line imply an increase, points below it a decrease. Numbers indicate stands as follows: 1, 1200 m; 2, 1130 m; 3, 975 m; 4, 885 m; 5, 800 m; 6, 690 m. (Original data.)

decline is inversely related to its current basal area and density; the less red beech present, the more immediate does its demise appear.

The results illustrate some of the problems of interpreting matrices derived from data on the relative distribution patterns of canopy trees and their sub-canopy 'replacements'. No account has been taken of the differential growth rates and survivorships of the different beech species under gap and shade conditions. Abundance in the sub-canopy may not fairly reflect the potential of a species to capture a gap. Moreover, if the potential canopy trees among the juveniles of the red beech population are generally situated in large canopy gaps, then taking an arbitrary radius of 5 m from each sample tree may bias the replacements in favour of the more shade-tolerant silver beech. However, our inventory of the gaps did not reveal hoards of red beech saplings, so we are left with the predicted decline for this species.

In each stand a minimum of 10 trees of each species were cored and several seedlings and saplings sacrificed to provide some initial height-growth data. That the growth rate of red beech was least where the transition matrix predicted greatest decline seems to provide some independent verification, reflecting a senile population with few survivors, low growth rates and poor regeneration.

Average growth rates were ranked: mountain > red > silver. In all stands except one, red and mountain beech showed statistically significant ($P < .05$) correlations between diameter and estimated age. Silver beech showed no such correlations; moreover its growth rate had generally greater variance, both within individual trees with time, and between trees in a stand. Throughout these forests silver beech seedlings form small thickets in which mean height-growth rates decline as density increases. On average, seedlings take 23 years to reach 1 m in height, although those surviving to the sapling class are faster-growing. The less frequent red beech seedlings and saplings do not show this difference and average about 15 years to 1 m.

These differences in mature and seedling growth rates and variances provide insights into the behaviour of the two species in mixed stands. Red beech is a gap coloniser and highly dependent on safe seedling microsites. In such situations, height growth is rapid. On the Rangataua lava flow silver beech maintains much more extensive seedling populations. Saplings are recruited from these as small openings in the canopy allow, but growth rates vary considerably, and young trees may stagnate again if the light gap is reduced or closes over. Red beech apparently relies on occasional forest destruction, while silver beech succeeds by taking advantage of illuminated opportunities in the less disturbed forest. This interpretation also holds for the relationship between mountain and silver beech, although here the roles are more clearly differentiated.

At the highest stand studied (1200 m) the canopy was composed entirely of mountain and silver beech. The transition matrix suggested some reciprocity in replacement relationships but predicted a shift towards more silver beech in

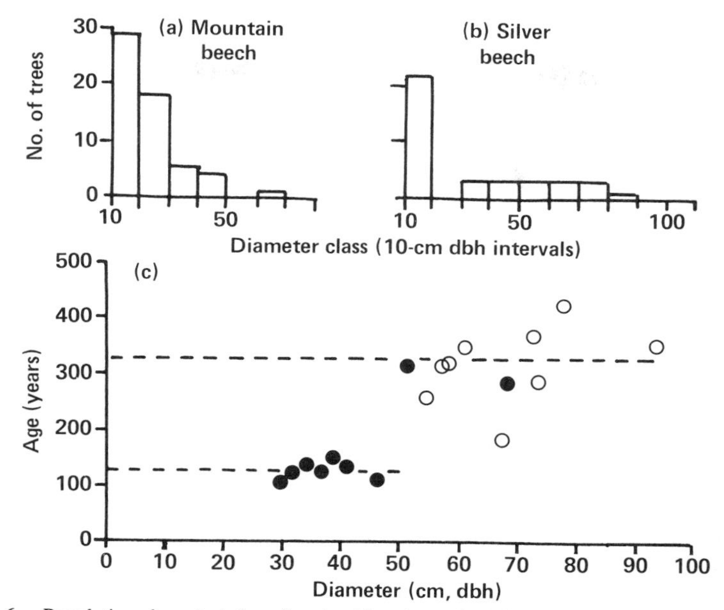

Fig. 6. Population characteristics of a mixed beech stand at 1200 m on the Rangataua lava flow. a, Size–frequency distribution of mountain beech. b, Size–frequency distribution of silver beech. c, Age–diameter relationships of cored trees in larger size classes of each species. Black circles: mountain beech; open circles: silver beech. Dashed lines indicate approximate cohort establishment dates. (Original data.)

future. The size–frequency distributions of the two species suggested that mountain beech was the younger of the two populations, and this was confirmed by the increment cores (Fig. 6). The results indicate that a majority of the mountain beech trees on the site belong to a single cohort with a characteristic skewed structure. This population developed sometime after 1820 within a pre-existing forest, probably composed predominantly of silver beech but with an unknown number of mountain beech, represented now by a few senile trees c. 300 years old, their maximum longevity (Wardle 1970b). Fallen logs and a solitary red beech pole suggest that partial canopy destruction initiated mountain beech 'invasion'. The high proportion of silver beech poles (10–20 cm dbh) is in agreement with the matrix prediction (based on the abundance of saplings) of an increase in this species.

This stand illustrates how a reconstruction of past history can illuminate the predictions of a simplistic transition matrix. The high living basal area of the stand (103 m^2 ha^{-1}) indicates that future thinning of the mountain beech cohort is to be expected (cf. Wardle 1970b). Most of the gaps created will be captured by saplings or poles of the longer-lived silver beech, so that, barring future disaster, the stand will move towards silver beech dominance once again. A similar disturbance-mediated relationship between these two species has been noted by Wardle (1984) in South Island forests. Pure-stand basal areas, relative longevity, shade tolerance and seed production characteristics all support the view that silver beech will dominate undisturbed mixtures.

IV. CONCLUSION

Reconstructions of forest history have been largely responsible for reappraisal of the usefulness of 'climax vegetation' as an organising paradigm in plant ecology. The evidence suggests that, on a realistic sampling scale of a few hectares or less, most forests are not stable in composition or in the demographic characteristics of their constituent species: rather they are in various stages of recovery from past disturbances (reviewed by White 1979). Random sampling strategies suitable for "stand description" are quite inappropriate for a study of the dynamics of a "pre-emptive crazy quilt" (Horn 1981), but a description of this quilt may indeed provide clues to the dynamics of the system (Harper 1977 p. 635).

Several species in the genera *Agathis, Podocarpus, Dacrydium* and *Libocedrus* live for 800–1000 years and have even-aged populations which dominate the site for several centuries without much regeneration, then senesce more or less synchronously. Some of the gaps created may be recolonised, so that secondary blurred waves are created (Veblen & Stewart 1982; Morton *et al.* 1984). The size, frequency and persistence of tree-fall gaps may determine the rate of decline of the local population, but the recurrence time of the rejuvenating disturbance may be the selective force determining longevity and dispersal characteristics.

Thus I present a picture of long-lived species populations which wait or migrate in space and time in the variable environment created by glacial–interglacial oscillations. Such species exceed the time scales of normal "ecology" and force us to retrospective techniques such as palynology and dendrochronology. I suggest that by combining such techniques with predictive modelling a more complete understanding of their dynamics may be reached.

ACKNOWLEDGMENTS

I am grateful to many students, park rangers, Forest Service employees and others who have helped me in the various field programmes alluded to in this paper. They are too numerous to mention by name. The work has been financially supported by Massey University, the Australian National University and the University of Auckland.

REFERENCES

Ahmed, M. (1984). *Ecological and dendrochronological studies on* Agathis australis *Salisb. (kauri).* Ph.D. thesis, University of Auckland, Auckland, New Zealand.

Daubenmire, R. (1968). *Plant Communities: A Textbook of Synecology.* Harper & Row, New York.

Ecroyd, C. E. (1982). Biological flora of New Zealand, 8, *Agathis australis* (D. Don) Lindl. (Araucariaceae) kauri. *New Zealand Journal of Botany,* **20,** 17–36.

Enright, N. J. & Ogden, J. (1979). Applications of transition matrix models in forest dynamics: *Araucaria* in Papua New Guinea and *Nothofagus* in New Zealand. *Australian Journal of Ecology,* **4,** 3–23.

Fritts, H. C. (1976). *Tree Rings and Climate.* Academic Press, London.

Grubb, P. J. (1977). The maintenance of species richness in plant communities. The importance of the regeneration niche. *Biological Reviews of the Cambridge Philosophical Society,* **52,** 107–145.

Harper, J. L. (1977). *Population Biology of Plants.* Academic Press, London.

Harper, J. L. & White, J. (1974). The demography of plants. *Annual Review of Ecology and Systematics,* **5,** 419–463.

Harper, J. L., Williams, J. T. & Sagar, G. R. (1965). The behaviour of seeds in soil. Part 1. The heterogeneity of soil surfaces and its role in determining the establishment of plants from seed. *Journal of Ecology,* **53,** 273–286.

Henry, J. D. & Swan, J. M. A. (1974). Reconstructing forest history from live and dead plant material—an approach to the study of forest succession in southwestern New Hampshire. *Ecology,* **55,** 772–783.

Horn, H. S. (1975). Markovian properties of forest succession. *Ecology and the Evolution of Communities* (Ed. by M. L. Cody & J. M. Diamond), pp. 196–211. Belknap Press, Cambridge, Massachusetts and London.

Horn, H. S. (1981). Some causes of variety in patterns of secondary succession. *Forest Succession: Concepts and Application* (Ed. by D. C. West, H. H. Shugart & D. B. Botkin), pp. 25–35. Springer, New York.

Jackson, W. D. (1968). Fire, air, water and earth—an elemental ecology of Tasmania. *Proceedings of the Ecological Society of Australia,* **3,** 9–16.

June, S. R. (1974). *The germination, growth and survival of red beech seedlings in relation to forest regeneration.* M.Sc. thesis, Massey University, Palmerston North, New Zealand.

June, S. R. & Ogden, J. (1975). Studies on the vegetation of Mount Colenso, New Zealand. 3. The population dynamics of red beech seedlings. *Proceedings of the New Zealand Ecological Society,* **22,** 61–66.

June, S. R. & Ogden, J. (1978). Studies on the vegetation of Mount Colenso, New Zealand. 4. An assessment of the processes of canopy maintenance and regeneration strategy in a red beech (*Nothofagus fusca*) forest. *New Zealand Jounral of Ecology,* **1,** 7–15.

Kirkland, A. (1961). Preliminary notes on seeding and seedlings in red and hard beech forests of north Westland and silvicultural implications. *New Zealand Journal of Forestry,* **3,** 482–497.

LaMarche, V. C., Jr., Holmes, R. L., Dunwiddie, P. W. & Drew, L. G. (1979). *Tree Ring Chronologies of the Southern Hemisphere.* 4. Australia. Chronology Series. V. Laboratory of Tree-Ring Research, University of Arizona, Tuscon.

Lloyd, R. C. (1960). Growth study of regenerated kauri and podocarps in Russell Forest. *New Zealand Journal of Forestry,* **8,** 355–361.

Mirams, R. V. (1957). Aspects of the natural regeneration of kauri (*Agathis australis* Salisb.). *Transactions of the Royal Society of New Zealand,* **84,** 661–680.

Morton, J., Ogden, J., Hughes, T. & MacDonald, I. (1984). *To Save a Forest—Whirinaki.* David Bateman Ltd., Auckland, New Zealand.

Ogden, J. (1971a). Studies on the vegetation of Mount Colenso, New Zealand. 1. The Forest Continuum. *Proceedings of the New Zealand Ecological Society,* **18,** 58–65.

Ogden, J. (1971b). Studies on the vegetation of Mount Colenso, New Zealand. 2. The population dynamics of red beech. *Proceedings of the New Zealand Ecological Society,* **18,** 66–75.

Ogden, J. (1978a). Investigations of the dendrochronology of the genus *Athrotaxis* D. Don (Taxodiaceae) in Tasmania. *Tree-Ring Bulletin,* **38,** 1–13.

Ogden, J. (1978b). On the diameter growth rates of red beech (*Nothofagus fusca*) in different parts of New Zealand. *New Zealand Journal of Ecology,* **1,** 16–18.

Ogden, J. & Powell, J. A. (1979). A quantitative description of the forest vegetation on an altitudinal gradient in the Mount Field National Park, Tasmania, and a discussion of its history and dynamics. *Australian Journal of Ecology,* **4,** 293–325.

Veblen, T. T. & Stewart, G. H. (1982). On the conifer regeneration gap in New Zealand: The dynamics of *Libocedrus bidwillii* stands on South Island. *Journal of Ecology,* **70,** 413–436.

Veblen, T. T., Schlegel, F. M. & Escobar, B. R. (1980). Structure and dynamics of old-growth *Nothofagus* forests in the Valdivian Andes, Chile. *Journal of Ecology,* **68,** 1–31.

Wardle, J. (1970a). The ecology of *Nothofagus solandri.* 3. Regeneration. *New Zealand Journal of Botany,* **8,** 571–608.

Wardle, J. (1970b). The ecology of *Nothofagus solandri.* 4. Growth, and general discussion to parts 1 to 4. *New Zealand Journal of Botany,* **8,** 609–646.

Wardle, J. A. (1984). *The New Zealand Beeches. Ecology, Utilisation and Management.* The New Zealand Forest Service, Christchurch, New Zealand.

White, J. (1980). Demographic factors in populations of plants. *Demography and Evolution in Plant Populations* (Ed. by O. T. Solbrig), pp. 21–48. Blackwell Scientific Publications, Oxford.

White, P. S. (1979). Pattern, process, and natural disturbance in vegetation. *Botanical Review,* **45,** 229–299.

2

Plant Demography: A Community-level Interpretation

José Sarukhán

Daniel Piñero

Departamento de Ecología
Instituto de Biología
Universidad Nacional Autonoma de México
Mexico City, Mexico

Miguel Martínez-Ramos[1]

Estación de Biología Tropical Los Tuxtlas
Instituto de Biología
Universidad Nacional Autónoma de Méxio
San Andrés Tuxtla, Verracruz, Mexico

I. INTRODUCTION

As with many other approaches which focus only on parts of the population biology of organisms, demography has been sometimes considered as an end in itself in the study of plant populations. Besides describing and explaining populational behaviour of plants, demographic studies must constitute a dual avenue to study phenomena at two levels of organization: at the individual (or subpopulational) and at the community (or supra-populational) levels. Long-term demographic studies may be used to spot and interpret individual-level variability concerning those components of individual Darwinian fitness such as differential survivorship, growth and fecundity (Sarukhán, Martínez-Ramos & Piñero 1984). Typically plant size (especially when it estimates closely the amount of photosynthetic or reserve tissue) is positively correlated with survivorship, growth and fecundity, rather independently of age. When individuals of the same age encounter a patchier environment, the variance of individual behaviour increases correspondingly (Sarukhán, Martínez-Ramos & Piñero

[1]Present address: Departamento de Ecología, Instituto de Biología, Universidad Nacional Autonoma de México, 0410 Mexico City, Mexico.

STUDIES ON PLANT DEMOGRAPHY:
A FESTSCHRIFT FOR JOHN L. HARPER

1984). The studies that render themselves suitable to such analyses show that individual-level variability has important implications, through genetic hierarchy, on the genetic structure of the population, at least in the short term (Sarukhán, Martínez-Ramos & Piñero 1984; Bullock 1982; Piñero & Sarukhán 1982). It is impossible at the moment, with the available information on individual-level variability of demographic parameters, to find any discernible patterns by which such variation in plant populations may be attributed to genetic or environmental factors. A better understanding of what constitutes the relevant environment for individuals of a population is needed to help determine the causes of individual variation. This is particularly complicated when individuals may sample differences in competitive interactions with neighbours and when as a result of growth they also sample different parts of the environment of the community they live in.

The aim of this essay is to discuss some of the community-level consequences that may be derived from demographic information about one component of the community and so gain insight into some of its synecological processes (in the sense of Anglo-American ecologists such as Whittaker 1978). This second avenue of application of actuarial information about a plant population has been seldom explored, and few examples exist that give to synecological studies a dimension of populations and individuals.

II. PLANT DEMOGRAPHY AND THE STUDY OF MOSAIC REGENERATION IN TROPICAL FORESTS

Forest regeneration in the tropics occurs in gaps formed by tree-falls, by the filling or 'healing' of the gaps through succession of different communities of species, both pioneer, colonizing taxa and light-demanding, suppressed, late secondary or primary forest species. Many authors (Sarukhán 1968; Whitmore 1974, 1975; Hartshorn 1978, 1980; Denslow 1980; Brokaw 1982) have described the importance which secondary succession and the gap-forming processes in the tropical forests have on the composition and dynamics of the forest as well as on its diversity. Gap-forming tree-falls are caused usually during storms (cyclones, hurricanes) when combined conditions of water saturation of the soil and very strong winds bring down the most exposed trees in the forest canopy, forming cleared areas of up to 600–700 m^2. Gaps can of course be smaller, if only a small tree or the large limbs of a tree fall.

A mosaic of different regenerative phases thus comprises what is often colloquially spoken of as 'virgin' or untouched tropical forest. Various descriptive studies of such patchwork of vegetation have been carried out (e.g. Oldeman 1978).

Lang and Knight (1983) have summarized the problems that beset the study of the dynamics of tropical forests, specifically of the late successionary stages; there are generally poor records on the history of forest disturbance, there is a widespread lack of time-recording structures in tropical plants, such as annual growth rings, and there is a paucity of long-term observations of forest dynamics. Of these, the first two may be solved, at least partially, by information and knowledge coming from studies at the population level. Our research on the detailed population dynamics of *Astrocaryum mexicanum* has allowed us to bridge the levels of population and community studies. The capacity to record the passage of time rather accurately, by means of the scars left by successive leaf cohorts on the trunk of the palm, has been discussed in detail (Sarukhán 1978, 1980; Piñero, Martínez-Ramos & Sarukhán 1984) and provides a vital time axis not only for demographic purposes, but also to date environmental events that affect the community in which the palm lives.

In addition to recording the passage of time, a special feature of the behaviour of the palm has proved to be an asset in allowing the understanding of forest dynamics. *Astrocaryum mexicanum* palms possess a single erect trunk which never produces suckers since it has one, terminal leaf bud. Severe damage to this will result in the death of the individual. The main source of mortality of mature palms (over 1.5 m tall) is an accidental hit by large, falling branches and entire trees blown over in the process of gap formation in the forest. However, when the hits by large falling objects are not direct or do not kill the terminal bud, palms are very often bent downwards, and frequently all the way to the forest floor without becoming uprooted, since they have an excellent anchoring root system. On average, over a quarter of all the adult individuals in the forests we have studied (corresponding to c. 360 palms per hectare) have been hit and bent at least once during their lifetime.

If the palm survives the hit and the bending, it will do two interesting things (see Fig. 1). First, the terminal meristem will start growing upwards, producing a well-defined kink that remains permanently in the trunk. Second, if the palm has been totally bent, it will normally change the morphology of its adult leaves to that of an immature individual and cease reproduction if it already was a reproductive adult. Very often surviving palms lose much of their leaf crown; this affects for some years their growth and reproduction. Mendoza (1981) has experimentally produced defoliation and observed the resulting effects on survivorship, growth and reproduction in palms of all ages.

Of these two behaviours, the one relevant for the present discussion is the turning upwards of the bent trunk within a year or two of the accidental event. By counting the leaf scars below the kink (l_1 in Fig. 1) we can estimate the age at which an individual was hit and brought downwards, and by estimating the age above the kink (l_2 in Fig. 1), we know how long ago was the palm brought down. The presence, in one spot of the forest, of several palms bent at the same time is considered as a strong suggestion that a disturbing event, severe enough as to have formed a gap in the above canopy, has occurred.

Besides the permanent observation plots, where most of the studies on the demography of *A. mexicanum* have been carried out at Los Tuxtlas (Piñero,

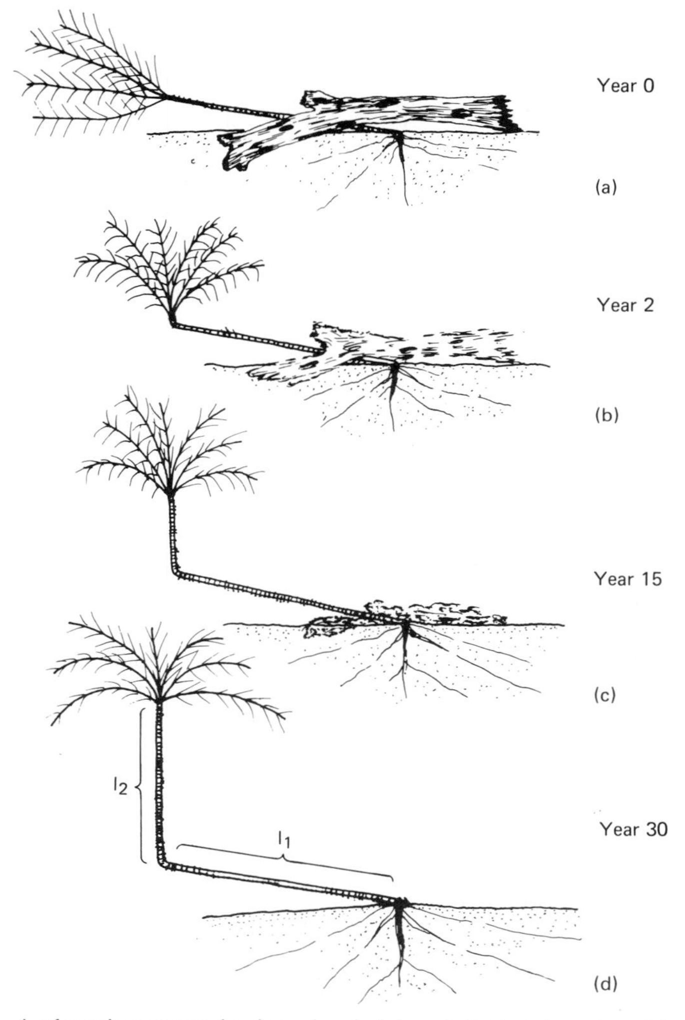

Fig. 1. A schematic sequence showing a, knocked-down individual; b, up-turn of the trunk 1–2 years afterwards; c, growth of the up-turned portion of the trunk; and d, the usual form in which bent-down palms are found in the forest. See text for explanation of l_1 and l_2.

Sarukhán & González 1977; Sarukhán 1978; Pĩnero, Martínez-Ramos & Sarukhán 1984), a 5-hectare permanent forest plot has been set up where all trees 1 cm diameter at breast height (dbh) and larger have been recorded in a detailed map of the site, measuring their girth, height, crown cover and crown depth. These include all *A. mexicanum* adult individuals, whether they are erect or bent. A total of approximately 3482 adult individuals (≥1 m in trunk height) of the palm have been recorded and measured in the 5-ha plot, of which c. 1800 show a distinctively bent trunk. Several important pieces of information are derived from these observations.

The first is that sizable disruptions of the forest canopy, sufficient to produce light gaps of differing sizes, occur regularly in time and space. Because of the frequency with which *A. mexicanum* palms occur in the forest and their longevity, we can estimate for each spot of the forest where several individuals have been bent, how long ago it was last subjected to a gap-forming event. The 5-ha site has been divided in a 5 × 5-m grid; for each 25-m^2 subsite all the palms showing a bent trunk have been selected, counting how many years back each individual was brought down. The modal frequency of the ages of all the upturned trunks represent the age after disturbance for that subsite. A map of the mosaics of gap stages of different age, from recently opened gaps to very old successional sites, is then made (M. Martínez-Ramos, J. Sarukhán & E. Alvarez-Buylla, unpublished). A considerable degree of heterogeneity at a very small scale (25 m^2) is apparent in the mosaic of the forest (Fig. 2). We have also observed at the site a gradient of younger to older stages from an exposed forest edge to the central, more protected part of it. This is also true for the steepest slopes within the forest, which support a more disturbance-prone mosaic of younger, successional gap phases.

A second aspect is that we can detect periods at which the forest as a whole

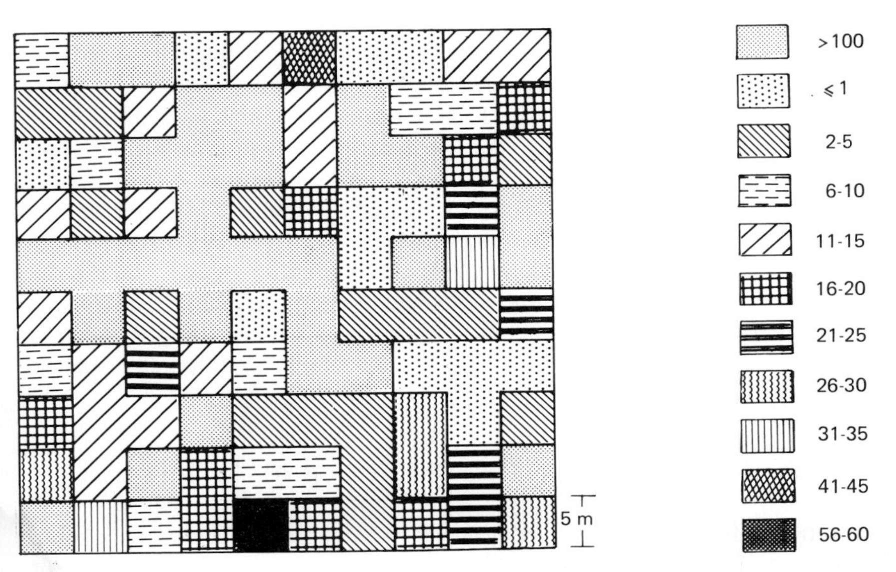

Fig. 2. A portion of 2500 m^2 (50 × 50 m) of the 5-ha permanent plot of observation at the tropical rain forest in Los Tuxtlas, Veracruz, Mexico. Each 5 × 5-m subsite shows the age of that particular spot of the forest. The most frequent forest ages are over 100 years, 29%; 2–5 years, 15%; and 11–15 years, 14%.

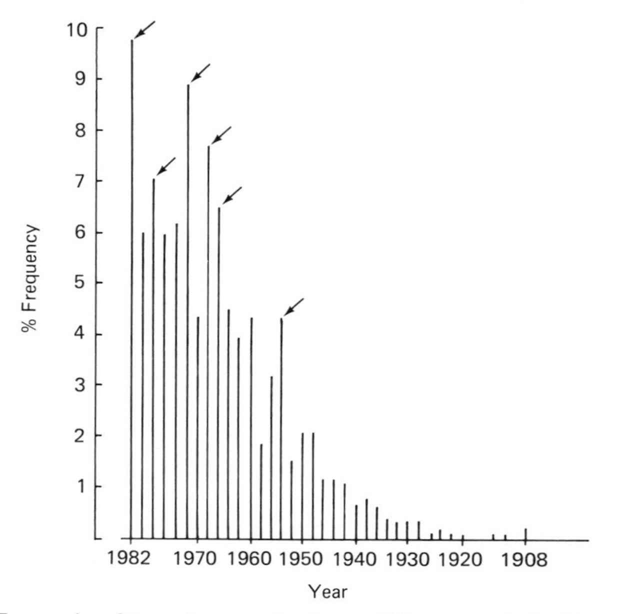

Fig. 3. Frequencies of *A. mexicanum* palms bent at different years in the 5-ha permanent plot at the tropical rain forest at Los Tuxtlas, Veracruz, Mexico. Arrows point at years of extensive perturbation by tree falls. $N = 1800$ individuals ≥ 1 m tall.

was subjected to unusually high rates of gap formation. In Fig. 3, periods of severe perturbation can be spotted at 2, 12, 16–18 and maybe 24 and 30 years prior to the recording date in 1982. Because of the dwindling numbers of older categories of palms, it is difficult to detect clearly generalized events of perturbation for all the forest earlier than 1950.

Besides using *A. mexicanum* palms as 'recording devices' of important environmental events in the regenerative processes of the tropical rain forest, it has been possible to relate the structure and floristic composition of every spot of the forest by means of correlating these to the age of each gap phase in the forest, down to a scale of some 100 m^2. In consequence, we are able to assemble together isolated 'frames' of successive ages of spatially and temporally isolated events of gap formation and regeneration for the same forest.

III. DETERMINATION OF FOREST TURNOVER RATES

The possibility of using the detailed understanding of a species' population dynamics to interpret events at the synecological level has proved to be of importance. The use of *A. mexicanum* palms as indicators of time of disturbance has allowed us to assess with great precision the turnover rates for different parts of the forest; thus, we calculate for the tropical rain forest at Los Tuxtlas a turnover rate of 144 years for the more stable portions and of only 24 years for the forest edges. The rates for the stable portions differ very little from the values found for forests in Barro Colorado, Panama (137 years, Lang & Knight 1983), and La Selva, Costa Rica (135 years, Hartshorn 1978). Reviews on the subject by Martínez-Ramos (1985) and Brokaw (1985) coincide fairly closely with the former estimates. The rates for the forest edges bear importance from the conservation point of view, since long edges in forest reserves will mean large areas which are very susceptible to being naturally disturbed with greater frequency. Whether this high rate of turnover can be buffered and contained within a certain width of the forest, or whether it has a considerable domino effect, is unknown. Floristic composition in gap phases at the forest edge could be considerably changed by the replacement of many long-lived upper and mid-storey species by shorter-lived, high-turnover, pioneer species.

IV. THE STUDY OF THE DYNAMICS OF FOREST STRUCTURE AND COMPOSITION

A logical consequence of being able to age the patches composing a forest mosaic is to achieve a fairly rapid understanding of many aspects of the structural dynamics and floristic composition of the forest. So far, most descriptions of the dynamics have been based on somewhat vague estimations of age of forest phases or on information from distinct forest plots, distant in either space or time, or both.

The distributions of diameter classes for two different sets of species (pioneers and upper-canopy 'nomads') in three different gap phases, aged by means of upturned trunks, of *A. mexicanum* are shown in Fig. 4. The first group of species require light gaps for their seeds to germinate and include *Cecropia obtusifolia, Heliocarpus appendiculatus, Bellotia campbellii, Trema micrantha, Carica papaya* and *Cnidoscolus multilobus*. They show a transition from a reverse J-shaped distribution of diameter classes (gaps 1–10 years old) to one in which small-diameter classes are almost absent (20- to 30-year-old gaps). Of course, a given gap may experience one or more subsequent disturbances which renew parts of the gap; this results in the presence of pioneers of a size (or age) that would not be expected had the gap undergone only one sucessional sequence; this, in fact, is the case for Fig. 4c. The second group of species are those which will not germinate in light gaps, but their seedlings and saplings will be 're-leased' from a state of light suppression and will grow very actively, sometimes dominating a portion of the gap with saplings and small trees. The data in Fig. 4d, e and f show a clear tendency for greater representation of larger-diameter classes of this group of species as the succession proceeds.

An immediate consequence of the preceding is the generation of tree growth curves for each species as a function of age. A schematic representation for such

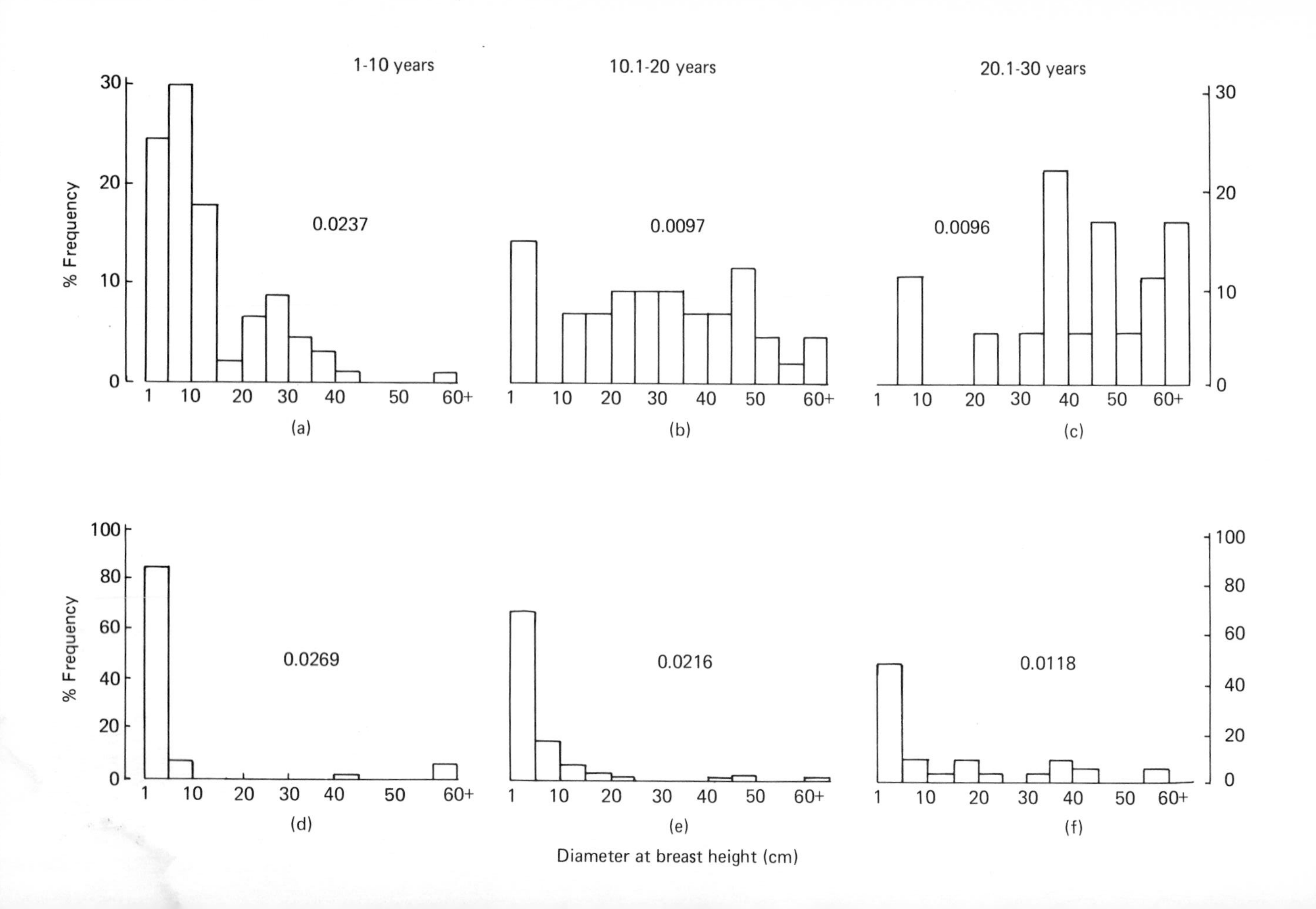

1-10 years
10.1-20 years
20.1-30 years
% Frequency
0.0237
0.0097
0.0096
(a)
(b)
(c)
0.0269
0.0216
0.0118
(d)
(e)
(f)
Diameter at breast height (cm)

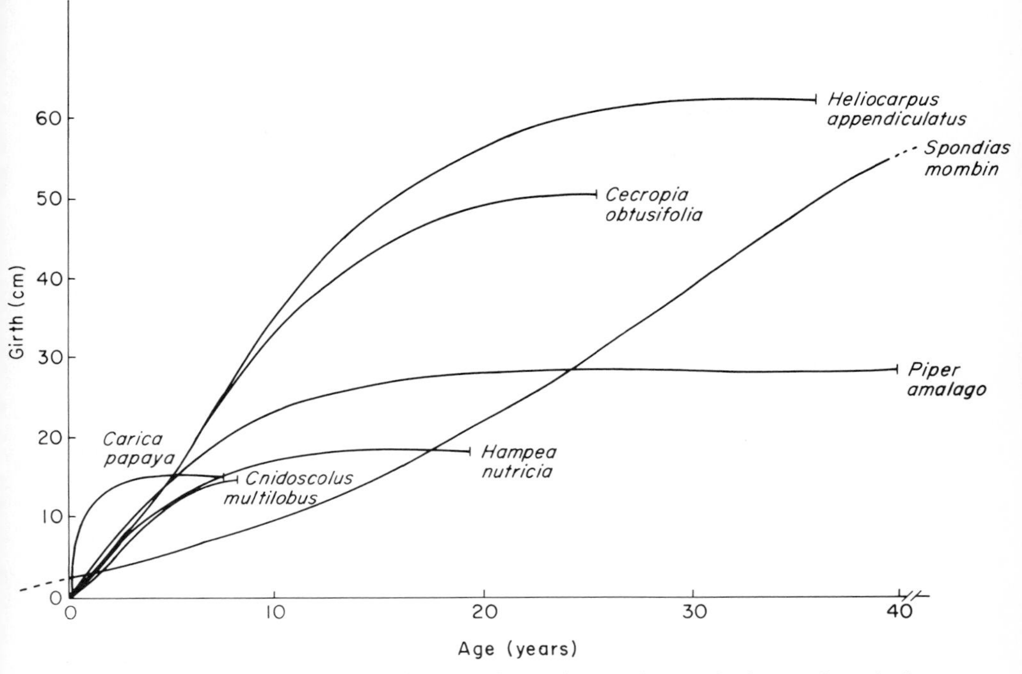

Fig. 5. Schematic growth curves for selected secondary species growing in gap phases in the tropical rain forest at Los Tuxtlas, Veracruz, Mexico. *Spondias mombin* was recorded with already established individuals in the newly opened gap.

curves obtained by M. Martínez-Ramos (unpublished) is shown in Fig. 5 for some of the light-demanding pioneer and nomad species at Los Tuxtlas, based on measurements of individuals present in gaps of different ages, using *A. mexicanum* individuals as estimators of gap age. *Spondias mombin* (a nomad), which started with already established individuals, *Heliocarpus appendiculatus* and *Piper amalago,* appear as the longest-lived, whereas *Carica papaya* and *Cnidoscolus multilobus* have maximum life span of 9 years and *Hampea nutricia* extends its life to nearly 16 years. This method provides very detailed information of growth rates and life spans for many of the important species participating

---

Fig. 4. Distributions of diameter classes for groups of pioneer (top row) and upper-canopy nomad (lower row) species in three different gap phases (1–10, 10.1–20 and 20.1–30 years) present in the tropical rain forest at Los Tuxtlas, Veracruz, Mexico. Pioneer species: *Cecropia obtusifolia* (a, b, c), *Heliocarpus appendiculatus* (a,b,c), *Bellotia campbellii* (a,b,c), *Trema micrantha* (a,b,c), *Carica papaya* (a) and *Cnidoscolus multilobus* (a). Nomad species: *Spondias mombin* (d,e,f), *Omphalea oleifera* (d,e,f), *Sapium lateriflorum* (d,e,f), *Bernoullia flammea* (f), *Bursera simaruba* (d,e,f) and *Robinsonella mirandae* (d,e,f). Letters in brackets indicate which species are involved in each histogram. Population density (m^{-2}) is indicated above each histogram.

in the regeneration of light gaps, without having to follow regenerative processes for long periods of time.

V. THE INFLUENCE OF *ASTROCARYUM MEXICANUM* ON FOREST DIVERSITY

The high diversity of tropical rain forests is probably one of the main biological features of this type of vegetation; woody (particularly arboreal) species are usually represented by only a few individuals per hectare, many of them by even lower densities. In certain forests, however, where this situation holds true for arboreal species, some species, not normally counted as trees either because of their form or their relatively small girths, may be exceedingly abundant and may play an important role in influencing the structure and composition of their communities. They may do this by exerting great influence on the environmental conditions of the forest floor, where all future recruitments for all the species of the forest are established. One of the most obvious influences is through the modification of light regimes, which in turn determine the germination behaviour (Vázquez-Yanes 1976; Valio & Joly 1979; Vázquez-Yanes & Orozco-Segovia 1982; Vázquez-Yanes & Smith 1982; Aminuddin and Ng 1982). Such is the case for *A. mexicanum* in the tropical rain forest of Los Tuxtlas, and indeed in much of its area of distribution in tropical Mexico. There is a marked negative relationship between the abundance of *A. mexicanum* palms and the diversity of the specific patch of forest in which it may be found. Figure 6 shows such a relationship, in which total leaf cover of the palm is compared to the number of the other species present in the forest. This dominating role of a species such as *A. mexicanum* is confirmed by the fact that in certain areas of the forest where *A. mexicanum* is not abundant, another species of very similar size (*Faramea occidentalis,* Rubiaceae) fills the niche left by the palm and attains similar values of importance (in terms of number of individuals, basal area and crown cover). If one combines values of these two species to the diversity of forest patches, the negative correlation becomes even stronger (Martínez-Ramos 1980). Palms and other small trees typical of and restricted to the understorey (up to 7–9 m tall) may be playing a controlling role in certain tropical rain forests. This is not fully realized, since normally the attention of plant ecologists and foresters is caught by the large canopy and emergent trees.

An additional evidence of the important role played by *Astrocaryum mexicanum* in the community it lives in is its effect on both the pattern of distribution of its own recruitments in the forest and the probability of survivorship of seedlings and juvenile palms. The permanent plots, where demographic observations on populations of this tropical palm are being carried out, represent different densities of palms (Piñero and Sarukhán 1982), an equivalent of 1100–2800 individuals per hectare.

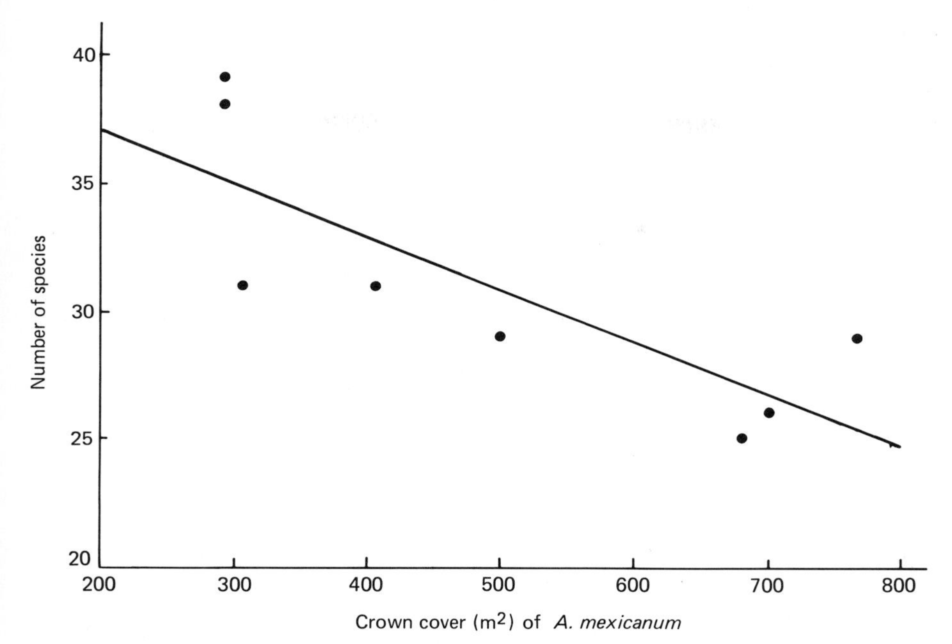

Fig. 6. The relationship between the importance (estimated as crown cover) of *Astrocaryum mexicanum* in a site and the number of species with individuals $\geq$1 cm dbh present at the same site in the tropical rain forest at Los Tuxtlas, Veracruz, Mexico ($p < .005$, $r = .8147$).

One of the first interesting observations about distribution patterns is that they change markedly with age from a highly aggregated one at the seedling stage in high-density sites to a far less aggregated and almost regular distribution for the reproducing matures in low-density sites (D. Piñero, unpublished data). But of greater relevance is the detailed analysis of how the density of *A. mexicanum* plays an important role in defining not only pattern, but also, and very precisely, spatial fixation of individual leaf crowns and of new recruitment to the population. As described elsewhere (Piñero, Sarukhán & González 1977), exact distribution maps of every palm individual in each of six 600-m^2 permanent sites have allowed us to make detailed analyses of individual distributions. Populations of *A. mexicanum* establish a strict hierarchical distribution, more easily detectable in high-density sites. If the leaf-crown sizes of all the mature individuals are projected into the forest floor, two features become obvious. Crown projections of the mature individuals very seldom superimpose, so that in very high densities, one obtains a crowding of circles of different sizes, touching the neighbouring ones only tangentially, but never overlapping, despite the fact that crowns of palms of different age may be placed at different heights above the forest floor (see Fig. 7), sometimes up to 6–7 m in difference. It is as if each mature palm cast an invisible, untrespassable vertical cylinder, producing together something like the tubing of an enormous pipe organ.

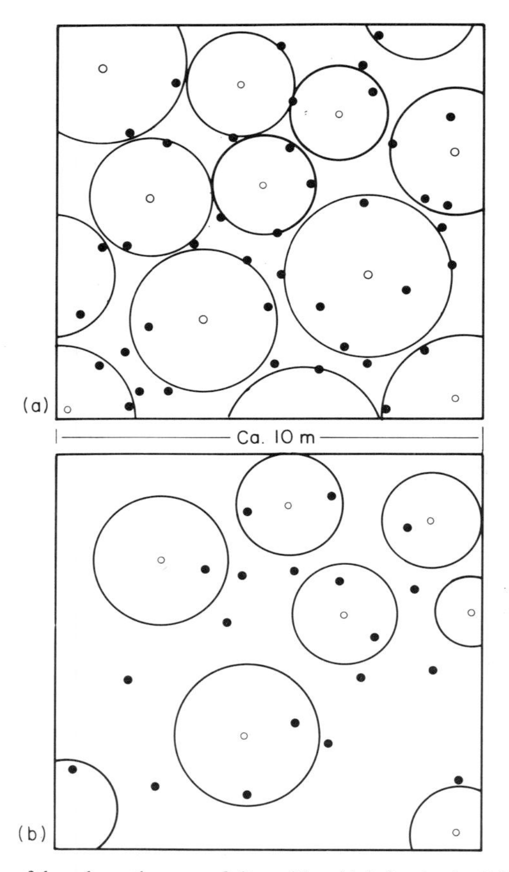

Fig. 7. Portions of the schematic maps of sites with a, high density (= 2,800 ind. ha^{-1}); b, low density (= 1100 ind. ha^{-1}) of *A. mexicanum* in the tropical rain forest at Los Tuxtlas, Veracruz, Mexico. Large circles represent the projection on the ground of the leaf crown of mature individuals (over 1.5 m tall) of *A. mexicanum;* small circles mark the position of the trunk. Black dots mark the position of each seedling and juvenile individual of the palm.

The positions and consequently the chances of survivorship of seedlings and juveniles, are strictly determined by the crown projections of their mature conspecifics. An analysis of the location of these two younger categories of plants in sites of high density (Fig. 7a) shows that there is a probability of .85 that a seedling or juvenile of *A. mexicanum* will occur within a band 0.5 m at each side of the circular projection of the crown of mature individuals. This constriction in the location of seedlings and juveniles is loosened or even lost in sites of lower density of mature palms (Fig. 7b), where the probabilities for an individual of these two categories of appearing within or outside the 1-m band around the crown projection are no more than those expected by chance.

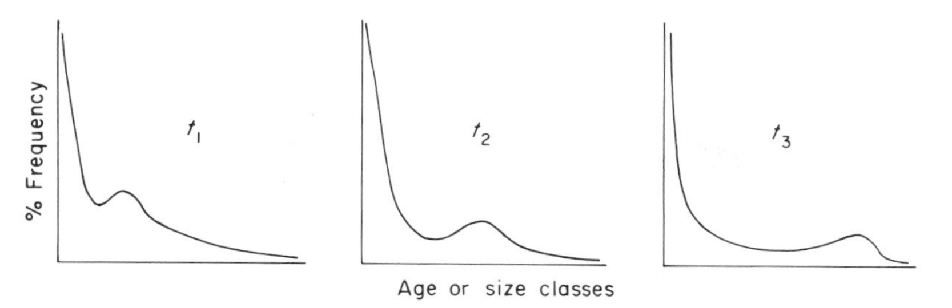

Fig. 8. The progression of a single demic wave caused by a gap formation influencing demographic components such as fruit production, seed and seedling survivorship and growth rate of seedlings and juvenile palms. There may be more than one hump simultaneously in a given population of *A. mexicanum*.

Quite clearly, it is not only the density of mature individuals of *A. mexicanum* that may affect the distribution of its own recruitment, and indeed of seedlings and saplings of many other species, but also the age structure of its mature individuals. Sequences of different age structures have been observed for *A. mexicanum,* having originated mostly as a result of the opening of large gaps in the forest; these gaps usually affect groups in the vicinity of 100 mature palm individuals and their seedling and juvenile phases, creating, by virtue of differential survivorship and transition probabilities, a so-called demic wave that displaces itself along time in the population until it disappears by the death of the last individuals belonging to it. Figure 8 is a schematic representation of one such 'demic wave' sequence, based on data obtained at our stable forest and gap-phase sites at Los Tuxtlas, Veracruz. The increased relative densities of varying age classes determine the greater abundance of individuals with leaf crowns placed at certain heights, and therefore exerting different spatial influences from one spot of the forest to the next and also along time for the same area.

VI. CONCLUDING REMARKS

John L. Harper started his Presidential Address to the British Ecological Society by pointing out the contrast between the approach of the 'vegetationalists', with their concern to describe and interpret areas of land, and Darwin's preoccupation with numbers, when he asked questions about the plants and vegetation in his back garden (Harper 1967). This depicted the neglect by many plant ecologists of dynamic, short-term changes in populations, and of plant demography, which by that time had not gained its present momentum. Now, 18 years later, we may ask ourselves if we can make an effort to link population ecology (and plant demography) with the vegetational approach that (in the words of Harper) had been brought to adulthood by Schimper and Warming. This would indeed be one very fruitful closing of a generation gap!

ACKNOWLEDGMENTS

There are events in the academic life which influence permanently much of what one achieves as a scientist; sometimes, this influence transcends the individual and affects a whole group. The stay of one of us (J. S.) at the School of Plant Biology, University College of North Wales, under John L. Harper's guidance had such an effect. We, and much of Mexican plant ecology, are in great and permanent debt to him.

Parts of the research reported here have been supported by various grants from CONACYT, Mexico.

REFERENCES

Aminuddin, M. & Ng, F. S. P. (1982). Influence of light on germination of *Pinus caribaea, Gmelina arborea, Sapium baccatum* and *Vitex pinnata. Malaysian Forester,* **45,** 62–68.

Brokaw, N. V. L. (1982). Treefalls: Frequency, timing, and consequences. *The Ecology of a Tropical Forest* (Ed. by E. G. Leigh Jr., A. S. Rand & D. M. Windsor), pp. 101–108. Smithsonian Institution Press, Washington, D.C.

Brokaw, N. V. L. (1985). Treefalls, regrowth, and community structure in tropical forests. *Natural Disturbance: An Evolutionary Perspective* (Ed. by S. T. A. Pickett & P. S. White), pp. 53–69. Academic Press, New York.

Bullock, S. H. (1982). Population structure and reproduction in the Neotropical tree *Compsoneura sprucei. Oecologia,* **55,** 238–242.

Denslow, J. S. (1980). Patterns of plant species diversity during succession under different disturbance regimes. *Oecologia,* **46,** 18–21.

Hartshorn, G. S. (1978). Treefalls and tropical forest dynamics. *Tropical trees as Living Systems* (Ed. by P. B. Tomlinson & M. Zimmermann), pp. 617–638. Cambridge University Press, London.

Hartshorn, G. S. (1980). Neotropical forest dynamics. *Biotropica,* **2,** Suppl., 23–30.

Lang, G. E. & Knight, D. H. (1983). Tree growth, mortality, recruitment, and canopy gap formation during a 10-year period in a tropical moist forest. *Ecology,* **64,** 1075–1080.

Martínez-Ramos, M. (1980). *Aspectos sinecológícos del proceso de renovación natural de una selva alta perennifolia.* B.Sc. thesis, Facultad de Ciencias, U.N.A.M. México.

Martínez-Ramos, M. (1985). Claros, ciclos vitales de los árboles tropicales y la regeneración natural de las selvas altas perennifolias. *Regeneración de Selvas.* II (Ed. by A. Gómez-Pompa & S. del Amo), pp. 191–239. Ed. Alhalambra, Mexicana, Mexico.

Mendoza, A. (1981). *Modificaciones del equilibrio foliar y sus efectos en el comportamiento reproductivo y vegetativo en* Astrocaryum mexicanum. B.Sc. thesis, Facultad de Ciencias, U.N.A. M. México.

Oldeman, R. A. A. (1978). Architecture and energy exchange of dicotyledonous trees in the forest. *Tropical Trees as Living Systems* (Ed. by P. B. Tomlinson & M. Zimmermann), pp. 535–560. Cambridge University Press, London.

Piñero, D. & Sarukhán, J. (1982). Reproductive behaviour and its individual variability in a tropical palm. *Astrocaryum mexicanum. Journal of Ecology,* **70,** 461–472.

Piñero, D., Sarukhán, J. & González, E. (1977). Estudios demográficos en plantas: *Astrocaryum mexicanum* Liebm. I. Estructura de las poblaciones. *Boletín de la Sociedad Botánica de México,* **37,** 69–118.

Piñero, D., Martínez-Ramos, M. & Sarukhán, J. (1984). A population model of *Astrocaryum mexicanum* and a sensitivity analysis of its finite rate of increase. *Journal of Ecology,* **72,** 977–991.

Sarukhán, J. (1968). *Análisis sinecológico de las selvas de* Terminalia amazonia *en la planicie costera del Golfo de México.* M.Sc. thesis, Colegio de Postgraduados, Chapingo, México.

Sarukhán, J. (1978). Demography of tropical trees. *Tropical Trees as Living Systems* (Ed. by P. B. Tomlinson & M. Zimmermann), pp. 163–186. Cambridge University Press, London.

Sarukhán, J. (1980). Demographic problems in tropical systems. *Demography and Evolution in Plant Populations* (Ed. by O. T. Solbrig), pp. 168–188. University of California Press, Berkeley.

Sarukhán, J., Martínez-Ramos, M. & Piñero, D. (1984). The analysis of demographic variability at the individual level and its populational consequences. *Perspectives on Plant Population Ecology* (Ed. by R. Dirzo & J. Sarukhán), pp. 83–106. Sinauer, Associates, Sunderland, Massachusetts.

Valio, I. F. M. & Joly, C. A. (1979). Light sensitivity of the seeds on the distribution of *Cecropia glaziovi* Snethlage (Moraceae). *Zeitschrift für Pflanzenphysiologie,* **91,** 371–376.

Vázquez-Yanes, C. (1976). Seed dormancy and germination in secondary vegetation tropical plants: the role of light. *Comparative Physiology and Ecology,* **1,** 30–34.

Vázquez-Yanes, C. & Orozco-Segovia, A. (1982). Seed germination of a tropical rain forest pioneer tree (*Heliocarpus donnell-smithii*) in response to diurnal fluctuations of temperature. *Physiologia Plantarum,* **56,** 295–298.

Vázquez-Yanes, C. & Smith, H. (1982). Phytochrome control of seed germination in the tropical rain forest pioneer trees *Cecropia obtusifolia* and *Piper auritum* and its ecological significance. *New Phytologist,* **92,** 477–485.

Whitmore, T. C. (1974). Change with time and the role of cyclones in tropical rain forest on Kolombangara, Solomon Islands. *Commonwealth Forestry Institute Paper,* **46.**

Whitmore, T. C. (1975). *Tropical Rain Forests of the Far East.* Oxford University Press (Clarendon), London and New York.

Whittaker, R. H. (Ed.) (1978). *Classification of Plant Communities.* Dr. W. Junk, The Hague.

3

Fires and Emus: The Population Ecology of Some Woody Plants in Arid Australia

James C. Noble

Division of Wildlife and Rangelands Research
CSIRO
Deniliquin, N.S.W., Australia

I. INTRODUCTION

The principal feature distinguishing arid ecosystems from their more mesic counterparts is extreme rainfall variability. Periods of protracted drought are punctuated by pulses of abundant, often super-abundant rainfall. These episodes create the paradoxical situation whereby water is the dominant erosion agent in most arid ecosystems of the world. Several different types of drought are recognized, including meteorologic drought, hydrologic drought and agricultural drought, the latter having some direct biological connotations. As Whalley (1973) contends, these definitions have little relevance in semi-arid and arid pastoral regions where drought is the norm and the native vegetation copes in various ways with extreme moisture depletion.

Infrequent wet episodes are probably as important as drought in terms of population regulation. High rainfall results in heavy seed production and also seedling recruitment if it continues for a second season; but in localized situations waterlogging may cause substantial plant mortality. Probably the most significant effect of high rainfall in semi-arid areas is the abundant growth that results in the form of grasses such as variable speargrass (*Stipa variabilis*) thus providing fuel for extensive wildfires which commonly follow these wet seasons (Noble, Smith & Leslie 1980).

The degree to which fire interacts with grazing and rainfall, both before and after a fire, probably has equal, if not greater, significance than the effects of fire alone. Rainfall patterns following both fire and drought will determine what species germinate and establish, and these in turn interact with herbivores at-

STUDIES ON PLANT DEMOGRAPHY:
A FESTSCHRIFT FOR JOHN L. HARPER

tracted to regenerating areas producing palatable, new growth. Grazing may be due to a combination of domestic livestock (amenable to management), native and feral animals (less easily managed) and invertebrates (uneconomic to control). Not all animal interactions are necessarily deleterious to the vegetation, since seedlings of some undesirable shrub species may be significantly reduced by selective grazing.

Because of the infrequent, episodic nature of regeneration events in arid ecosystems, special importance is attached to maintaining vigorous, reproductive populations of perennial plant species. This applies especially to the more useful forage species such as the perennial grasses and low shrubs but also to other woody growth forms which may be valued for their topfeed (i.e. canopies to be utilized during droughts) or which maintain sufficient cover to reduce soil erosion and provide some habitat diversity for ephemeral herbage species.

In this paper two groups of non-forage, woody plants will be compared: one growing on heavy-textured soils and another on light-textured soils or dunefields. Their distributions extend well into the semi-arid zone of southern Australia and both have strongly contrasting reproductive biologies.

II. THE SPECIES AND THEIR DISTRIBUTION

A. Nitre Bush (*Nitraria billardieri*—Zygophyllaceae)

Nitre bush is a procumbent shrub growing to a height of 2 m and a diameter of 6 m. It occurs on sands, loams and clays which are generally saline (Noble & Whalley 1978a). The genus has an unusual disjunct world distribution. In the Southern Hemisphere it occurs only in Australia, where *Nitraria billardieri,* a tetraploid species ($2n = 48$), is the sole species; the remaining species, all diploid ($2n = 24$), are found in Eurasia, North Africa and the Middle East (Bobrov 1965).

Because *Nitraria* tends to become a conspicuous component of heavily grazed areas, particularly in originally saltbush- (*Atriplex* spp.-) dominant pastures growing on heavy-textured clays, such changes in abundance can be used to monitor changes in condition of these chenopod communities. Dominance by *Nitraria* is especially evident on stock routes which have had a long history of heavy grazing by sheep and cattle as they were moved on foot from one district to another, particularly during drought. This heavy grazing is recorded in the words of that most Australian of literary works, *Such Is Life* by J. Furphy (alias Tom Collins) (1944, p. 206), first published in Sydney in 1903.

> The mile-wide stock-route from Wilcannia to Hay was strewn with carcases of travelling sheep along the whole two hundred and fifty miles. . . . I remember once, in passing along

> the fifty-mile stretch of that route which bisects the One Tree Plain, that, taking no account of sheep, I never was out of sight of dying cattle and horses, let alone the dead ones. The famine was sore in the land. To use the expression of men deeply interested in the matter, you could flog a flea from the Murrumbidgee to the Darling (river).

B. Mallee (*Eucalyptus* spp.—Myrtaceae)

The mallee shrublands of southern Australia are dominated structurally and floristically by several species of *Eucalyptus*. This particular vegetation type has traditionally been regarded as occupying a transitional zone between the more humid sclerophyll woodlands and forests to the south and the arid or Eremaean flora to the north (Wood 1929). Probably the two most widespread species are pointed mallee (*Eucalyptus socialis*) and congoo mallee (*E. dumosa*). Other common species include yorrell (*E. gracilis*), red mallee (*E. oleosa*), lerp mallee (*E. incrassata*) and slender-leaf mallee (*E. foecunda*).

Although today *mallee* is generally regarded as a term describing semi-arid populations of eucalypts with a multi-stemmed habit, the word was originally derived from the indigenous Aboriginal tribes. These people used the term *mali* in a much narrower context when referring to a specific individual within a population whose surface roots contained a readily accessible and valuable source of drinking water (Fig. 1). The explorer E. J. Eyre published the earliest account of Aborigines collecting water from mallee roots (1845, Vol. 1, p. 349): "when wandering among the scrubs, and by means of which they are enabled to remain out almost any length of time, in a country quite destitute of surface water . . . the quantity of water contained in a good root, would probably fill two-thirds of a pint".

Pastoralists in the mallee areas generally refer to larger individuals growing up to 10 m in height with only 1–3 stems (c. 25 cm in diameter) as "bull mallee", these tending to be confined to the heavier-textured loams found in the interdune swales. Smaller shrubs only grow to 3 m, and because of their tendency to produce up to 20 thin stems (c. 2.5 cm diameter), they are called "whipstick mallee". This latter growth form is characteristically dominant on the lighter-textured sands and sandy loams found on mallee dunes.

Fire has had an important influence on the development of mallee communities, which are regarded as some of the most inflammable to be found in semi-arid southern Australia (Noble 1982). Such flammability has been attributed to substantial quantities of litter fuel that accumulate (up to 12 tonnes/ha) beneath the canopy. When ignited, this fuel creates an intense localised fire, which in turn causes the canopy to ignite aided by the essential oils found in the green leaves of most eucalypts. Partially burnt embers of leaves, together with fragments of burning bark that formerly hung in ribbons from the branches, are lifted up by strong convection currents to be carried forward by wind, thereby causing

Fig. 1. Free water running out of four 1-m segments cut from surface lateral roots of a large water mallee (*Eucalyptus socialis*) near Yalata, South Australia. These yielded 250 ml of clear, tasteless water after 2 min.

spot fires in front of the main fires. This spotting process enables the fire to travel quickly although the ground fuel is often discontinuous; this condition makes fire suppression extremely difficult and hazardous.

The disposition of mallee communities to burn is further enhanced by the highly inflammable nature of under-storey species such as porcupine grass (*Triodia irritans*). This grass creates its own highly aerated fuel bed after it has grown for more than 10 years, with flammability further enhanced by resin and lipid contents up to 35 and 17%, respectively (Leigh & Noble 1981). Porcupine grass is one of the few under-storey species which has close affinities with mallee, but only when growing on light-textured soils usually associated with dunes.

On the heavier-textured loams of the inter-dune swales, the principal grass fuel is variable speargrass, a short-lived perennial (1–3 years) which only occurs in substantial quantities after above-average rainfall seasons. Given sufficient rainfall it is possible for more than one fire to occur within a 5-year period with lasting effects on the structure and species composition of subsequent communities.

This readiness to burn, often frequently, under certain conditions has led ecologists to speculate that much of the remaining dense mallee shrublands in semi-arid Australia may have been quite different in appearance prior to the arrival of Europeans. Once these areas were settled by pastoralists, every attempt was made to suppress fires which previously may have burnt for months over large areas as a result of either Aboriginal burning or lightning strikes.

III. PHENOLOGY AND REPRODUCTIVE BIOLOGY

A. Nitre Bush

The annual growth cycle of *Nitraria,* particularly in inland Australia, is characterised by almost total leaf abscission during the winter. This leaf abscission is thought to be due to a combination of factors, including infestation by leaf mites, increasing salt accumulation as leaves age, and low temperatures. This winter defoliation is of interest because it occurs when rainfall is more likely to be effective than during the summer. This has led some botanists (e.g. Specht & Rayson 1957) to suggest that such perennials whose main growing period occurs during the summer are growing "out-of-phase" with the Mediterranean-type climate prevailing across southern Australia. It may well be that shoot growth of these species is in fact "in-phase" if it is mediated primarily by temperature (Nix 1982).

Nitraria plants have two types of branches: brachyblasts and dilochoblasts, the latter ending in spines and the former bearing flowering shoots. Although isolated flowers have been recorded on bushes during the late summer–early autumn period, usually after significant summer rainfall events, the main flowering period occurs during spring from mid-October to mid-November.

Nitraria plants flowering near Deniliquin, N.S.W., were found to be hermaphroditic, or andromonoecious, carrying both male and bisexual flowers. The breeding ratios of the population (n = 1350) were determined as 1 female to 2.3 males to 0.5 hermaphrodites (Noble & Whalley 1978a).

Pollination of *Nitraria* flowers is entomophilous, and a wide range of insects have been observed feeding on flowers near Deniliquin. Even at night small insects have been observed in the flowers. Excessive rainfall during flowering has an adverse effect on pollination, presumably because it interferes with insect feeding, and heavy fruit production always follows a dry spring.

The mature fruit of nitre bush is a small drupe up to 20 mm long and 10 mm wide, varying in colour when fully ripe from dark red to yellow. The outer pericarp encloses the stone or putamen, which tapers to a point with small round or oval depressions at the opposite end. During the peak fruiting period in December, up to 1730 seeds per m^2 (projected canopy basal area) can be produced on individual female bushes.

B. Mallee

Like nitre bush, vegetative growth by mallee eucalypts occurs primarily during the late spring and summer, although this is the hottest and driest part of the year. Some winter growth of mallee coppicing after fire has been observed (Noble 1982); however, this has largely been attributed to a lack of the insect predators which normally attack the vulnerable 'naked' buds; such insect populations are slow to recover from the fire.

Like *Nitraria,* the mallee eucalypts can flower at different times of the year, depending on rainfall events, although major flowering periods, usually lasting around 2 months for most species, occur during the summer. Often flowering of different mallee species is quite clearly stratified, thus preventing large-scale hybridization where species are compatible. Simultaneous floral bud initiation of five mallee species was observed in the autumn near Pooncarie, N.S.W., after the stand had been burnt by a wildfire 4 years earlier (Noble 1984). The first species (*E. gracilis*) commenced flowering 4 months later, although the last (*E. foecunda*) did not commence flowering until 21 months had elapsed. Similar variability exists between eucalypt species for the time taken for full seed maturation, although it is likely in the case of mallee species that seed would be fully ripe within 1 year following flowering (K. W. Cremer, personal communication).

Mallee eucalypts have bisexual flowers whose pollination is effected by a range of insects, birds (especially honeyeaters) and possibly bats. The flowers of some species such as *E. incrassata* are rich in nectar and pollen and are actively sought by commercial apiarists whenever there is good nectar flow after favourable rainfall. This may occur as frequently as every 2–5 years.

On reaching maturity the bulk of the seed, which is quite small (c. 1 mm long), is retained in capsules that remain attached within the canopy for several years.

IV. SEED DISPERSAL AND ESTABLISHMENT BIOLOGY

A. Nitre Bush

The mature fruit of nitre bush has many of the characteristics regarded as typical of species whose seeds are dispersed primarily by birds. These include

Fig. 2. Emus being sampled on ''Barratta'' Station near Deniliquin, N.S.W., in heavily grazed saltbush country now dominated by nitre bush shown in the middle distance.

such features as an attractive edible part, signalling colours when mature, no smell, semi-permanent attachment and inner protection of the seed against digestion. Certainly the ripening fruit of *Nitraria* is preferentially selected by one particular bird, the emu (*Dromaius novaehollandiae*), which is commonly found on the Riverine Plain (Fig. 2). Although emus are normally observed only in small groups, often only as individuals or breeding pairs for much of the year, they congregate into large flocks (up to 80) whilst feeding on *Nitraria* fruit. Examination of the gut contents of mature emus weighing from 30 to 40 kg (Davies 1967) indicated that *Nitraria* fruit and seed comprised 96% of the birds' crop contents. Substantial quantities of seed (up to 7800 seeds and weighing c. 470 g) have been found during these peak feeding periods, 60% being held in the gizzard (Fig. 3).

Faecal droppings analyzed during the 3-month period when the *Nitraria* fruit was ripening also indicated their concentration on this food source. Individual deposits contained up to 1350 seeds. Seed collected from emu faeces showed a marked capacity to germinate quickly compared with the extremely low, and slow, germination of seed obtained from fruit collected by hand from individual bushes. After 24 days, 62% of the seed collected from emu droppings had germinated under controlled laboratory conditions, compared with only 6% of hand collected seed (Noble 1975).

The increased germination capacity of seed voided by emus was attributed to the removal during digestion of the outer layers of the pericarp of the *Nitraria*

Fig. 3. Emu dissection on "Barratta" showing the gizzard filled with nitre bush stones.

fruit. Chemical analysis indicated high concentrations of both sodium (0.4% dry weight) and chloride (3.5%) in the epicarp and mesocarp, which would create an adverse osmotic situation, thereby restricting imbibition by the embryo. Subsequent germination experiments confirmed that leachate derived from soaking hand-picked seed in distilled water severely reduced germination of seed collected from emu faeces (Noble 1975).

A number of other animals eat the mature fruit of *Nitraria,* including sheep, kangaroos, and rabbits. Although they operate to some extent as dispersal agents, none appears to have the same effect as emus in stimulating germination (Noble & Whalley 1978a). Even Aboriginals may have acted as dispersal agents since they consumed the entire fruit without bothering to spit out the stone. C. Wilhelmi (1860), who was the assistant to the Port Lincoln (South Australia) Protector of Aborigines, wrote that the native name for nitre bush was *karambi.* "In December and January the bushes are so full of fruit, that the natives lie down on their backs under them, strip off the fruit with both hands and do not rise until the whole bush has been cleared of its load."

Enhanced germination of emu-ingested seed was observed in the field near Deniliquin in February 1973, when heavy rains (102 mm over 2 days) followed a drought of 10 months during which only 188 mm of rain had fallen (Fig. 4). It was apparent that seedlings were emerging at very high densities. Given that

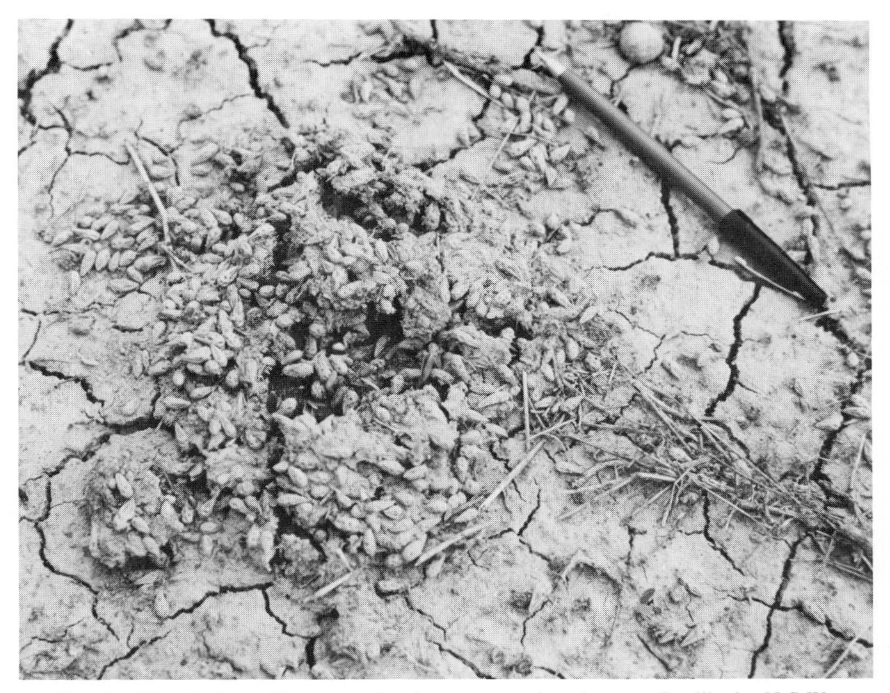

Fig. 4. Nitre bush seedlings emerging from an emu dropping near Deniliquin, N.S.W.

there may be approximately 1000 *Nitraria* seeds in an individual emu dropping of c. 15 cm diameter, even if only 50% germinated the density would still be close to 3 seedlings per cm^2.

Four months after germination commenced, it was clear that interspecific interference from cool-season annual species was also having a significant effect on *Nitraria* seedling survival. By July, annual herbage dry matter was estimated at 1600 kg per ha, of which 90% comprised annual medics (*Medicago* spp.) (Noble & Whalley 1978b). As shown in Fig. 5, hand weeding of all the annuals had a marked effect on *Nitraria* seedling survivorship with 8% of the original population surviving after 8 years despite 50% being lost after 26 months. In contrast, 50% of the unweeded controls had died after only 2 months, and these populations had virtually disappeared after 3 years. Waterlogging in small depressions was also a factor influencing seedling survival but only to a limited extent.

B. Mallee

The seeds of mallee eucalypts, because of their small size, are capable of wind dispersal, although it is unlikely that such dispersal occurs over any significant

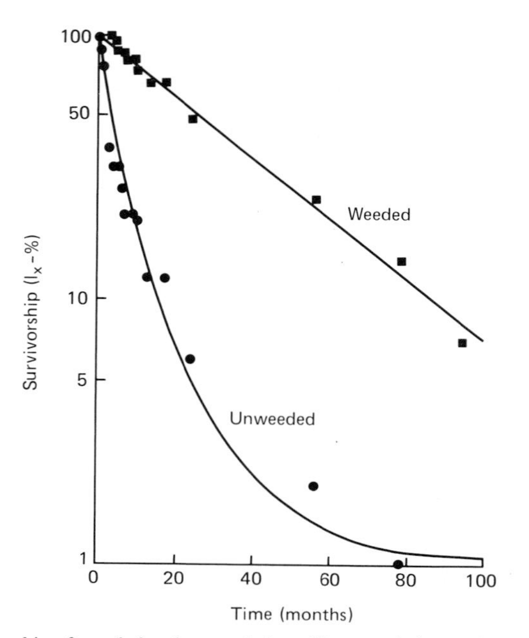

Fig. 5. Survivorship of weeded and unweeded seedling populations (1973 cohorts) of nitre bush on "Barratta" Station, Deniliquin, N.S.W. The curves drawn are defined as follows: Weeded: $\log y = 2.0 - 0.01x$ ($r^2 = .99$). Unweeded: $\log y = 1.95e^{-0.04x}$ ($r^2 = .94$).

distance (Cremer 1977). Occasionally there may be local redistribution of seed on the ground, where there is surface flow of rainwater. Losses of seed retained within eucalypt capsules may be extremely heavy as a result of fungal attack (Drake 1975) and predation by birds (Ashton 1975) and insects (Drake 1975).

In a study of one mallee species, *E. incrassata,* growing in northwestern Victoria, Wellington (1981) found that seed was released intermittently throughout the year at around 70 seeds per square metre per year under fecund canopies, although rates of shedding increased during dry periods. He also found that similar individuals could release up to 300 seeds per square metre after experimental induction of dehiscence. The estimated total seed yield of a large fecund individual of *E. incrassata* on an individual stem basis is shown in Table I.

Wellington (1981) has also shown that predation of fallen seed by harvester ants can result in significant losses to underground caches unless there is "predator satiation" following synchronous release of all canopy-stored seed. He calculated the half-life of a seed of *E. incrassata* on the surface of an unburnt stand at c. 1.7 days. Following fire, the simultaneous release of seed from the aerial 'bank' resulted in soil seed reserves lasting up to periods of at least 200 days with little loss in viability.

Seeds from most eucalypts are shed intermittently over a 2- to 4-year period following ripening, but fire greatly accelerates dehiscence (Mount 1969). This

TABLE I.

Estimates of the Total Seed Load of a Large Fecund Individual of Lerp Mallee (*E. incrassata*) during January 1980[a]

Stem	No. of capsules	Estimated total no. of seeds ± (S.D.)	Mean no. of seeds per capsule
1	0	0	0
2	995	2749 ± 452	2.8
3	1329	3610 ± 585	2.7
4	442	2782 ± 451	6.3
5	1013	1133 ± 184	1.1
6	1384	3055 ± 495	2.2
Total	5164	13367 ± 2166	2.6

[a]After Wellington (1981).

enhanced 'seed rain' is sufficient to offset the heavy predation by seed-harvesting ants described earlier, thus ensuring sufficient seed remains for germination should there be sufficient rainfall. Germination of fallen seed is further enhanced by fire-induced soil changes, particularly those resulting from heating and ash addition (Wellington 1981).

Burning treatments using replicate 20-ha plots were applied during autumn, winter and spring to mallee with a *Triodia* under-storey growing near Pooncarie in southwestern New South Wales. Although quite high fire intensities were recorded during the autumn (April), and winter (July) burns, the highest intensities were recorded during the spring (October) burns (Noble 1982). The hot fires in the spring-burnt treatments resulted in extensive seedling recruitment (c. 3350 seedlings per ha), far exceeding that recorded in the autumn-burnt plots (c. 140 seedlings per ha). No germination was observed on plots burnt during the winter, and this lack was attributed largely to low soil temperatures inhibiting germination of those seeds escaping predation.

Very heavy seedling mortality (c. 50%) was recorded in the spring-burnt treatments during 1 month (mid-December to mid-January) of the first summer following germination while the seedlings were obviously susceptible to moisture shortage. Wellington (1981) measured leaf water potential and stomatal conductance in mallee seedlings and confirmed that much of the early seedling mortality could be directly attributed to low moisture content even though mature individuals coppicing after fire survived.

Six years after the initial establishment event, the surviving 10% of the original seedling cohort appear to be well established, suggesting that gaps were present in the pre-burn mallee population. Obviously in these arid environments, carrying capacity is a highly dynamic state varying widely between droughts and pluvial periods. Nonetheless it is apparent that many mallee seedlings are still capable of surviving indefinitely as 'advanced growth' seedlings that appear to make no growth, suggesting that simple density measurements may not adequately define carrying capacity. Whether or not these suppressed populations remain static by retaining their leaves, or alternatively, producing new leaves and losing old ones at the same rate, is one of the many interesting physiological and demographic aspects of these populations waiting to be resolved. In an arid context, however, these genets are obviously extremely fit and represent an important regenerative bank whereby advanced meristem populations can rapidly exploit any gaps that may develop, especially in the absence of germination events.

V. MALLEE COPPICE DYNAMICS

The tolerance of mallee eucalypts to episodic fire is based primarily on their ability to produce epicormic shoots from meristematic accessory strands in the lignotuber (Cremer 1972), which, by virtue of their subterranean location, are protected from lethal fire temperatures. Nonetheless multiple fire events in these populations, e.g. more than two fires in 10 years, can have a dramatic effect on population stability depending on the season of burn. After three consecutive autumn burns at Pooncarie, N.S.W., only 18% of the original mallee population still survived, whereas 93% was still alive after three consecutive spring burns (Noble 1982). A mallee genet prior to burning may possess only three to four stems, for example, but after decapitation by fire may ultimately produce clusters or fascicles of epicormic shoots from as many as 40 active meristematic sites on the lignotuber, each site having the potential to produce one dominant stem (see Fig. 6). A simple analogy has been made between a tillering grass and a coppicing mallee which is producing lignotillers (Noble 1982). The ability of mallee plants to coppice rapidly after fire enables rapid canopy refoliation at ground level, thus balancing any respiratory burden imposed by underground organs, particularly roots. The high root water content may also be an important factor enabling rapid coppicing after mid-summer fires. As in tillering grasses, ultimately some of these lignotillers assume ascendancy, presumably through correlative inhibition.

An individual mallee plant displays the classic population dichotomy where flux may occur at either the genet (N) or ramet (η) level, the latter represented by individual stems or stem fascicles. Demographic studies of fascicle populations in mallee coppicing after fire have shown that events occurring early in ramet life history can have a persistent effect on age structure of these populations. Subterranean epicormic shoots, some emerging from as deep as 20 cm, closely resemble populations of emerging seedlings and presumably are subject to pre- and post-emergence hazards. Removing the earth surrounding the lignotubers of recently burnt mallee genets, and therefore eliminating most of these pre-emergence hazards, doubled the average number of visibly active lignotuber

Fig. 6. A large mallee coppicing vigorously after a prescribed burn applied 2 years earlier on "Birdwood" Station, Pooncarie, N.S.W. The number of branchlets in just one shoot fascicle emerging from a single meristem site on the lignotuber is clearly evident.

meristems (Fig. 7a). This resulted in a relatively even-aged ramet population dominated by two fascicle cohorts emerging 2 and 3 months after the fire (Fig. 7b). This age profile contrasted strongly with the control population, where stem ramet recruitment was still occurring 10 months after burning. Subsequent survivorship data will provide some insight into the self-thinning of mallee stem populations of contrasting age structure.

VI. CONCLUDING REMARKS

This chapter has attempted to describe how subtle, but highly irregular, interactions between abiotic and biotic factors are critical in determining the success or failure of perennial shrub regeneration in arid ecosystems. The complexity of these interactions is accentuated by the phenomenon that reproductive success, especially for the species used in these case studies, depends on quite different seasonal events. Seasonal conditions are of paramount importance in ensuring an adequate supply of viable seed in the first place. In the case of nitre bush, dry spring conditions are required to ensure adequate pollination, whereas a wet spring is essential to ensure adequate mallee seed production.

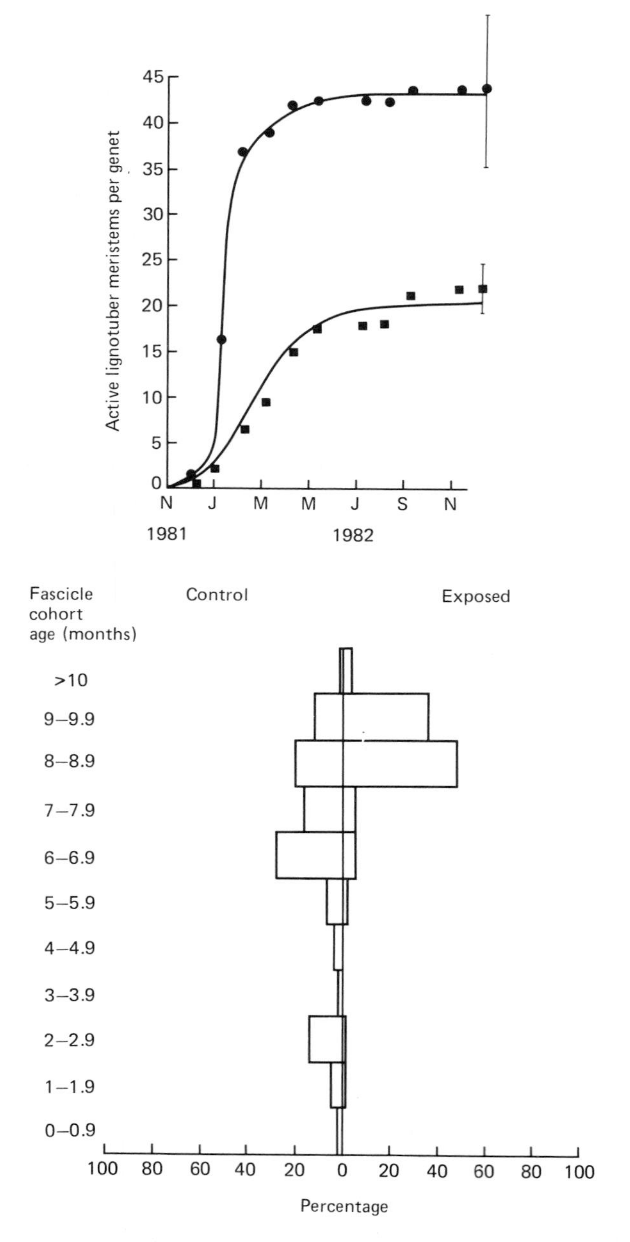

Fig. 7. The response to post-fire soil removal by mallee stem fascicle populations in terms of both (a) visibly active lignotuber meristems per genet and (b) the subsequent effect on age structure of these ramet populations. Circles represent exposed populations and squares, unexposed populations.

During the 2–4 years following flowering, when most of this seed is retained in the mallee 'canopy bank', further substantial rainfall is required to produce sufficient grass fuel to carry fire. Dry weather is then required to cure the fuel before it will burn with sufficient intensity to kill mallee stems, thereby releasing more than enough seed to compensate for losses to seed harvesters. Still more rain is then required after the fire to ensure adequate seed germination and seedling establishment at a time when mature mallee individuals are at their least competitive following decapitation by fire.

Both the rainfall regime antecedent to such perturbations as fire and that consequential to perturbations are therefore critical in determining the development of a plant population as outlined schematically by Harper and White (1974). The former regime controls the sexual reproductive process, and the latter largely dictates the success or failure of germination (birth) and seedling establishment (juvenile period).

The importance of biotic interactions with rainfall regime has already been referred to in respect of insect pollination of both species and mallee seed predation by harvester ants. However, rarely is the biology of an arid plant population apparently so dependent on animal interaction as in the case of nitre bush. Whether there has been co-evolution at both trophic levels is unknown but what is certain is that on heavy soils such as those found on the Riverine Plain, widespread recruitment of nitre bush is unlikely to occur in the absence of emus.

Because of the rarity of major recruitment events of most woody perennials in these arid ecosystems, demography is often only relevant when directed at the sub-population or modular level, particularly during periods shorter than the genet's lifetime. In the absence of any recruitment events, subterranean gaps are more likely to be exploited through either accelerated lateral root growth by adjoining survivors or, if they have a clonal capacity, by lateral root extension combined with adventitious shoot growth as in the case of mature nitre bush. Mallee eucalypts have only a limited clonal capacity as old lignotubers age and decay.

Ultimately, studies of resource capture in arid ecosystems must concentrate primarily on how soil water is captured and utilized by different plant populations. Only by undertaking appropriate studies based on demographic analyses and experimental perturbations (Harper 1977) can more sophisticated management systems be devised for these complex, and as yet poorly understood, environments.

ACKNOWLEDGMENTS

The author acknowledges his debt to John L. Harper for providing a considerable intellectual stimulus while he was studying plant population biology at Bangor. The technical support provided by a number of people based at Deniliquin over past years is gratefully acknowledged. Particular mention must be made, however, of the support provided by Peter A. Mills during the *Nitraria* studies and Ronald A. Bawden during the mallee studies. The assistance provided by colleagues who read drafts of this chapter is also appreciated. Bruce Wellington kindly allowed unpublished material to be incorporated from his thesis.

REFERENCES

Ashton, D. H. (1975). Studies of flowering behaviour in *Eucalyptus regnans* F. Muell. *Australian Journal of Botany,* **23,** 399–411.

Bobrov, E. G. (1965). On the origin of the flora of the Old World deserts as illustrated by *Nitraria. Botanicheskii Zhurnal (Leningrad),* **50,** 1053–1067.

Cremer, K. W. (1972). Morphology and development of the primary and accessory buds of *Eucalyptus regnans. Australian Journal of Botany,* **20,** 175–195.

Cremer, K. W. (1977). Distance of seed dispersal in eucalypts estimated from seed weights. *Australian Forest Research,* **7,** 225–228.

Davies, S. J. J. F. (1967). Sexual dimorphism in the emu. *Emu,* **67,** 23–26.

Drake, D. W. (1975). Seed abortion in some species and interspecific hybrids of *Eucalyptus* from Australia. *Australian Journal of Botany,* **23,** 991–995.

Eyre, E. J. (1845). *Journals of Expeditions of Discovery into Central Australia, and Overland from Adelaide to King George's Sound (1840–41).* T & W. Boone, London.

Furphy, J. (1944). *Such is Life. Being Certain Extracts from the Diary of Tom Collins.* Angus & Robertson, Sydney.

Harper, J. L. (1977). *Population Biology of Plants.* Academic Press, London.

Harper, J. L. & White, J. (1974). The demography of plants. *Annual Review of Ecology and Systematics,* **5,** 419–463.

Leigh, J. H. & Noble, J. C. (1981). The role of fire in the management of rangelands in Australia. *Fire and the Australian Biota* (Ed. by A. M. Gill, R. H. Groves & I. R. Noble), pp. 471–495. Australian Academy of Science, Canberra.

Mount, A. B. (1969). Eucalypt ecology as related to fire. *Proceedings of the Tall Timbers Fire Ecology Conference,* **9,** 75–108.

Nix, H. (1982). Environmental determinants of biogeography and evolution in Terra Australis. *Evolution of the Flora and Fauna of Arid Australia* (Ed. by W. R. Barker & P. J. M. Greenslande), pp. 47–66. Peacock Publications, Adelaide.

Noble, J. C. (1975). The effects of emus (*Dromaius novaehollandiae* Latham) on the distribution of the nitre bush (*Nitraria billardieri* DC.). *Journal of Ecology,* **63,** 979–984.

Noble, J. C. (1982). The significance of fire in the biology and evolutionary ecology of mallee *Eucalyptus* populations. *Evolution of the Flora and Fauna of Arid Australia* (Ed. by W. R. Barker & P. J. M. Greenslade), pp. 153–159. Peacock Publications, Adelaide.

Noble, J. C. (1984). Mallee. *Management of Australia's Rangelands* (Ed. by G. N. Harrington, A. D. Wilson & M. D. Young), pp. 223–240. CSIRO, Melbourne.

Noble, J. C. & Whalley, R. D. B. (1978a). The biology and autecology of *Nitraria* L. in Australia. I. Distribution, morphology and potential utilization. *Australian Journal of Ecology.* **3,** 141–163.

Noble, J. C. & Whalley, R. D. B. (1978b). The biology and autecology of *Nitraria* L. in Australia. II. Seed germination, seedling establishment and response to salinity. *Australian Journal of Ecology,* **3,** 165–177.

Noble, J. C., Smith, A. W. & Leslie, H. W. (1980). Fire in the mallee shrublands of western New South Wales. *Australian Rangeland Journal,* **2,** 104–114.

Specht, R. L. & Rayson, P. (1957). Dark Island heath (Ninety-Mile Plain, South Australia). I. Definition of the ecosystem. *Australian Journal of Botany,* **5,** 52–85.

Wellington, A. B. (1981). *A study of the population dynamics of the mallee* Eucalyptus incrassata *Labill.* Ph.D. thesis, Australian National University.

Whalley, R. D. B. (1973). Drought and vegetation. *The Environmental, Economic and Social Significance of Drought* (Ed. by J. V. Lovett), pp. 81–98. Angus & Robertson, Sydney.

Wilhelmi, C. (1860). Manners and customs of the Australian natives, in particular of the Port Lincoln District (S.A.). *Proceedings of the Royal Society of Victoria,* **5,** 164–203.

Wood, J. G. (1929). Floristics and ecology of the mallee. *Transactions of the Royal Society of South Australia,* **53,** 359–378.

4

Differences in Life Histories between Two Ecotypes of Plantago lanceolata *L.*

J. M. van Groenendael[1]

Institute for Ecological Research
Department of Dune Research Weever's Duin
Oostvoorne, The Netherlands

La seule façon d'etre suivi,
c'est de courir plus vite
que les autres

F. Picabia

I. INTRODUCTION

It has been argued by Harper (1982) that taxonomic categories fail as descriptors of ecological entities because of the bias in taxonomy towards stable, conservative characters that can be used to define a taxon. Those characters that enable a plant to survive in a specific habitat and that are of interest to an ecologist vary as a reaction to the variable conditions in which that plant grows. Such variability could be the result of plasticity of the individual, as well as the genetic differences between the individuals. Habitat-related genetic differences between populations (ecotypes) have been an object of study for a long time (Heslop-Harrison 1964; Langlet 1971). All sorts of characters in plants, including life history traits, have been reported to show ecotypic variation; in general, these have resulted from strong and readily identifiable selective forces (Turesson 1925; Antonovics, Bradshaw & Turner 1971; Gadgil & Solbrig 1972; Law, Bradshaw & Putwain 1977; Warwick 1980).

Usually the genetic variation between ecotypes is established by growing them under uniform conditions in the greenhouse or experimental garden and testing them in experiments with the selective force as a variable. When dealing with

[1]Present address: Department of Vegetation Science, Plant Ecology and Weed Science, Agricultural University, 6708 PD Wageningen, The Netherlands.

STUDIES ON PLANT DEMOGRAPHY:
A FESTSCHRIFT FOR JOHN L. HARPER

complex life history traits like survival or fecundity it can be especially difficult to estimate the genetic variation (Lewontin 1974; Primack & Antonovics 1981). Furthermore, in an evolutionary context it is not sufficient to establish the presence of genetic variation; it is also necessary to demonstrate fitness differences as a result of the genetic differences in the traits under consideration. A powerful test in this respect, especially when working with ecotypes, consists of reciprocal transplanting and measuring the actual fitness differences (Antonovics & Primack 1982). An indirect test is based on the assumption that strong selection on important traits must result in relatively low levels of additive genetic variance (Lewontin 1965; Stearns 1977; Schmidt & Lawlor 1983).

An important tool when dealing with life history traits is the sensitivity analysis developed by Caswell (1978). This analysis is applied in matrix projection models describing the growth of a population that were originally developed by Lewis (1942) and Leslie (1945). It calculates the sensitivity of the population growth rate to small changes in the model parameters. Since this population growth rate can be used as a measure of the fitness of a population (Fisher 1958; Emlen 1973) and since the model parameters are life history traits, it is possible to make a hypothesis about the relative importance of life history traits for the fitness (Schmidt & Lawlor 1983). The more realistic the model is, the more accurate the prediction of the effect of an important trait on the fitness will be.

In this paper two ecotypes of *Plantago lanceolata* L. will be used to illustrate these ideas. Two main questions will be addressed:

1. What are the differences in life histories between two ecotypes of *P. lanceolata* from contrasting habitats? Hypotheses on the importance of several life history traits will be generated by using a matrix model and sensitivity analysis.
2. Are these differences really important in the field? Two tests will be used: an indirect test involving the genetics of life history traits and a direct one based on reciprocal transplanting.

II. THE DIFFERENCES IN LIFE HISTORIES OF ECOTYPES

Plantago lanceolata is a rather short-lived perennial rosette plant which produces long-stalked inflorescences from axillary meristems and can also give rise to side rosettes (Van Groenendael 1985a). The flowers are born in spikes, each flower containing two ovules that need cross-fertilisation. The plant is known to form distinct types in the field (Böcher 1943; Primack 1976; Teramura 1978), and these differences are maintained under uniform conditions, showing substantial genetic control (Warwick & Briggs 1979; Primack & Antonovics 1981, 1982; Slim & Van der Toorn 1982).

Table I.

Characteristic Differences between Two Populations of *Plantago lanceolata* from a Dry Dune Grassland (DG) and a Wet Meadow (WM), Respectively

	WM mean	CV%	DG mean	CV%
No. of leaves	3.8	25	6.6	27
Length of longest leaf (mm)	203	24	34	24
No. of ears	1.7	47	9.2	69
Length of scape (mm)	380	29	54	61
Length of ear (mm)	17.1	34	7.3	31
No. of seeds/ear	57.17	53	15.74	75
Seed weight (mg)	1.91	39	0.73	75
No. of side rosettes	0		3.6	
Percentage of adults flowering	20		31	
Percentage of adults with side rosettes	0		12	
No. of seeds in the soil per m^2 (1-11-1979)	500		1700	
No. of seedlings per m^2 1979	1158		779	
1980	465		486	
Percentage germinating in spring 1979	44		68	
1980	46		75	
Percentage of seedlings surviving 1979 → 1980	26		59	
1980 → 1981	28		45	
Half-life of adults in months	46		20	

In the Netherlands *P. lanceolata* is found in a variety of grassland habitats. Two habitats were selected that were strongly contrasting. Each of them has been more or less the same for at least a hundred years; this has probably provided a continuity in selective forces peculiar to each. The choice was based on the species composition of the habitats, which is the most reliable indicator of habitat conditions (Westhoff & Van der Maarel 1978). Ordination of all communities where *P. lanceolata* is found confirmed this choice and provided an objective measure of 'ecological distance' between the two habitats (Haeck *et al.* 1981). The first site is a dry and open dune grassland, grazed as commonage by cattle and horses for several hundred years (Noë & Blom 1982). Occasionally and unpredictably the grassland suffers from catastrophic droughts in spring and summer. The second site is a permanently waterlogged, closed hay field, situated in a former river bed now filled with a thick peat layer. The vegetation is mown once a year in July.

The population from the dry site is the shorter-lived of the two. It has a seedbank, its juveniles and adults share mortality risks equally and individuals show a tendency towards monocarpy. The other population is longer lived, juveniles carry the greatest mortality risk, there is almost no seedbank and

												Σ
B_1 0.10	.	.	B_{2a} 0.13	B_{2s} 0.02	B_{2z} 0.13	B_3 0.42	B_{3z} 0.22	B_4 0.39	B_{4z} 0.12	B_5 0.28	B_6 0.12	N_{1a} 1.93
.	B_1 0.17	.	B_{2a} 0.21	B_{2s} 0.03	B_{2z} 0.23	B_3 0.72	B_{3z} 0.37	B_4 0.66	B_{4z} 0.21	B_5 0.47	B_6 0.20	N_{1s} 3.27
.	.	.	Z_2 0.07	Z_2 0.17	.	Z_3 0.69	Z_3 0.14	Z_4 0.33	Z_4 0.06	Z_5 0.17	Z_6 0.07	N_{1z} 1.68
P_{1a} 1.61	.	.	.	.	.	.	.	.	.	.	.	N_{2a} 1.61
.	P_{1s} 3.08	.	.	.	.	.	.	.	.	.	.	N_{2s} 3.08
.	.	P_{1z} 1.68	.	.	.	.	.	.	.	.	.	N_{2z} 1.68
.	.	.	P_{2a} 1.43	P_{2a} 2.86	.	.	.	.	.	.	.	N_3 4.29
.	.	.	.	.	P_{2z} 1.32	.	.	.	.	.	.	N_{3z} 1.32
.	.	.	.	.	.	P_3 2.46	.	.	.	.	.	N_4 2.46
.	.	.	.	.	.	.	P_{3z} 0.59	.	.	.	.	N_{4z} 0.59
.	.	.	.	.	.	.	.	P_4 1.10	P_{4z} 0.21	.	.	N_5 1.31
.	.	.	.	.	.	.	.	.	.	P_5 0.38	.	N_6 0.38
											Σ	23.60

Fig. 1. Goodman transition matrix for a population of *Plantago lanceolata* in a dry dune grassland. Each matrix element represents an age category and is in itself a matrix of size categories, not shown. Figures indicate relative sensitivities (see text). *B*, number of seeds produced per age category; *Z*, number of side rosettes produced per age category; *P*, transition probabilities between age categories; *N*, number of individuals per age category. Subscripts refer to years: 1, year of germination; 2, first year; 3, second year, etc.; and to cohorts: *a*, autumn cohort; *s*, spring cohort; *z*, cohort of side rosettes.

individuals are definitely iterocarpic (Table I). Both life histories seem to be a response to the environment, being hazardous and unpredictable in the dry site and more stable in the wet site.

Detailed field observations in permanent quadrats with mapped individuals over a 3-year period provided information on age and size-dependent mortality and fertility schedules. This large set of data has been summarised in a comprehensive matrix model (J. M. van Groenendael & P. Slim, unpublished). In the case of an iterocarpic perennial, both size and age are important categories to describe the life cycle, and therefore a modified version of the model proposed by Law (1983) was used (Figs. 1 and 2). The main structure of the matrix is based on age, divided into years. The maximum number of years considered is three times the half-life of the population. Special categories have been created for autumn and spring cohorts of seedlings and their juvenile survival because of

																		Σ
.	.	.	.	.	.	B_{4a} 0.02	B_{4s} 0.01	B_{5a} 0.10	B_{5s} 0.05	B_{6a} 0.11	B_{6s} 0.05	B_7 0.17	B_8 0.16	B_9 0.13	B_{10} 0.11	B_{11} 0.09	B_{12} 0.03	N_{1a} 1.05
.	.	.	.	.	.	B_{4a} 0.01	B_{4s} 0.01	B_{5a} 0.05	B_{5s} 0.03	B_{6a} 0.06	B_{6s} 0.03	B_7 0.10	B_8 0.09	B_9 0.08	B_{10} 0.06	B_{11} 0.05	B_{12} 0.02	N_{1s} 0.59
P_{1a} 1.05	.	.	.	.	.	.	.	.	.	.	.	.	.	.	.	.	.	N_{2a} 1.05
.	P_{1s} 0.57	.	.	.	.	.	.	.	.	.	.	.	.	.	.	.	.	N_{2s} 0.57
.	.	P_{2a} 0.95	.	.	.	.	.	.	.	.	.	.	.	.	.	.	.	N_{3a} 0.95
.	.	.	P_{2s} 0.59	.	.	.	.	.	.	.	.	.	.	.	.	.	.	N_{3s} 0.59
.	.	.	.	P_{3a} 1.04	.	.	.	.	.	.	.	.	.	.	.	.	.	N_{4a} 1.04
.	.	.	.	.	P_{3s} 0.59	.	.	.	.	.	.	.	.	.	.	.	.	N_{4s} 0.59
.	.	.	.	.	.	P_{4a} 1.00	.	.	.	.	.	.	.	.	.	.	.	N_{5a} 1.00
.	.	.	.	.	.	.	P_{4s} 0.57	.	.	.	.	.	.	.	.	.	.	N_{5s} 0.57
.	.	.	.	.	.	.	.	P_{5a} 0.85	.	.	.	.	.	.	.	.	.	N_{6a} 0.85
.	.	.	.	.	.	.	.	.	P_{5s} 0.47	.	.	.	.	.	.	.	.	N_{6s} 0.47
.	.	.	.	.	.	.	.	.	.	P_{6a} 0.68	P_{6s} 0.42	.	.	.	.	.	.	N_7 1.10
.	.	.	.	.	.	.	.	.	.	.	.	7 0.82	.	.	.	.	.	N_8 0.82
.	.	.	.	.	.	.	.	.	.	.	.	.	P_8 0.57	.	.	.	.	N_9 0.57
.	.	.	.	.	.	.	.	.	.	.	.	.	.	P_9 0.37	.	.	.	N_{10} 0.37
.	.	.	.	.	.	.	.	.	.	.	.	.	.	.	P_{10} 0.20	.	.	N_{11} 0.20
.	.	.	.	.	.	.	.	.	.	.	.	.	.	.	.	P_{11} 0.06	.	N_{12} 0.06
																	Σ	12.44

Fig. 2. Goodman transition matrix for a population of *Plantago lanceolata* in a wet meadow. Structure and symbols as in Fig. 1.

strong seasonal influences. Side rosettes are also treated separately, until the effects of season or mode of birth have subsided. Each matrix element is in itself a small matrix based on five size categories. All B elements contain mean numbers of offspring produced per age and per size category. All P elements contain survival probabilities per age and size, beginning with the probability of surviving from the moment of germination until the first census date, taken as the first of July. The numerical values for each element older than 3 years have been

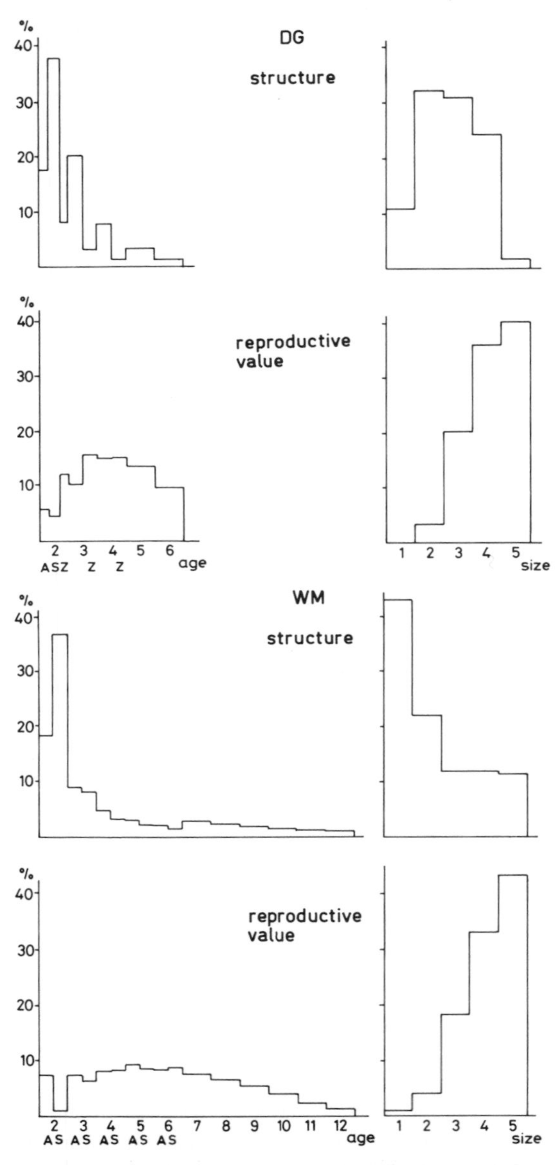

Fig. 3. Limit properties of the Goodman transition matrices for the population of a dry dune grassland (DG) and for the population of a wet meadow (WM). Stable age and size structure and the reproductive value, given age and size, are shown. Letters refer to different cohorts within 1 year, until they are merged (*A* = autumn cohort, *S* = spring cohort, and *Z* = cohort of side rosettes). Mean length of generation and population growth rate were 791 and 1.9310, respectively, in the dry and 495 and 1.0353 for the wet population.

inferred from the fate of adult individuals present at the beginning of the field observations. The other data come from cohorts of seedlings followed over time.

The matrices were solved numerically. The dominant eigenvalue λ is the population growth rate, the right eigenvector **W** gives the stable age–size distribution, and the left eigenvector **V** gives the reproductive value of each age–size category. The scalar product of both vectors $\langle \mathbf{V}, \mathbf{W} \rangle$ is a measure of the mean length of generation (Leslie 1966). For the two populations under study this demographic information is given in Fig. 3. Comparison between calculated and actual age/size distributions in July in the field gave a satisfactory fit (J. M. van Groenendael & P. Slim, unpublished). There are more small and young individuals in the wet site with a greater chance of dying than in the dry site. This results in an unexpectedly low mean length of generation for the population from the wet site. Field observations had led to the opposite impression, based on the longevity of adults. Judging from the reproductive values (Fig. 3), size is more important in the wet than in the dry site: there is a greater reward for growing into size category 5. The same holds for growing old. The spring cohort in the wet site becomes gradually more important because of a better survival over time than the autumn cohort, which faces heavy winter mortality when still small. The autumn cohort is more important in the dry site because it survives the summer droughts better. In this dune grassland population side rosettes contribute significantly to the reproductive value, especially when young.

A more comprehensive impression is given by the sensitivity analysis. The effect of small changes in any of the matrix elements a on the population growth rate can be expressed as follows:

$$\frac{\delta\lambda}{\delta a_{ij}} = \frac{v_i \, w_j}{\langle \mathbf{V}, \mathbf{W} \rangle} \cdot a_{ij}$$

A slightly different normalisation has been used in comparison with the original sensitivity measure of Caswell (1978). This brings out better the biological implications (H. de Kroon, A. Plaisier & J. M. van Groenendael unpublished).

The results for both populations were calculated, summed over the size matrices, and are presented in Fig. 1 and 2. The sensitivities for size-dependant seed production and survival are omitted for clarity (full details are given by J. M. van Groenendael & P. Slim, unpublished). The following hypotheses can now be formulated, based on the model's properties and the sensitivity analysis, using Fisher's theorem:

Judged by sensitivity levels the selective pressure on life history traits is greatest in the dry site. This should then result in lower additive genetic variance for this population in life history traits and in other characters directly influencing those traits.

Germination and establishment are important phases in both life histories, but especially in the dry site. Staying alive as an adult is relatively more important in the wet site. Considering that the greatest mortality occurs in seed and seedling phases in the wet site, there is surprisingly small selective pressure in these phases. The death of a few adults seems as important as the death of many seedlings (see also Antonovics & Primack 1982).

The effect of a seedbank in the dry site is rather unimportant, but side rosettes are an important source of new recruits; their importance is mainly limited to the first year.

Immediate germination in autumn is most important in the wet site; delayed germination in spring is most important in the dry site. This is confirmed by the actual germination pattern in the field. It should be noted, however, that the other cohorts have relatively high reproductive values, which may account for the existence of the two germination flushes.

III. TESTS OF LIFE HISTORY HYPOTHESES

Two kinds of experiments have been proposed to test the hypotheses given in the previous section. Reciprocal transplantation of ecotypes has been strongly advocated (Antonovics & Primack 1982) because it allows direct comparison of different life histories in terms of fitness, by recording fitness parameters such as survival. In such transplantation experiments, the various phases of the life cycle should be investigated as completely as possible. Three phases of the life cycle of *P. lanceolata* were tested: the survival of seeds in the soil, the germination and subsequent survival of seeds and seedlings, and the growth and survival of young adults.

The other test used is the more common one of growing plants from the different sites under uniform conditions to establish the genetic basis of the differences between various morphological features which may be related to the life history traits we are interested in.

A. Reciprocal Transplantation

The first phase tested by reciprocal transplantation was the fate of seeds in the soil. After seeds were buried in nylon mesh tubes in the autumn of 1979, survival was tested by retrieving a set of tubes every 3 months and testing the surviving seeds for viability. Apart from site and origin, seed size and depth of burial were explicitly taken into account. A total of 4800 seeds were tested this way. Time, depth of burial and relative seed size are major determinants of the survival of seeds in the soil, but there were minor but significant site effects and effects of origin (Van Groenendael 1985b). Seeds from both populations survive better in the dry site, and seeds from the dry site do better in both habitats (Fig. 5a). The

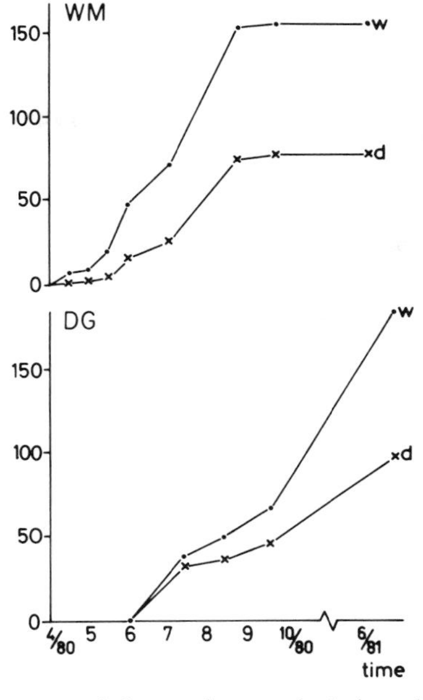

Fig. 4. Germination of seeds from dry grassland (d) or wet meadow (w) transplanted into the dry grassland (DG) or into the wet meadow (WM).

interaction between site and origin, which suggests local adaptation, is significant only after taking into account the effects of size and depth of burial. The contrast is mainly a result of high mortalities in the wet site for seeds from the dry site when buried deep or when small. The seeds from the wet site are less affected by the contrast in environment. They show a more uniform survival pattern.

In the second phase germination and establishment were investigated. Marked seeds were sown in replicate plots in a centimetre grid in the spring of 1980. After germination, seedlings were followed over time. The survivors were measured at regular intervals. In total 3200 seeds were sown. All germination takes place in one season in the wet site, whereas there is delayed germination in the dry site (Fig. 4). The timing of the germination, represented by the shape of the curves, shows a strong environmental control, regardless of origin, except for a somewhat retarded start in the wet site for seeds from the dry site. This can be explained by a greater sensitivity to lower temperatures in germination for seeds from this site, found under standard laboratory conditions (Van Groenendael 1985b). The total number of seeds that germinate is strongly affected by the origin of the seeds and to a lesser extent by site effects, but there is no significant

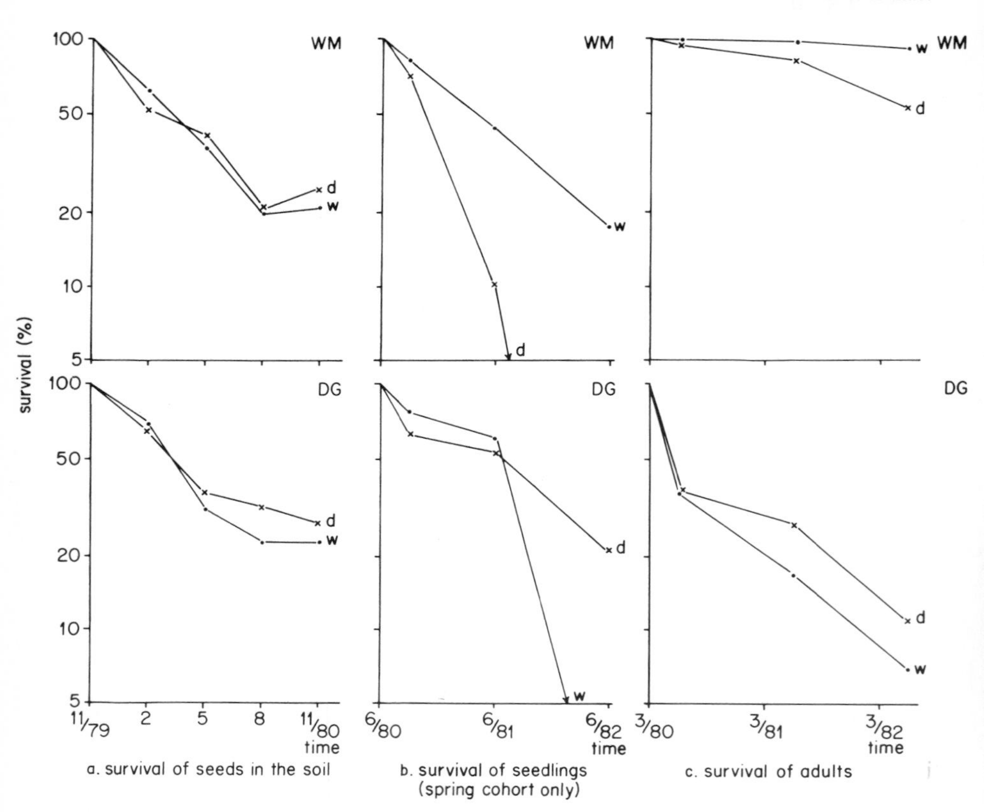

Fig. 5. Survivorship curves of various experimental cohorts (a, b, c) of populations from dry grassland (d) or wet meadow (w) transplanted into the dry grassland (DG) or into the wet meadow (WM).

interaction term in the analysis of variance (Van Groenendael 1985b). This points at a rigid control over the level of germination. After germination the seedlings from the wet site do establish significantly better than the seedlings from the dry population in both habitats (Fig. 5b). There is weaker, but still significant site effect, due to better establishment in the dry site. However, it is not until the period of drought in the spring of the second year that interaction becomes evident, with each ecotype performing best in its own habitat and the strongest contrast for the young adults from the wet site.

The last part of the life cycle that was investigated was juvenile and adult survivorship, using shoots propagated vegetatively from leaves (Wu & Antonovics 1975) collected from flowering individuals in the field. Each site was represented by 10 genotypes, and 20 shoots per genotype were used for reciprocal transplanting in a randomised block design in the spring of 1980 in both habitats. Survival was recorded at regular intervals over a 2-year period. Strong site effects dominate the first few months, as a result of spring drought (Fig. 5c). After the first year interaction became significant, again with the strongest contrast in survival for plants from the wet site. Comparison of survival of individual gentoypes between sites, using a χ^2 analysis, showed no significant differences between genotypes from the wet site. They all reacted very uniformly on the contrast between the sites, indicating genetic homogeneity. Between the genotypes from the dry site, however, there were significant differences in survival pattern between the sites, suggesting stronger genetic heterogeneity (Van Groenendael 1985b).

When comparing these results with the predictions from the model one can see similarities. When there is strong selection pressure, local specialisation is expected, and this is reflected in the analyses of variance in strong interaction terms in the survival parameters, the ultimate test for the fitness in these experiments. A strong contrast is found for adult survival in plants from the wet site, indicating strong selection pressure, also predicted in the model. No strong selection was apparent in survival of seeds in the soil, confirming the model's prediction. The weak interaction terms, however, point in the direction of a stronger contrast for the seeds from the dry site. When the seedbank is considered to be a way to increase the establishment of new recruits, this again is in accordance with the model.

There are, however, two drawbacks in this reciprocal transplantation test. The first is that only in the long run is the test really reciprocal, after all the hazards of life have been encountered. In the short term there might be a severe imbalance in the reciprocality. For instance, the big seeds from the wet site germinate readily and, without competition, grow fast in the dry site. At the same time seedlings from the dry site suffer from shading in the wet habitat. Full reciprocality is achieved only after periods of drought. Until then there is a wrong estimation of local adaptation. The second point is that it is usually not clear how this local adaptation in survival is achieved, whether by genetic canalisation or

TABLE II.

Variability in Two Traits under Controlled Conditions and in the Field for Both Populations of *Plantago lanceolata*

Variable	Greenhouse		Field	
	Mean	CV%	Mean	CV%
Seed weight DG	1.32	46	0.73	75
Seed weight WM	1.71	29	1.91	39
Length of cotyledon DG	25.70	32	13.17	37
Length of cotyledon WM	42.10	24	23.40	53

by plasticity, or through what traits. As an example, the effects of seed size will be discussed. The seeds from the dry site are small and very variable (Table II). They are produced under unfavorable conditions. When grown in the laboratory there is a substantial gain in seed weight and a reduction in variability, but with a considerable amount of variability still left. In the field there is a significant negative correlation between the number of seeds per mm of spike and the weight of the seed ($r = -0.4490$, $p < .01$; $n = 28$). This suggests a trade-off between number of seeds and the individual size of a seed. Seed size is far more strictly regulated in the wet meadow population, showing the opposite tendencies and no significant correlation ($r = 0.2135$, n.s.; $n = 30$).

Seed size in turn determines the length of the cotyledons, being bigger for the wet meadow population. The cotyledons in this population are also more variable in the field. This variability is mainly phenotypic and strongly reduced under laboratory conditions (Table II). It is easy to see the advantage of such variability, because it enables these seedlings to reach the light in a dense vegetation. There is a significant positive correlation between the length of the cotyledons and the growth in the first three months in the field, irrespective of the habitat ($r = 0.6368$ in the wet site and $r = 0.6407$ in the dry site, $p < .01$; $n = 20$). No such correlation exists for the dry population ($r = 0.3456$, $n = 17$ and $r = 0.0421$, $n = 19$, respectively, both non-significant). The conclusion is that variability in seed size, and relative constancy in the seedlings developing from them, permits the dry population to establish enough seedlings. In the wet habitat such variability is fatal. Constancy in seed size and plasticity in seedlings, on the other hand, give the seedlings from the wet population at least a chance in a competitive environment. In a non-competitive environment and in combination with a high and constant growth rate, plants grow too big to withstand periods of drought. (The relative growth rates after 6 weeks were 155 and 145 mg/g/day for the wet and dry site populations, respectively; the coefficients of variation were 0.28 and 0.51, respectively.)

B. Laboratory Test

To elaborate the point of selective pressure on the variability of traits somewhat further, the following experiment has been done. Three plantlets from the clonally propagated genotypes from the last experiment were saved and grown to flowering in the greenhouse. Several morphological traits were measured and analysed in a nested analysis of variance. This procedure permits not only a comparison between origins but also a comparison between and within 'mothers' (see Table III). The higher the genetic variance, the higher the value in the F statistic. The coefficient of variation in combination with the F value indicates how variable a trait is under controlled conditions and whether this variability is genetically controlled or results from plasticity. At the population level there are

TABLE III.

Mean, Coefficient of Variation and Variance Ratio between versus within Clones for Several Morphological Traits of Two *Plantago lanceolata* Populations, Measured in Clonally Propagated Offspring from 10 Flowering Adults from Each Habitat Grown in the Greenhouse

	DG[c]				WM[d]			
	Mean	CV%	*F*	*p*	Mean	CV%	*F*	*p*
Length of longest leaf (cm)[a]	12.5	28	4.55	<.01	20.1	23	11.30	<.001
Width of longest leaf (mm)[a]	13.8	42	2.82	<.05	20.6	32	4.49	<.01
Length of longest scape (cm)[a]	39.2	20	6.88	<.001	53.2	16	4.46	<.01
Estimated biomass[a]	242	82	2.07	<.1	647	50	4.26	<.01
No. of leaves[b]	12.9	31	2.87	<.05	15.0	18	2.06	<.1
No. of spikes[b]	28.3	44	1.76	ns	6.3	63	11.50	<.001
No. of side rosettes[b]	9.2	48	5.86	<.001	1.9	118	3.82	<.01
Length of longest spike (mm)[b]	28.0	25	9.07	<.001	39.8	32	6.46	<.001
No. of seeds per spike[b]	52.9	40	12.15	<.001	92.3	36	1.45	ns

[a] Traits related to size.
[b] Traits related to production of offspring.
[c] Population from dry grassland.
[d] Population from wet meadow.

significant differences in all 9 traits measured, which confirm the distinctness of the two ecotypes. In general on the genotype level, more traits show significant genetic variance in the wet than in the dry site. When there is significant genetic variance, the variability in those traits is lower in the dry than in the wet population. According to the model, overall selective pressure was highest in the dry site, and, as a consequence, low levels of additive genetic variance were predicted. Although the additive genetic variance was not measured here, the low levels of total genetic variance (of which the additive genetic variance forms a part) in the dry population seem to suggest the accuracy of the predicted trend.

Looking in more detail and grouping traits into two categories, the one related to the production of offspring, and the other related to the size of an adult (labels *b* and *a* in Table III, respectively), it becomes clear that there is more genetic control over adult size characteristics than over traits related to reproduction in the wet population, and the reverse is true for the dry population. Also when there is a significant genetic control, the variability is relatively lower in size-related traits than in traits related to reproduction in the wet population; again the reverse is true but less markedly so for the dry population. These observations on the total genetic variance are not conclusive with respect to the predictions of the model, but they are certainly along the lines predicted for the additive genetic variance. Apparently traits are under more strict genetic control in the wet site than in the dry site and have low variabilities when traits related to adult size are considered. Exactly those traits were expected to be under selective pressure according to the model. The variability in traits in the dry site was more plastic, especially in size-related traits. There was a higher genetic control in traits related to reproduction and the variability was also somewhat lower in these traits, where, according to the model, the higher selective pressure was expected.

IV. CONCLUDING REMARKS

The conclusion from the previous section is that the differences between the two populations have a genetic basis and that the differences seem to be adaptive. The predictions from the model are borne out quite well, confirming the usefulness of such a matrix model in an evolutionary context.

On the basis of the model, certain life history traits have been identified that could be responsible for these differences in fitness. Adult survival was the most important trait for the population from the wet site. In this phase the population proved to be most sensitive to a change in the environment because of narrow genetic specialisation, directed at maintaining adults in a competitive environment. Adults from the population of the dry site proved to be more heterogeneous and less affected by the contrast. This heterogeneity is necessary to survive in a hazardous environment. Juvenile survival, however, was more important for this population, which is geared towards producing enough offspring, regardless of the size of individuals, and releasing them slowly into the environment. This variability prevents establishment in a competitive environment. Seed and seedlings from the wet site proved to be more homogeneous and their establishment was hardly affected by the contrast in the environments.

It is clear that the whole of the life history traits determines the fitness of a population. Such a combination of life history traits has been described as a 'life history tactic' (Stearns 1976) or as a 'plant strategy' (Grime 1979) and is defined as a set of co-adapted traits, that are inherited as such. Although there is some proof for the existence of genetic correlation for life history traits (Etges 1982), it remains difficult to see how such correlation has been generated by selection. One solution could be that there exists a more simple and more general underlying structure which regulates a set of traits at the same time, and on which selection can operate. A possible candidate for such a general, underlying regulator is the way the growth of a plant is controlled.

Recently, it has become sufficiently clear that the shoot system of a plant consists of discrete construction units or modules (Hallé, Oldeman & Tomlinson 1978; Harper & Bell 1979; White 1979, 1984). It is possible to express the optimal design of a plant, optimal in terms of future offspring, in the use a plant makes of its meristems that produce these modules (Smith 1984; Watson 1984). With regard to the two ecotypes of *P. lanceolata* used in this study, the major differences in life history tactics can also be related to differences in modular growth (Van Groenendael 1985a). In the wet site it is important to maintain a large main rosette, to be able to compete with other tall grasses and forbs. In terms of modular growth this means a slow initiation of new modules (short internode, with a leaf attached to it and axillary meristem) in the main rosette so that each unit can grow big enough. It also means a tight control over the use of the axillary meristems, so that no side rosettes are formed and only a few ears, that again can grow big, are formed. In the dry site it is important to produce many offspring. This can be achieved by a rapid production of small leaves in the main rosette. The axillary meristems associated with the leaves then produce side rosettes, causing a rapid proliferation of growing points and many inflorescences in the main as well as in side rosettes.

A large part of the differences noticed in the life histories of the two populations of *P. lanceolata* from contrasting habitats can be reduced to two fundamental controls: control over the speed of initiation of new modules or plastrochron time in the main rosette and control over the induction of second-order meristems. These two controls define the form of the plant and the way its resources will be allocated. It offers a way of understanding life history tactics not only in terms of co-adaptation of traits but also on a deeper level of regulation, affecting several life history traits at once.

ACKNOWLEDGMENTS

The following people commented on the text: C. Blom, S. ter Borg, J. van Damme, H. de Kroon, M. van Mansfeld, V. Westhoff and J. Woldendorp. Their contributions are gratefully acknowledged. A. Ormel typed various versions of the manuscript, and H. Klees did the drawings.

REFERENCES

Antonovics, J. & Primack, R. B. (1982). Experimental ecological genetics in *Plantago*. VI. The demography of seedling transplants of *P. lanceolata*. *Journal of Ecology*, **70**, 55–75.

Antonovics, J., Bradshaw, A. D. & Turner, R. G. (1971). Heavy metal tolerance in plants. *Advances in Ecological Research*, **7**, 1–85.

Böcher, T. W. (1943). Studies on variation and biology in *Plantago lanceolata* L. *Dansk Botanisk Arkiv*, **11**, 1–18.

Caswell, H. (1978). A general formula for the sensitivity of population growth rate to changes in life history parameters. *Theoretical Population Biology*, **14**, 215–230.

Emlen, J. M. (1973). *Ecology: An Evolutionary Approach*. Addison-Wesley, Reading, Massachusetts.

Etges, W. J. (1982). "A new view of life-history evolution"?—A response. *Oikos*, **38**, 118–122.

Fisher, R. A. (1958). *The Genetical Theory of Natural Selection*. Dover, New York.

Gadgil, M. & Solbrig, O. T. (1972). The concept of r- and K- selection: Evidence from wild flowers and some theoretical considerations. *American Naturalist,* **106,** 14–31.
Grime, J. P. (1979). *Plant Strategies and Vegetation Processes.* Wiley, New York.
Groenendael, J. M. van (1985a). Teratology and metameric plant construction. *New Phytologist,* **99,** 171–178.
Groenendael, J. M. van (1985b). *Selection for different life histories in* Plantago lanceolata *L.* Ph.D. thesis, Catholic University, Nijmegen, The Netherlands.
Haeck, J., Aart, P. J. M., van der, Dorenbosch, H., Maarel, E. van der & Tongeren, O. van der (1981). The occurrence of *Plantago* species. *Verhandelingen der Koninklijke Nederlandse Akademie van Wetenschappen, Afdeling Natuurkunde, Reeks 2,* **81,** 26–33.
Hallé, F., Oldeman, R. A. A. & Tomlinson, P. B. (1978). *Tropical Trees and Forests: An Architectural analysis.* Springer-Verlag, Berlin and New York.
Harper, J. L. (1982). After description. *The Plant Community as a Working Mechanism* (Ed. by E. I. Newman), pp. 11–25. Blackwell Scientific Publications, Oxford.
Harper, J. L. & Bell, A. D. (1979). The population dynamics of growth form in organisms with modular construction. *Population Dynamics,* (Ed. by R. M. Anderson, B. D. Turner & L. Taylor), pp. 29–53. Symposium of the British Ecological Society, 20. Blackwell Scientific Publications, Oxford.
Heslop-Harrison, J. (1964). Forty years of genecology. *Advances in Ecological Research,* **2,** 159–240.
Langlet, O. (1971). Two hundred years of genecology. *Taxon,* **20,** 653–722.
Law, R. (1983). A model for the dynamics of a plant population containing individuals classified by age and size. *Ecology,* **64,** 224–231.
Law, R., Bradshaw, A. D. & Putwain, P. D. (1977). Life-history variation in *Poa annua. Evolution,* **31,** 233–246.
Leslie, P. H. (1945). On the use of matrices in certain population mathematics. *Biometrika,* **33,** 183–212.
Leslie, P. H. (1966). The intrinsic rate of increase and the overlap of successive generations in a population of guillemots (*Uria aalge* Pont.). *Journal of Animal Ecology,* **35,** 291–301.
Lewis, E. G. (1942). On the generation and growth of a population. *Sankhyá,* **6,** 93–96.
Lewontin, R. C. (1965). Selection for colonizing ability. *The Genetics of Colonizing Species* (Ed. by H. G. Baker & G. L. Stebbins), pp. 77–94. Academic Press, New York.
Lewontin, R. C. (1974). *The Genetic Basis of Evolutionary Change.* Columbia University Press, New York.
Noë, R. & Blom, C. W. P. M. (1982). Occurrence of three *Plantago* species in coastal dune grasslands in relation to pore-volume and organic matter content of the soil. *Journal of Applied Ecology,* **19,** 177–182.
Primack, R. B. (1976). *The evolutionary basis of population dynamics in the genus* Plantago. Ph.D. thesis, Duke University, Durham, North Carolina.
Primack, R. B. & Antonovics, J. (1981). Experimental ecological genetics in *Plantago.* V. Components of seed yield in the ribwort plantain *Plantago lanceolata* L. *Evolution,* **35,** 1059–1079.
Primack, R. B. & Antonovics, J. (1982). Experimental ecological genetics in *Plantago.* VII. Reproductive effort in populations of *P. lanceolata* L. *Evolution,* **36,** 742–752.
Schmidt, K. P. & Lawlor, L. R. (1983). Growth rate projection and life history sensitivity for annual plants with a seed bank. *American Naturalist,* **121,** 525–539.
Slim, P. & Toorn, J. van der (1982). Variability in morphological characteristics of *Plantago lanceolata. Verhandelingen der Koninklijke Nederlandse Akademie van Wetenschappen, Afdeling Natuurkunde, Reeks 2,* **81,** 36–41.
Smith, B. H. (1984). The optimal design of a herbaceous body. *American Naturalist,* **123,** 197–211.

Stearns, S. C. (1976). Life-history tactics: A review of the ideas. *Quarterly Review of Biology,* **51,** 3–47.

Stearns, S. C. (1977). The evolution of life history traits: A critique of the theory and a review of the data. *Annual Review of Ecology and Systematics,* **8,** 145–171.

Teramura, A. H. (1978). *Localised ecotypic differentiation in three contrasting populations of* P. lanceolata. Ph.D. thesis, Duke University, Durham, North Carolina.

Turesson, G. (1925). The plant species in relation to habitat and climate. Contribution to the knowledge of genecological units. *Hereditas,* **6,** 147.

Warwick, S. I. (1980). The genecology of lawn weeds. VII. The response of different growth forms of *Plantago major* L. and *Poa annua* L. to simulated trampling. *New Phytologist,* **85,** 461–469.

Warwick, S. I. & Briggs, D. (1979). The genecology of lawn weeds. III. Cultivation experiments with *Achillea millefolium* L., *Bellis perennis* L., *Plantago lanceolata* L., *Plantago major* L. and *Prunella vulgaris* L. collected from lawns and contrasting grassland habitats. *New Phytologist,* **83,** 509–536.

Watson, M. A. (1984). Developmental constraints: Effect on population growth and patterns of resource allocation in a clonal plant. *American Naturalist,* **123,** 411–426.

Westhoff, V. & Maarel, E. van der (1978). The Braun-Blanquet approach. *Classification of Plant Communities* (Ed. by R. H. Whittaker), pp. 287–399. Dr. W. Junk, The Hague.

White, J. (1979). The plant as a metapopulation. *Annual Review of Ecology and Systematics,* **10,** 109–145.

White, J. (1984). Plant metamerism. *Perspectives on Plant Population Ecology* (Ed. by R. Dirzo & J. Sarukhán), pp. 15–47. Sinauer Associates, Sunderland, Massachusetts.

Wu, L. & Antonovics, J. (1975). Experimental ecological genetics in *Plantago* I. Induction of leaf shoots and roots for large scale vegetative propagation and tolerance testing in *P. lanceolata. New Phytologist,* **75,** 277–282.

5

Variation and Differentiation in Populations of Trifolium repens *in Permanent Pastures*

Roy Turkington

Botany Department
University of British Columbia
Vancouver, British Columbia, Canada

I. INTRODUCTION

In 1756, the Count de Buffon wrote (in French) "and if each type is further examined in different regions, varieties will be found responding in terms of size and form; to varying degrees all take on a colour from the region" (in Langlet 1971). This is one of the earliest clear descriptions of intraspecific differentiation in plant populations. For over two centuries there have been many studies describing patterns of differentiation in plant populations but the processes generating the patterns have either been ignored or referred to only speculatively; only a few researchers have given serious consideration to processes (e.g. Antonovics 1978, 1984; Bradshaw 1984). In most of this work the studies were dominated by what Harper (1977) has called a 'Wallacian attitude': questions were asked about situations in which the observed *abiotic* conditions varied markedly and were shown to be the prime determinants of the observed patterns of differentiation. These were important studies demonstrating clearly that micro-evolution occurred over very short distances, often in the presence of extensive gene flow, and over relatively short time intervals. In contrast, a 'Darwinian attitude' asks questions about *biotic* factors in the environment. This Darwinian attitude was revived by Harper (1967); in 1977 he wrote, "it is therefore rather odd that although it is Darwin's influence that is usually acknowledged in the development of evolutionary theory, it is a Wallacian attitude that has dominated the

STUDIES ON PLANT DEMOGRAPHY:
A FESTSCHRIFT FOR JOHN L. HARPER

study of adaptation in plants''. Further,

> much of the early science of plant ecology sought for correlations between vegetation and physical, not biotic, factors in the environment, most particularly temperature, water supply and soil types . . . they represent those agents of natural selection that were of more concern to Wallace than to Darwin in accounting for how organisms are as they are and behave as they do. The role of biotic (Darwinian) forces in determining the distribution and abundance of species was largely neglected except by token reference to grazing animals (Harper 1983).

A Darwinian might expect micro-evolution to be dominated by biotic interactions, and in those studies that have considered micro-evolutionary patterns of differentiation in this context, biotic factors have often been shown to be prime determinants of the observed patterns (see review by Turkington & Aarssen 1984).

In this essay I will focus my attention on some aspects of the research on the ecology of white clover, *Trifolium repens,* from our own and from Professor Harper's laboratories. I will discuss some of the small-scale patterns of differentiation and population variation that have been documented with *Trifolium repens* and then speculate on a few of the many possible processes that may account for the observed patterns. Most of the work described from North Wales was done in a single small field of permanent grassland, at Aber, near Bangor (described in Turkington & Harper 1979) and affectionately (and hereafter) referred to as 'the field'. The work originating from our own laboratory has been done on the Chard dairy farm about 40 miles from Vancouver, British Columbia (described in Aarssen & Turkington, 1985b, c), where there are four adjacent pastures which were planted in 1977, 1958, 1939 and approximately 1914.

II. POPULATION DIFFERENTIATION

''A single genet of *Trifolium repens* lives long and during its life the parts may wander through a sward as disconnected stolon fragments (like a terrestrial *Lemna!*)'' (Harper 1978). As *T. repens* spreads through a sward it continually encounters a mosaic of new neighbourhoods and with up to 50 different species of varying age, size and habit, the local environment may change dramatically over short distances and over short periods of time. On this basis it is reasonable to postulate that this highly diverse biotic environment imposes selection and evolution on a micro-scale. There is considerable evidence that different grasses impose different constraints on the growth of *T. repens* (Chestnutt & Lowe 1970). A single clone of *T. repens* expresses quite different growth forms when grown in the presence of different species of grass (Turkington 1983a, b) so there is every reason to expect that different species of grass may exert different selection forces on genotypes of white clover. This question was investigated by Turkington and Harper (1979) in 'the field'. Clones of *T. repens* were collected from within patches dominated by *Agrostis tenuis, Cynosurus cristatus, Holcus lanatus* and *Lolium perenne*. The clones were propagated in a glasshouse and then transplanted back into 'the field' in all possible 16 combinations of site of clone origin with site of transplanting. One year after transplanting, the clones that had been transplanted back into their sites of origin outyielded clones transplanted into other sites in 'the field' ($p < .001$). In a second experiment the four *T. repens* populations were grown in all possible combinations with the four species of grass in a glasshouse experiment, and in this case the relatively minor variations in substrate that undoubtedly occurred in the field were eliminated by using a standard potting compost throughout the experiment. With the exception of *T. repens* from the *H. lanatus*-dominated swards, each clover type grew best in association with the grass from which it had originally been sampled ($p < .001$) (Fig. 1). This is a fine-scale level of differentiation within the clover population, selected by different forces imposed by different grass neighbours. This leads to the hypothesis that the mosaic of neighbours, mostly grasses, with which *T. repens* coexists, constitutes a prominent element of biotic heterogeneity that sorts genotypes of *T. repens* on the basis of neighbour-specific compatibilities. This hypothesis has been investigated by Richard Evans (unpublished data) in studies of the genetic component of variation in a population of *T. repens* in a pasture in British Columbia sown in 1939. One hundred individuals were collected from each of four neighbourhoods dominated by *Dactylis glomerata, Holcus lanatus, Lolium perenne* and *Poa compressa* and transplanted to common garden conditions, where they were later individually scored for a number of morphological

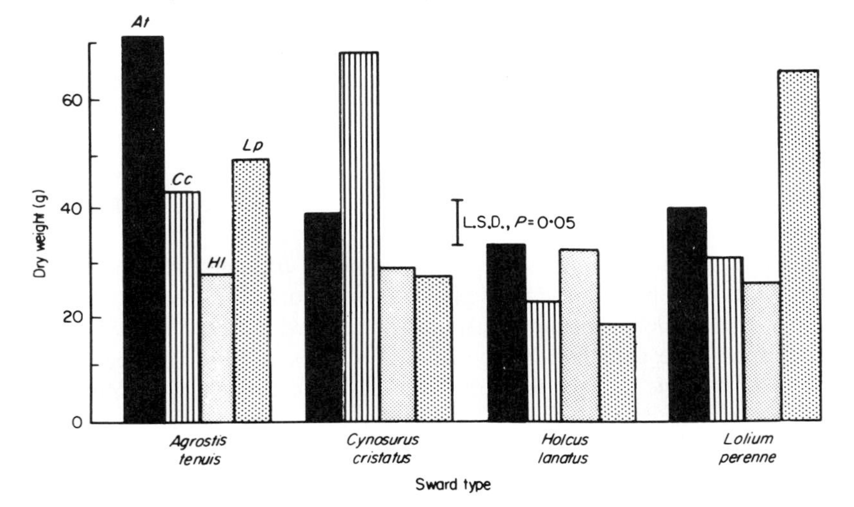

Fig. 1. The dry weight of plants of *Trifolium repens* from a permanent pasture, sampled from patches dominated by four different grasses and grown in all combinations of mixture with the four grass species. Clover 'types': At, *Agrostis tenuis;* Cc, *Cynosurus cristatus;* Hl, *Holcus lanatus;* Lp, *Lolium perenne*. [From Turkington & Harper (1979).]

TABLE I.

Summary of Analysis of Variance for Individuals of *Trifolium repens* Collected from Four Neighbourhoods Dominated by *Lolium perenne* (L), *Dactylis glomerata* (D), *Holcus lanatus* (H), and *Poa compressa* (P)[a]

Character	% Variation accounted for		Multiple range tests
Root weight	7.2	**	LD HP
Shoot weight	6.3	**	LD HP
Total weight	7.1	**	LD HP
Petiole length	18.9	**	LD HP
Primary stolon length	1.3	NS	
Secondary stolon length	1.9	NS	
Internode length	2.7	**	LD HP
Internode number	2.2	NS	
Primary stolon number	9.5	**	DL HP
Total stolon number	5.2	**	DL HP
Leaf mark		NS	

[a] Values represent the percentage variation accounted for by the variable 'neighbour'; characters for which the among-neighbour variance component is significant ($p < .01$) are designated by **. The right-hand column is Duncan's multiple range tests on the means for clovers classified by neighbourhood type (L,D,H,P) from which they were collected. Means are ranked from smallest to largest, and underlined sets of means are not significantly different ($p < .05$).

characteristics. Analysis of variance showed that for 7 out of the 11 characters a significant proportion of the variation was accounted for by the neighbouring grass species with which the individual of *T. repens* had been growing in the pasture (Table I). Results of Duncan's multiple range tests showed in addition that individuals of *T. repens* collected from *Lolium*- and *Dactylis*-dominated neighbourhoods were significantly differentiated from individuals collected from *Holcus*- and *Poa*-dominated neighbourhoods.

At a finer scale of investigation, Aarssen and Turkington (1985a) collected genets from four different pairs of physical neighbors of *T. repens* and *Lolium perenne,* from widely separated locations in this same pasture, from areas where both species were abundant. Individuals of *T. repens* were designated T and those of *L. perenne* L. Each individual was assigned a number (1–4) to identify the area where it was collected. When both species have the same number this indicates that they were growing together as physical neighbours in the pasture. All material was cloned, and the clovers and grasses were grown together for 1 year in pots in all of the 16 possible interspecific mixtures. Each clover genet had its highest yield when grown with its natural grass neighbour, but each grass genet had its lowest yield when grown with its natural clover neighbour. Com-

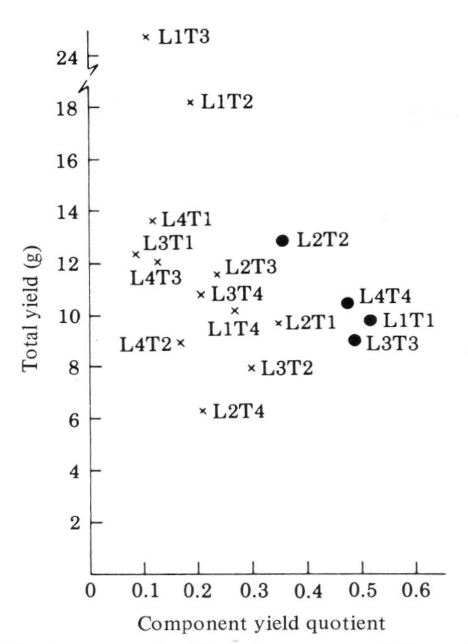

Fig. 2. Relationship between total yield and component yield quotient (yield of *T. repens* divided by yield of *L. perenne)* for different genet type combinations of *Lolium perenne* (L) and *Trifolium repens* (T). Naturally neighbouring genet pairs are circled and non-neighbour combinations are marked x. [From Turkington & Aarssen (1984).]

bined yield of natural neighbours was no greater than that of mixtures involving non-neighbours (Fig. 2). However, mixtures of natural neighbours had the highest component yield quotient, indicating biotic specialization in *T. repens* at the genotype level. Similarly, Trathan (1983) sampled *T. repens* and *L. perenne* from seven sites within 'the field', and polyacrylamide-gel electrophoresis revealed that both species were divided into distinct subpopulations with high numbers of unique genotypes. A greenhouse experiment showed that genotypes of *T. repens* which had been sampled close to genotypes of *L. perenne* produced a higher dry weight on average when grown with the *L. perenne* genotype with which it had been sampled than did any other genotype of *T. repens* (Trathan 1983). All of the studies outlined here are evidence of precise specialization of *T. repens* clones in their ability to grow in association with particular grass neighbours, and particular genotypes of *L. perenne*. It suggests that the mosaic of grass neighbours and genotypes acts as a powerful force of diversifying selection on these white clover populations, giving each individual clover a unique ecological identity.

A POSSIBLE MECHANISM

In many of the studies outlined above, competitive interactions have been invoked as the major process involved in generating the observed patterns. The many reviews in the literature attest to the volume of studies on plant competition, yet most of these studies have been largely descriptive, and only a few have paid close attention to the actual process of competition. From short-term descriptive and quantitative studies, ecologists often detect patterns but are forced to speculate about the processes that generated those patterns. Few testable hypotheses have emerged. This is not to imply that these studies are any less useful, but rather that they should only be considered as a necessary first step in the bigger question of understanding processes and mechanisms that generate patterns. This is not an easy task, and currently we are investigating one avenue that may increase our understanding of the processes underlying the close clover–grass relationships previously described. These investigations are being carried out by John Thompson (Botany Department) as part of a collaborative program with Brian Holl and Chris Chanway (Plant Sciences Department), and much of the following section was written in collaboration with them.

Biological nitrogen fixation is one of the most important factors determining long-term productivity in clover–grass pastures. Consequently, in searching for the processes at work in clover–grass interactions, the third major partner of the association, the nodule bacterium *Rhizobium trifolii,* must be considered. Williams (1970) recognized this question when he stated that "Another major gap in our understanding . . . is the precise relationship between the (white) clover plant together with its symbiont, and co-habiting grass species in a mixed sward". However, the relationship between *T. repens* and its microsymbiont *R. trifolii* is complex, even when the grass factor is ignored. There is abundant evidence of competition amongst *Rhizobium* strains in the rhizosphere and of selection of strains by the host plant (e.g. Masterson & Sherwood 1974; Jones & Hardarson 1979), but the processes governing these events are not understood.

In the analysis of variance of the genetic components of white clover–*Rhizobium* symbiosis, a non-additive plant genotype × *Rhizobium* strain interaction has been shown to account for 28% of the total variation in yield (Mytton 1975). Some authors have suggested that the most efficient plant–bacteria combination will necessarily result when *T. repens* is inoculated with a heterogeneous population of *R. trifolii* because selection by the host will occur (Masterson & Sherwood 1974; Mytton 1975) and result in a higher-yielding clover plant. Patterns of nodule formation and function are known to be influenced by acidity, calcium, cobalt, molybdenum, boron, light intensity, photoperiod, temperature, etc. The importance of these environmental factors on white clover productivity is recognized but still not adequately understood. The effectiveness of nodulation patterns as influenced by the nature of the host, the conditions of the rhizosphere, the strain of *Rhizobium,* and soil conditions have also been reasonably well described but not well understood.

There has been a tendency to concentrate on the influence of all these factors on white clover production. Because white clover is seldom grown in pure swards but as one species of a mixture in a pasture, many additional factors come into operation.

Grass species are known to influence the genetic composition of red and white clover populations (Charles 1968), and the grass–clover balance can be altered by changes in nitrogen (Vallis 1978). Consequently, careful analysis of the three-part system (grass–clover–*Rhizobium*) will be required to evaluate the importance of the individual components and to discover the causal mechanisms in their interrelationships.

Brian Holl and Chris Chanway are also investigating a possible fourth factor in the interaction. Although symbiotic dinitrogen fixation will be of major importance in assimilation of atmospheric nitrogen, diazotrophic rhizocoenosis (root-associated dinitrogen fixation) between grasses and free-living bacteria should not be ignored. Two developments suggest that grass-associated bacteria may play a significant role in the relationships described between individual clover and grass plants. The first is the demonstration that *Rhizobium* species are not limited to their leguminous hosts for induction of nitrogenase activity. Both free-living and non-legume root-associated *Rhizobium* have been shown to fix dinitrogen (Scowcroft & Gibson 1975; Gotz & Hess, 1980). Thus it may not be unreasonable to hypothesize that strains of *R. trifolii* can either associate with, or at least proliferate in, the rhizosphere of *L. perenne.* If this were the case, one could envisage a "tri-symbiosis", in which the clover and grass plants growing near one another are intimately associated with the local strain of *R. trifolii* found in their root systems. The second demonstration of importance suggests that a more conventional explanation may be found. Enhanced nodulation and plant performance have been observed in several species of legumes inoculated with mixed cultures of *Rhizobium* and *Azospirillum* or *Azotobacter.* Singh & Subba Rao (1979) correlated the presence of root-associated *Azospirillum brasilense* to increased nodulation and grain yield in soybean. El-Bahrawy (1983) has demonstrated similar trends in soybean by using *Azotobacter* and *R. japonicum.* Also, a *Rhizobium* × *Azospirillum* genotype–genotype interaction has been shown to be important in superior nodulation and dinitrogen fixation in winged bean and soybean (Iruthayathas, Gunasekaran & Vlassak 1983). These studies suggest that an intimate and, in some cases, highly specific relationship between *Azospirillum* or *Azotobacter* and *Rhizobium* exists. Consequently, it is possible that a relationship between free-living or *L. perenne* root-associated bacteria and strains of *R. trifolii* has evolved in these old pastures, such that symbiotic dinitrogen fixation is enhanced by the presence of the non-*Rhizobium* bacterial species.

III. POPULATION VARIATION

The patterns of differentiation just described are one component of population variation. *Trifolium repens* is a highly variable species commonly showing considerable differences within and between populations in a wide range of morphological and physiological characteristics (Turkington & Burdon 1983; Burdon 1983). The species has been shown to be polymorphic for cyanogenic glycosides (Dirzo & Harper 1982), leaf marks (Cahn & Harper 1976a, b; R. Evans, unpublished), leaf size (Harper 1977), relative growth rate (Burdon & Harper 1980; Trathan 1983), stolon length (Aarssen & Turkington 1985b), susceptibility to infection by *Rhizobium* (Mytton 1975), nitrogen fixation ability (Connolly, Masterson & Conniffe 1969), incompatibility systems (Atwood 1940), date of first flowering (Davies & Young 1967), flower production (Aarssen & Turkington 1985b), aggressiveness to different associated species of grass in the sward (Turkington & Harper 1979; Trathan 1983), and petiole length (Kerner von Marilaun 1895).

Burdon (1980a) investigated the intraspecific diversity of a population of *Trifolium repens* from 'the field' in North Wales. The vast majority of the 50 plants assessed differed from one another with respect to at least one character, and usually several characters; the clones on average differ significantly in 3.3 characters (Fig. 3). The level of diversity found within this single population for a range of characters was of a magnitude more commonly encountered in comparisons between populations taken from distinctly different environments. In this same pasture, Cahn & Harper (1976a) identified an average of between 3 and 4 clones per 10 × 10-cm quadrat; more recently, Trathan (1983), using isoenzyme analysis, identified 48 to 50 genotypes of *T. repens* per square meter.

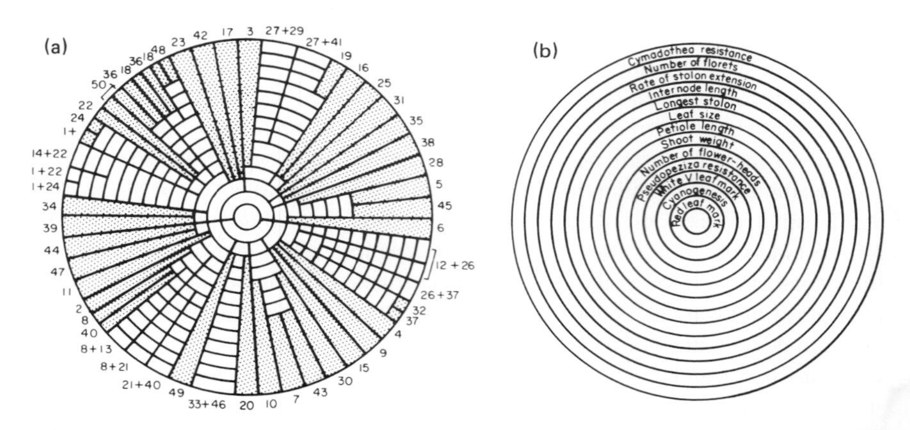

Fig. 3. a, Individuality diagram for 50 clones of *Trifolium repens* sampled from a permanent pasture. b, The 13 characters used to construct the individuality diagram. [From Burdon (1980a).]

In a similar study in British Columbia, Canada, Aarssen and Turkington (1985b) investigated the nature and extent of diversity within single natural populations of *Holcus lanatus, Lolium perenne* and *T. repens* collected from four adjacent pastures at the Chard dairy farm. The degree of variability within each population was large, but all three species showed significant differences in the quantity of variation present among the different-aged pastures. Many characters, especially for *L. perenne* and *T. repens,* showed a decline in mean values and variance with increasing pasture age (Fig. 4). As with the studies described from North Wales, this study also demonstrated an overall high level of variation within the *T. repens* population.

A POSSIBLE MECHANISM

The origin and maintenance of these high levels of diversity, especially in a clonal species, where there is the potential for a single, highly vigorous clone to dominate large areas by steadily eliminating less fit genotypes, is in question. Burdon (1980a) discounted selection and evolution on a micro-scale due to physical factors because previous studies had shown the physical environment of 'the field' to be quite uniform. Burdon argued that the local component of environmental variation is determined almost exclusively by biotic factors such as the identity, age and size of neighbours and the degree to which they compete.

> In a permanent pasture containing a large number of different plant species, the local environment clearly may change radically not only from place to place in the field but also from time to time. The tremendous temporal and spatial variation in the nature and intensity of competitive interactions which results from this patchiness represents a complex array of local directional selective forces which, in an otherwise uniform environment, are likely to dominate the immediate fitness of individuals and produce evolution on a micro-scale. . . . The tremendously variable nature of the local biotic environment may thus act to produce and maintain a high level of variation in the plant population (Burdon 1980a).

Although Aarssen and Turkington (1985b) confirm Burdon's demonstration of enormous variation in the *T. repens* population, their study also shows that population variation decreases with time. Variable local biotic environments may then be merely slowing down the rate of decrease of population variation. Both the Burdon (1980a) and Aarssen and Turkington (1985b) studies have demonstrated high levels of population variation, but they do not offer explanations of the origin and maintenance of the variation, i.e. they do not demonstrate a process or a mechanism to account for the observed patterns.

One very obvious process by which variation can be maintained in spite of the demonstrated genotype depletion is continuous input of new genotypes. New genotypes are continually introduced into both the Bangor field and the British Columbia pasture populations, but they apparently do not occur in the main sward-dominated areas (Turkington *et al.* 1979; R. Parish unpublished). Roberta

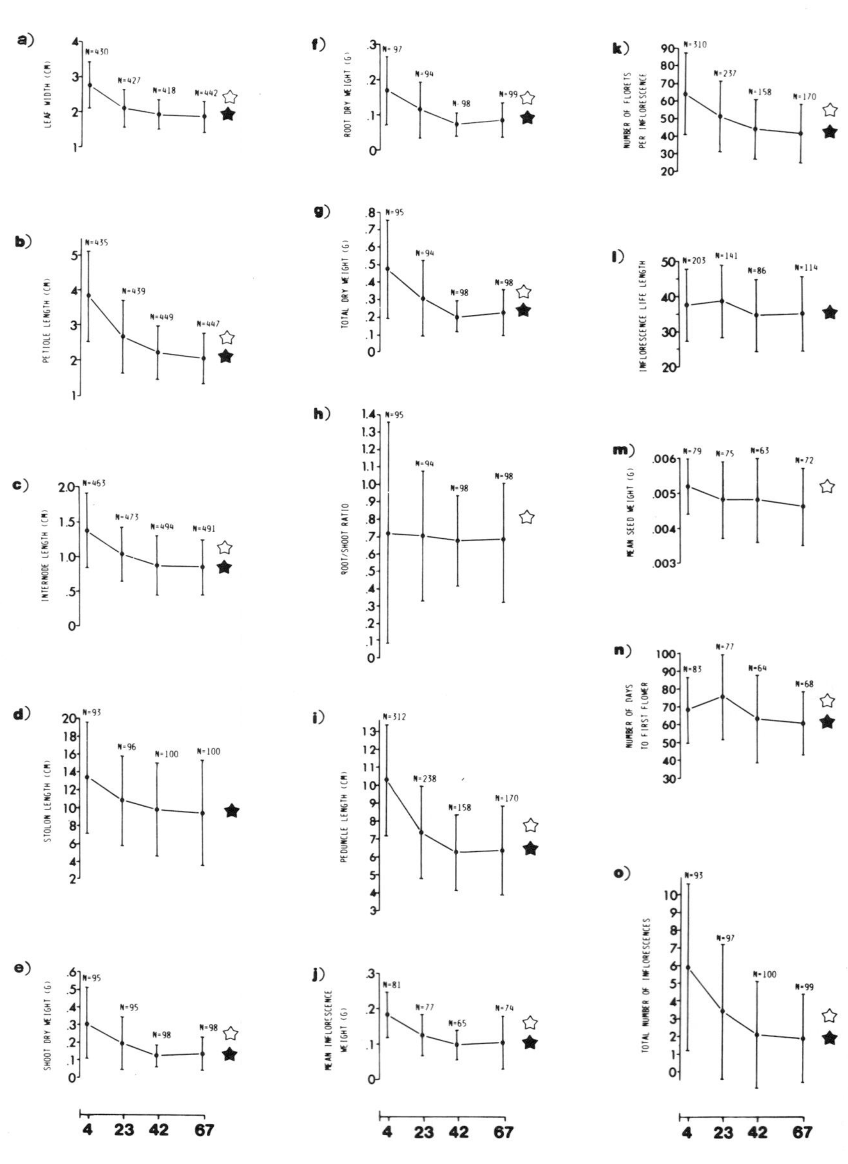

Fig. 4. The relationship between each recorded character for *Trifolium repens* and the age of the pasture from which plants were collected. Means and standard deviation bars are plotted. An open symbol (☆) indicates a significant difference among variances based on an F_{max} test. A closed symbol (★) indicates a significant difference among means based on a one-way analysis of variance.

Parish has shown that irregular recruitment of new *T. repens* seedlings occurs on mole hills and dung piles (Table II), but these still represent a small genotype input to the entire pasture. However, although there is a tendency to dismiss seedling input if it is at a very low level, Soane and Watkinson (1979), using a simulation model, have shown that in clonal species "it is clear that occasional establishment of seedlings (≅3% of the ramet population per year) is sufficient to maintain considerable genet variation within the population . . . consequently it may not always be necessary to search for disruptive forces to explain diversity within the field".

Burdon (1980a) pointed out, citing the work of Breese, Haywood and Thomas (1965), that significant differences have been shown to occur in *Lolium perenne* between individual plants derived by clonal multiplication from the same parent and suggests that some of the observed variation in the *T. repens* population may be somatic. From an ecological point of view, the significance of somatic mutations has been developed primarily by Whitham and his colleagues (e.g. Whitham & Slobodchikoff 1981; Whitham, Williams & Robinson 1984). Whitham & Slobodchikoff (1981) argue that somatic mutations may be an important source of heritable variation in plants and list examples of variation within progeny of asexually propagated plants. Because of the population-like (modular) structure of most plants (Harper 1977, 1981) the role of somatic mutations in modular and unitary organisms is fundamentally different. A beneficial somatic mutation that occurs in any apical meristem of a clover can spread, but if the mutation is deleterious, then the death of that single meristem will have little effect on the survival of the whole plant. In unitary organisms, somatic mutations have no known mechanisms of being preserved and passed to succeeding generations. "Because of these basic differences, somatic mutations may enable plants to be better than animals at 'fine tuning' themselves to local environments . . . it would seem that somatic mutations represent a reasonable

TABLE II.

The Number of Seedlings of *Trifolium repens* Observed on Disturbances (Mole Hills and Dung) in Three Different-aged Pastures in British Columbia, and a Calculated Number of Seedlings per m² after 3 Months

	Date of pasture sowing		
	1939	1958	1977
Number of seedlings per disturbance			
Initial density	20.2	5.6	10.6
After 3 months	6.5	1.9	3.0
Calculated number of seedlings per m^2, after 3 months	4.9	1.2	1.5

source of heritable variation that would permit adaptation to local conditions'' . . . (Whitham & Slobodchikoff 1981). An individual clone of *T. repens* may be long-lived, and a large clone with many stolons may place apical meristems into a wide array of neighbourhoods. If intraclonal (somatic) variation exists, such a clone may evolve into a mosiac of genotypes. If such evolution by an individual is possible, this could lead to the patterns of population variation and the fine-scale differentiation patterns described in this essay.

ACKNOWLEDGMENTS

I would like to express my appreciation to Roberta Parish, Richard Evans and John Thompson for evaluating an early draft of this paper and for permitting me to use material which has not yet appeared in thesis form, and to Brian Holl and Chris Chanway for their collaborative efforts on the *Rhizobium* project. I will always be grateful to Bill and Mary Chard, who have given us unrestricted access to their pastures.

REFERENCES

Aarssen, L. W. & Turkington, R. (1985a). Biotic specialization between neighbouring genotypes in *Lolium perenne* and *Trifolium repens* from a permanent pasture. *Journal of Ecology,* **73,** 605–614.

Aarssen, L. W. & Turkington, R. (1985b). Intraspecific diversity in natural populations of *Holcus lanatus, Lolium perenne* and *Trifolium repens* from four different aged pastures. *Journal of Ecology* (in press).

Aarssen, L. W. & Turkington, R. (1985c) Vegetation dynamics and neighbour associations in pasture-community evolution. *Journal of Ecology,* **73,** 585–603.

Antonovics, J. (1978). The population genetics of mixtures. *Plant Relations in Pastures* (Ed. by J. R. Wilson), pp. 233–252. CSIRO, East Melbourne, Australia.

Antonovics, J. (1984). Genetic variation within populations. *Perspectives on Plant Population Ecology* (Ed. by R. Dirzo & J. Sarukhán), pp. 229–241. Sinauer Associates, Sunderland, Massachusetts.

Atwood, S. S. (1940). Genetics of cross incompatibility among self incompatible plants of *Trifolium repens. Journal of the American Society of Agronomy,* **32,** 955–968.

Bradshaw, A. D. (1984). Ecological significance of genetic variation between populations. *Perspectives on Plant Population Ecology* (Ed. by R. Dirzo & J. Sarukhán), pp. 213–228. Sinauer Associates, Sunderland, Massachusetts.

Breese, E. L., Haywood, M. D. & Thomas, A. C. (1965). Somatic selection in perennial ryegrass. *Heredity,* **20,** 367–379.

Burdon, J. J. (1980a). Intra-specific diversity in a natural population of *Trifolium repens. Journal of Ecology,* **68,** 717–735.

Burdon, J. J. (1980b). Variation in disease-resistance within a population of *Trifolium repens. Journal of Ecology,* **68,** 735–744.

Burdon, J. J. (1983). Biological flora of the British Isles, *Trifolium repens. Journal of Ecology,* **71,** 307–330.

Burdon, J. J. & Harper, J. L. (1980). Relative growth rates of individual members of a plant population. *Journal of Ecology,* **68,** 953–957.

Cahn, M. A. & Harper, J. L. (1976a). The biology of the leaf mark polymorphism in *Trifolium repens* L. I. Distribution of phenotypes at a local scale. *Heredity,* **37,** 309–325.

Cahn, M. A. & Harper, J. L. (1976b). The biology of the leaf mark polymorphism in *Trifolium repens* L. II. Evidence for the selection of leaf marks by rumen fistulated sheep. *Heredity,* **37,** 327–333.

Charles, A. H. (1968). Some selective effects operating upon white and red clover. *Journal of the British Grassland Society,* **23,** 20–25.

Chestnutt, D. M. B. & Lowe, J. (1970). Agronomy of white clover/grass swards. *White Clover Research* (Ed. by J. Lowe), pp. 191–213. Occasional Symposium No. 6. British Grassland Society, Hurley.

Connolly, V., Masterson, C. L. & Conniffe, D. (1969). Some genetic aspects of the symbiotic relationship between white clover (*Trifolium repens* L.) and *Rhizobium trifolii. Journal of Theoretical and Applied Genetics,* **39,** 206–213.

Davies, W. & Young, N. R. (1967). Self fertility in *Trifolium frageriferum. Heredity,* **21,** 615–624.

Dirzo, R. & Harper, J. L. (1982). Experimental studies on slug-plant interactions. III. Differences in the acceptability of *Trifolium repens* to slugs and snails. *Journal of Ecology,* **70,** 101–118.

El-Bahrawy, S. A. (1983). Associative effects of mixed cultures of *Azotobacter* and different rhizosphere fungi with *Rhizobium japonicum* on nodulation of symbiotic nitrogen fixation of soybean. *Zentralblatt für Mikrobiologie (Jena),* **138,** 443–449.

Gotz, E. M. & Hess, D. (1980). Nitrogenase activity induced by wheat plants (Tricitum aestivum). Zeitschrift fuer Pflanzenphysiologie, **98,** 453–458.

Harper, J. L. (1967). A Darwinian approach to plant ecology. *Journal of Ecology,* **55,** 247–270.

Harper, J. L. (1977). *Population Biology of Plants.* Academic Press, London.

Harper, J. L. (1978). The demography of plants with clonal growth. *Structure and Functioning of Plant Populations* (Ed. by Freysen, A. H. J. & Woldendorp, J. W.), pp. 27–48. North-Holland, Amsterdam.

Harper, J. L. (1981). The concept of population in modular organisms. *Theoretical Ecology: Principles and Applications* (Ed. by R. M. May), pp. 53–77. Blackwell Scientific Publications, Oxford.

Harper, J. L. (1983). A Darwinian plant ecology. *Evolution from Molecules to Man* (Ed. by D. S. Bendall), pp. 323–325. Cambridge University Press, London.

Iruthayathas, E. E., Gunasekaran, S. & Vlassak, K. (1983). Effect of combined inoculation of *Azospirillum* and *Rhizobium* on nodulation and N_2-fixation of winged bean and soybean. *Scientia Horticulturae (Amsterdam),* **20,** 231–240.

Jones, D. G. & Hardarson, G. (1979). Variation within and between white clover varieties in their preference for strains of *Rhizobium trifolii. Annals of Applied Biology,* **92,** 221–228.

Kerner von Marilaun, A. (1895). *The Natural History of Plants.* Vol. 2. Blackie, London.

Langlet, O. (1971). Two hundred years genecology [sic]. *Taxon,* **20,** 653–722.

Masterson, C. L. & Sherwood, M. T. (1974). Selection of *Rhizobium trifolii* strains by white and subterranean clovers. *Irish Journal of Agricultural Research,* **13,** 91–99.

Mytton, L. R. (1975). Plant genotype × *Rhizobium* strain interactions in white clover. *Annals of Applied Biology,* **80,** 103–107.

Scowcroft, W. R. & Gibson, A. H. (1975). Nitrogen fixation by *Rhizobium* associated with tobacco and cowpea cell cultures. *Nature (London)* **253,** 351–352.

Singh, C. S. & Subba Rao, N. S. (1979). Associative effect of *Azospirillum brazilense* with *Rhizobium japonicum* on nodulation and yield of Soybean (*Glycine max*). *Plant and Soil,* **53,** 387–392.

Soane, I. D. & Watkinson, A. R. (1979). Clonal variation in populations of *Ranunculus repens. New Phytologist,* **82,** 557–573.

Trathan, P. (1983). Clonal interactions of *Trifolium repens* and *Lolium perenne*. Ph.D. thesis, University of Wales, Bangor.

Turkington, R. (1983a). Leaf and flower demography of *Trifolium repens* L. I. Growth in mixture with grasses. *New Phytologist,* **93,** 599–616.

Turkington, R. (1983b). Leaf and flower demography of *Trifolium repens* L. II. Locally differentiated populations. *New Phytologist,* **93,** 617–631.

Turkington, R. & Aarssen, L. W. (1984). Local-scale differentiation as a result of competitive interactions. *Perspective on Plant Population Ecology* (Ed. by R. Dirzo & J. Sarukhán), pp. 107–127. Sinauer Associates, Sunderland, Massachusetts.

Turkington, R. & Burdon, J. J. (1983). Biology of Canadian weeds. 57. *Trifolium repens* L. Canadian Journal of Plant Science, **63,** 243–266.

Turkington, R. & Harper, J. L. (1979). The growth, distribution and neighbour relationships of *Trifolium repens* in a permanent pasture. IV. Fine scale biotic differentiation. *Journal of Ecology,* **67,** 245–254.

Turkington, R., Cahn, M. A., Vardy, A. & Harper, J. L. (1979). The growth, distribution and neighbour relationships of *Trifolium repens* in a permanent pasture. III. The establishment and growth of *Trifolium repens* in natural and perturbed sites. *Journal of Ecology,* **67,** 231–243.

Vallis, I. (1978). Nitrogen relationships in grass/legume mixtures. *Plant Relations in Pastures* (Ed. by J. R. Wilson), pp. 190–201, CSIRO, East Melbourne, Australia.

Whitham, T. G. & Slobodchikoff, C. N. (1981). Evolution by individuals, plant-herbivore interactions, and mosaics of genetic variability: The adaptive significance of somatic mutations in plants. *Oecologia,* **49,** 287–292.

Whitham, T. G., Williams, A. G. & Robinson, A. M. (1984). The variation principle: Individual plants as temporal and spatial mosaics of resistance to rapidly evolving pests. *A New Ecology: Novel Approaches to Interactive Systems* (Ed. by P. W. Price, C. N. Slobodchikoff & W. S. Gaud), pp. 16–51. Wiley, New York.

Williams, W. (1970). White clover in British Agriculture. *White Clover Research* (Ed. by J. Lowe), pp. 1–10. Occasional Symposium No. 6. British Grassland Society, Hurley.

6

Disasters and Catastrophes in Populations of Halimione portulacoides

Ad H. L. Huiskes and Atie W. Stienstra[1]

Delta Institute for Hydrobiological Research
Yerseke, The Netherlands

I. INTRODUCTION

Going through (a part of) John Harper's numerous publications, the reader will notice that two statements always return in one way or another. The first is the statement that a population consists of a number of individuals and that it is the response of these individuals to certain stimuli that results in the dynamics of the population (e.g. Harper 1964, 1968, 1977). The second statement is that one has to look at the behaviour of individual plants and plant populations from a 'plant's eye view' and not anthropocentrically (e.g. Harper 1977, 1982).

This has to be kept in mind whilst studying the behaviour of populations of *Halimione portulacoides*.[2] This salt marsh plant seems totally unfit for the habitat in which it grows: it prefers aerated soils (Chapman 1937, 1950), although large parts of the salt marsh soils are anaerobic; it is sensitive to inundation (A. M. Groenendijk, personal communication), whereas the salt marshes are flooded regularly; it grows best at low salt concentrations (Jensen 1985; Stienstra & Markusse 1978), although salinity of the salt marsh soil is usually much higher; its frost hardiness is low (Beeftink *et al.* 1978); and it occurs in salt marshes of the temperate climate. And yet it grows under all these 'adverse' circumstances and may cover large parts of the salt marsh (Beeftink 1959; Beeftink *et al.* 1978; Henriksen & Jensen 1979). This paper deals with the influence of disasters and

[1]Present address: Department of Plant Ecology, Lange Nieuwstraat 106, 3512 PN Utrecht, The Netherlands.

[2]Nomenclature follows Van der Meyden *et al.* (1983).

STUDIES ON PLANT DEMOGRAPHY:
A FESTSCHRIFT FOR JOHN L. HARPER

catastrophes on populations of *H. portulacoides*. The terms *disaster* and *catastrophe* are used in the way Harper described them:

> A disaster recurs frequently enough for there to be reasonable expectation of occurrence within the life cycles of successive generations. . . . A catastrophe occurs sufficiently rarely that few of its selective consequences are relevant to the fitness of succeeding generations. The selective consequence of disasters is therefore likely to be to increase short-term fitness and the consequence of catastrophes is to decrease it (Harper 1977).

A disaster to individuals of the species is, for instance, a severe frost that kills large portions of the population. A form of catastrophe is, for instance, a major change in abiotic environmental factors.

H. portulacoides is a chamaephyte and is solely confined to the salt marsh habitat. It is usually found in salt marshes with frequent but short inundations (Chapman 1950); in the Netherlands it occurs in areas that are inundated between 100 and 250 times a year (Beeftink 1965). Several studies point out that the species needs a well-drained substrate to grow on (Beeftink 1965; Chapman 1950; Jensen 1985), which explains why the species is often found on the creek banks in the salt marsh, where the grain size of the substrate is usually coarser than elsewhere. As in most halophytes a certain amount of salt in the substrate does not affect the growth of *H. portulacoides*. Moreover, like some other species it seems to need a high internal level of electrolytes for optimal growth (Albert & Popp 1977; Flowers, Troke & Yeo 1977; Greenway 1968; Osmond, Björkman & Anderson 1980). Although it can withstand moderate frost despite the fact that it is wintergreen (Kappen 1969), large parts of the population in a certain area can be destroyed by severe frost. The population can be restored by successional processes after a number of years (Beeftink *et al.* 1978). The percentage cover of live *Halimione* plants after a frost can drop from 90 to 10%. In the present study attention is paid to individual plants, especially to the fate of the emerging seedlings.

To safeguard the low-lying parts of the Netherlands against flooding by the sea and to minimize the salt intrusion via the rivers, a number of estuarine branches of the rivers Rhine, Meuse and Scheldt are closed off. This means that the tidal salt marshes along these estuaries are no longer flooded, become desalinized and gradually lose their salt marsh characteristics. *Halimione portulacoides,* too, disappears within a number of years. Although this process is not a sudden event, the result will be a complete disappearance of the population and may qualify as a catastrophe. In this paper an attempt will be made to describe the recovery of a population after a severe frost (disaster) and to explain the disappearance of the species from a formerly tidal salt marsh after the cessation of the tidal influence (catastrophe). As the recovery of the population is a population-dynamic process, it will be analysed with demographic techniques. The disappearance of *Halimione* is of course a population-dynamic process, too, but brought about by physiological processes, which we choose to describe by a physiological approach.

II. MATERIALS AND METHODS

A. Disaster

The research on the restoration of the population after the winter of 1978–1979 was carried out on a salt marsh near Ellewoutsdijk along the Western Scheldt (1 in Fig. 1). Four plots of 50 × 50 cm^2 were marked out with wooden stakes in areas where most of the *Halimione* shrub had died following severe winter frosts. In plots *a, b* and *c* the dead shrubs were left in place. Around plot *c* (within 50 cm) a few adult *Halimione* plants were still alive. Around plots *a* and *b* no live adult plants were found within this distance. From plot *d* all the dead material was removed. In all four plots the emerging seedlings were marked with pieces of coloured wire. At the first recording the plots were mapped as well.

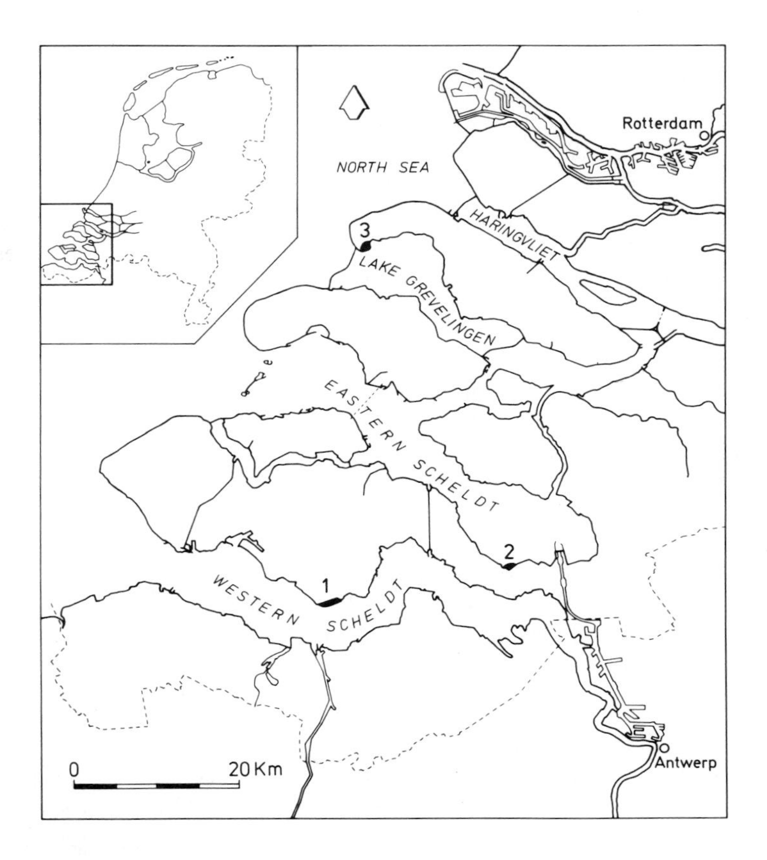

Fig. 1. Map of the southwest part of the Netherlands with the location of the study areas. 1 = Ellewoutsdijk marsh; 2 = Stroodorpepolder marsh (both tidal); 3 = Springersgors marsh (non-tidal).

B. Catastrophe

The influence of the termination of the inundation with salt water was studied by comparing the fate of the plants on tidal salt marsh (Stoodorpepolder marsh, 2 in Fig. 1) and on a former salt marsh behind one of the barrier dams in the process of desalinisation (Springersgors marsh, 3 in Fig. 1). As the main feature on this latter marsh was the change in soil properties, especially mineral levels and moisture content, it was decided to analyse this problem by looking at the uptake of minerals by the plants.

In these two marshes plots were marked out; from them leaf samples were taken at regular intervals during the growing season of 1975. Four plots were laid out in the tidal marsh: one on a creek bank in a closed *Halimione* vegetation (I), one a little away from the creek bank in the marsh where the *Halimione* does not form a closed sward (II), one in the marsh where *Halimione* grows between other species such as *Limonium vulgare* and *Puccinellia maritima* (III) and one on a creek bank where the species grows in a mixed vegetation with *Elymus pycnanthus* (=*Elytrigia pungens*) (IV). In the Springersgors marsh three plots were laid out: two of them in the marsh having small *Halimione* plants with a stunted growth in between a vegetation of various other species, mainly ruderal glycophytes (V and VI) and a third plot on a creek bank where *Halimone* also grew between other species but seemed to perform a little better (VII).

The leaves were sampled in four age classes: young, not fully developed leaves; fully grown leaves from the upper part of the branches; fully grown leaves from the lower part of the branches; yellow leaves from the base of the branches. Since a complete set of data could be obtained from the fully grown leaves from the upper part of the stem over the year, the discussion will especially focus on the findings in this age class. After being washed and dried at 80°C for 48 hours, the leaf samples were analysed for Na^+, K^+, Ca^{2+}, Mg^{2+}, P_2O_5, Cl^-, SO_4^{2-}, NO_3^- and total nitrogen. Sodium, potassium, calcium and magnesium were analysed by means of atomic absorption spectrophotometry, P_2O_5 colorimetrically in the same destruate, SO_4^{2-}, NO_3^- and Cl^- were measured in an acetic acid extract; sulphate and nitrate colorimetrically, chloride titrimetrically. Total nitrogen was determined by the Kjeldahl method.

Soil samples were taken at each collection of leaves (top 5 cm). The soil samples were analysed for moisture content, pH, silt content and the exchangeable amounts of the ions also measured in the plant samples.

III. RESULTS

A. Disaster

Figure 2 gives the fate of the seedlings of *H. portulacoides* over a 3-year period after the winter of 1978–1979. It is rather difficult to mark *H. por-*

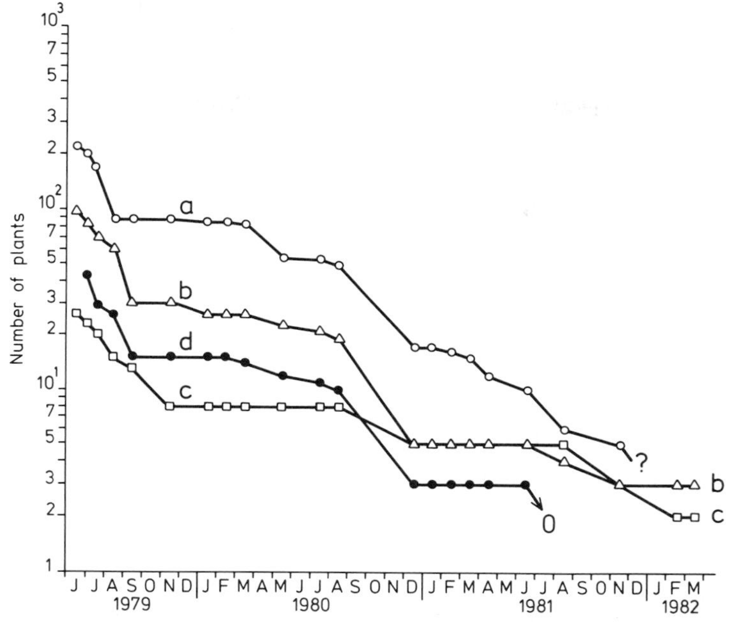

Fig. 2. The fate of cohorts of *Halimione portulacoides* seedlings in plots of 50 × 50-cm² in the salt marsh near Ellewoutsdijk. ○ = plot a; △ = plot b; □ = plot c; ● = plot d.

tulacoides as the stems lie in part flat on the ground and are likely to be silted over. This makes it difficult to distinguish the individual plants as they get older. After the first recording, marking and mapping of the seedlings, only the marked seedlings were noted during subsequent recordings. This means that plants that lost their marker one way or another were not taken into account although they were still there; the numbers of individuals recorded over the years may have been higher, therefore. But it is clear from Figure 2 that only a small portion of the seedlings reach the reproductive stage. No subsequent cohorts of seedlings were found after the first recording of the plots. Probably the seeds present all germinated in the same period, and no substantial input afterwards from elsewhere took place. Careful checks in the top layer of the soil of the plots did not reveal any seeds either. After 3 years plot *a* was covered with a mixed vegetation of *Halimione* and *E. pycnanthus*. Unfortunately no markers could be found any longer. In plot *d* no plants of *Halimione* were left after 3 years. The whole plot was covered with a dense stand of *E. pycnanthus*. The other two plots had a mixed vegetation of *E. pycnanthus* and *H. portulacoides*.

B. Catastrophe

Table I gives a survey of various soil parameters of the seven plots where the leaves were sampled. Only the soil samples taken at the first sampling date are

TABLE I.

Survey of Some Analyses on Soil Samples Taken on 26 June 1975 from Plots Marked Out in a Salt Marsh and a Former Salt Marsh in the Southwest Netherlands[a]

							Exchangeable cations (meq per 100 g dry soil)					Height above N.A.P.
Plot	Moisture (% dry soil)	NaCl (g/100 g dry soil)	NaCl (g/litre soil water)	pH (KCl)	Silt (% <16 μm)	Exch. P_2O_5 (mg/100 g dry soil)	Na^+	K^+	Ca^{2+}	Mg^{2+}	NH_4^+	(cm)
I	67.6	1.32	19.5	7.1	25	36	8.0	1.3	9.8	8.6	0.6	230
II	109.1	2.79	25.6	7.1	25	35	14.4	2.3	11.8	11.9	1.0	224
III	118.5	3.49	29.4	7.1	38	31	17.6	2.9	9.2	13.8	1.1	200
IV	57.1	1.27	22.3	7.3	20	40	6.7	1.3	8.0	8.2	0.5	235
V	49.2	0.08	1.7	7.2	30	63	2.0	4.0	22.0	12.1	1.1	
VI	62.5	0.03	0.4	7.1	40	67	1.6	4.7	20.0	12.5	1.3	
VII	33.9	0.01	0.4	7.1	32	49	0.6	1.9	10.3	5.2	0.4	

[a]The samples are taken from the top 5 cm of the soil. N.A.P. = Dutch Ordnance Level.

represented in the table, as the fluctuation over the growing season does not really contribute to the tenor of this paper.

In the Stroodorpepolder marsh plots II and III have a higher soil moisture content than the plots I and IV. The latter plots are on a creek bank and have a better drainage due to a coarser particle size. The amount of exchangeable ions is also lower on the creek banks, probably as a result of the same effect. In the Springersgors marsh the soils are much drier, especially in the summer months. The soils in this marsh have a low salinity as compared with the tidal marsh, which is of course not surprising after being embanked for 4 years. The pH does not differ very much, nor does the silt content. In the non-tidal marsh the levels of phosphate, calcium and potassium are higher than in the tidal marsh.

Figure 3 gives the results of the analyses of the major inorganic ions in the leaves of the upper half of the branches of *Halimione*. Only the results of the first sampling (late spring) are depicted. It is clear from the figure that the leaves have a high ion content. The levels change somewhat over the season, but the difference in ion levels between leaves sampled on the tidal salt marsh and on the former tidal salt marsh is more pronounced. There is also a marked difference in the ratio of analysed cations and anions.

The leaves contain high levels of Na^+ and Cl^- ions. In the late spring 65% of the analysed cations is Na^+, in summer and autumn 75%, for the leaves collected in the tidal marsh, and 55% and 65%, respectively, for the leaves of the former tidal marsh. For chloride the figures are even higher: more than 90% of the analysed anions is Cl^- in the tidal marsh; in the former tidal marsh it is 60–80%. Leaves sampled in the latter marsh also had higher nitrate and phosphate levels. In the leaves of the tidal marsh hardly any nitrate could be detected.

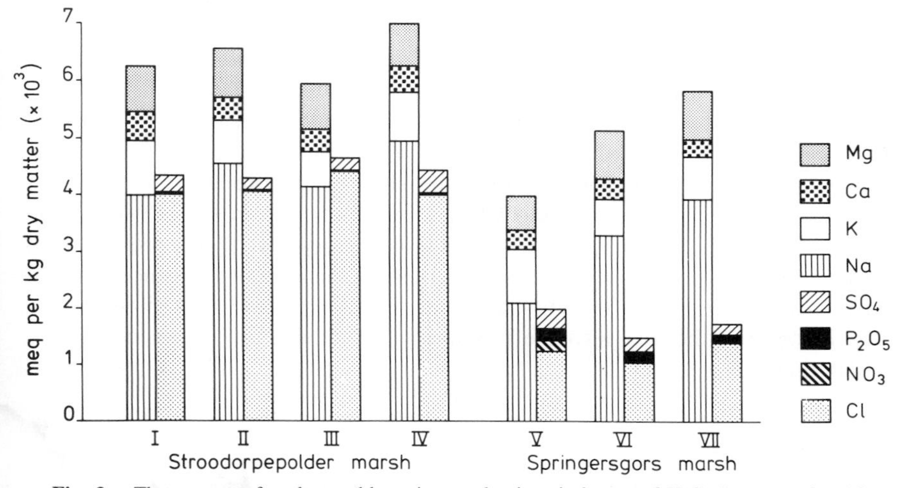

Fig. 3. The amount of exchangeable cations and anions in leaves of *Halimione portulacoides*, collected on a tidal (Stroodorpepolder) and non-tidal (Springersgors) former salt marsh.

The difference in calcium and magnesium levels of the leaves of the two marshes are small, but there is a difference in total amount of nitrogen.

IV. DISCUSSION

A. Disaster

The emergence of seedlings of *H. portulacoides* is not a rare event in itself. Every year seedlings can be found in the vicinity of adult plants (H. Schat, personal communication). But they are few, and a sudden emergence of numerous seedlings after the death of the old plants is certainly remarkable. What causes the sudden emergence is not exactly known. A possible explanation could be the change in the light, both in quantity and quality, that reaches the soil after the loss of the leaf canopy. This has been shown, for instance, for *Cirsium vulgare* (Pons 1983). Waisel (1972) mentions the light requirement for the germination of *Atriplex* species. It is also in accordance with the hypothesis of Cresswell and Grime (1981), who argue that seeds that have investing structures which retain chlorophyll throughout the maturation will contain phytochrome in its active form and need a light stimulus to germinate; they gave examples of other Chenopodiaceae with chlorophyll-containing maternal tissues envelopping the seed. *H. portulacoides* is also a Chenopod with green bracteoles around the seed and may therefore fit their hypothesis.

Although many of the plants die before they reach the adult stage, the few that remain are apparently able to colonize the area and to replace the plants that have died. Of course the clonal spread of the survivors plays an important role also. Beeftink *et al.* (1978) describe succession models in a *Halimione* vegetation, which show that a complete return to the original situation after a disaster may take 4 to 12 years, depending on the pathway of (and the species involved in) the succession and the kind of disaster. Three years of data were therefore far too limited to show a complete return to the situation before the frost, but the question whether seedlings would be able to play a role in the recolonisation could be answered in a positive way, although very few plants are likely to survive. From a population genetics point of view this means that genotypes that have proved to be fit for the habitat are spreading clonally and are kept within the population, but there is also the possibility of adding new genotypes to the populations.

B. Catastrophe

Quite in contrast with this restoration of the population after a disaster is the extinction of the population after a major environmental change. Beeftink (per-

TABLE II.

Chemical Analyses of Leaves of Some Glycophytes Growing in the Same Locations as *Halimione portulacoides* in Springersgors Marsh

	Chemical, meq. per kg dry matter								Total N, mmol per kg dry matter
Glycophyte	Cl^-	NO_3^-	P_2O_5	SO_4^{2-}	Na^+	K^+	Ca^{2+}	Mg^{2+}	
Solanum dulcamara	262	4	60	173	348	675	235	385	1621
Cirsium vulgare	1102	2	57	154	435	1013	655	574	1536
Sonchus arvensis	1291	56	81	506	2043	1008	605	451	1557
Epilobium angustifolium	333	1	76	—	478	497	720	738	1200

sonal communication) concluded on the results of his vegetation studies that it takes on average about 4 years for *Halimione* populations to disappear.

Four years after the cessation of the tidal influence on the Springersgors marsh the results show marked differences in soil characteristics between the two study areas (Table I). The Springersgors marsh has a much lower salt content and higher phosphate and calcium levels. And yet in the leaves of *Halimione* high levels of cations and anions are found, with sodium and chloride as the major ones. When the values are compared with those of other plants, they are higher than in most glycophyte species growing there (Table II).

Albert and Popp (1977) divided the halophytes into a number of 'physiotypes', whereby the Chenopodiaceae (and thus *Halimione*) were characterized by a high affinity for sodium and chloride ions and with low K^+/Na^+ ration. The halophytic chenopods grow normally in areas with low nitrate and phosphate levels. On the other hand, Chenopodiaceae are also known from ruderal areas accumulating large amounts of nitrate. Albert and Popp (1977) concluded that the accumulation of sodium and chloride in halophytic chenopods and the accumulation of nitrate in glycophytic chenopods were analogous processes. Osmond, Björkman and Anderson (1980), Kirkby and Knight (1977) and Jensen (1985) suggest an interaction between Cl^- and NO_3^-.

If the maintenance of a high electrolyte content, especially of chloride, is imperative for the plant to have an optimal growth, the plant needs a substrate that allows it to have a high uptake of chloride (Flowers, Troke & Yeo 1977; Yeo 1983; Jensen 1984). In the saline Stroodorpepolder marsh this is not a problem, but in the ever-more desalinizing Springersgors marsh, the plant is impaired in the uptake of sufficient electrolyte, which affects the growth in a negative way and causes the eventual disappearance of *Halimione*. However cogent this reasoning, it is still an assumption, not an explanation. What we ought to know is the electrolyte level in the leaves whereby the growth begins to be suboptimal. Only then is it possible to argue that the plant electrolyte level in the Springersgors marsh is too low for an optimal functioning of the plants. This paper therefore serves in that respect more as a hypothesis than a conclusion.

ACKNOWLEDGMENTS

The authors are indebted to Messrs. Markusse, Nieuwenhuize and Van Liere for collecting the samples and analysing them. They also wish to thank Ms. van Leerdam for typing the manuscript in its final form, Mr. Bolsius for preparing the drawings and Drs. Duursma and Beeftink for making valuable suggestions for improving the text.

REFERENCES

Albert, R. & Popp, M. (1977). Chemical composition of halophytes from the Neusiedler Lake region in Austria. *Oecologia,* **27,** 157–170.

Beeftink, W. G. (1959). Some notes on Skallingens salt marsh vegetation and its habitat. *Acta Botanica Neerlandica,* **8,** 449–472.

Beeftink, W. G. (1965). De zoutvegetatie van Z.W.-Nederland beschouwd in Europees verband. *Mededelingen Landbouwhogeschool Wageningen,* **65,** 1–167.

Beeftink, W. G., Daane, M. C., De Munck, W. & Nieuwenhuize, J. (1978). Aspects of population dynamics in *Halimione portulacoides* communities. *Vegetatio* **36,** 31–43.

Chapman, V. J. (1937). A note upon *Obione portulacoides* (L.) Gaertn. *Annals of Botany (London)* [New Series], **1,** 305–309.

Chapman, V. J. (1950). Biological flora of the British Isles. *Halimione portulacoides* (L.) Aell. *Journal of Ecology* **38,** 214–222.

Cresswell, E. G. & Grime, J. P. (1981). Induction of a light requirement during seed development and its ecological consequences. *Nature (London),* **291,** 583–585.

Flowers, T. J., Troke, P. F. & Yeo, A. R. (1977). The mechanism of salt tolerance in halophytes. *Annual Review of Plant Physiology,* **28,** 89–121.

Greenway, H. (1968). Growth stimulation by high chloride concentrations in halophytes. *Israel Journal of Botany,* **17,** 169–177.

Harper, J. L. (1964). The individual in the population. *Journal of Ecology,* **52,** Supplement, 149–158.

Harper, J. L. (1968). The regulation of numbers and mass in plant populations. *Population Biology and Evolution* (Ed. by R. C. Lewontin), pp. 139–158. Syracuse University Press, Syracuse.

Harper, J. L. (1977). *Population Biology of Plants.* Academic Press, London.

Harper, J. L. (1982). After description. *The Plant Community as a Working Organism* (Ed. by E. I. Newman), pp. 11–25. Blackwell Scientific Publications, Oxford.

Henriksen, K. & Jensen, A. (1979). Nitrogen mineralisation in a salt marsh ecosystem dominated by *Halimione portulacoides. Ecological Processes in Coastal Environments* (Ed. by R. L. Jefferies & A. J. Davy), pp. 373–384. Blackwell Scientific Publications, Oxford.

Jensen, A. (1985). On the ecophysiology of *Halimione portulacoides* (L.) Aellen. *Vegetatio,* **61,** 231–240.

Kappen, L. (1969). Frostresistenz einheimischer Halophyten in Beziehung zu ihrem Salz-, Zucker- und Wassergehalt im Sommer und Winter, *Flora (Jena), Abteilung B,* **158,** 232–260.

Kirkby, E. A. & Knight, A. H. (1977). Influence of the level of nitrate nutrition on ion uptake and assimilation organic acid accumulation and cation-anion balance in whole tomato plants. *Plant Physiology,* **60,** 349–353.

Osmond, C. B., Björkman, O. & Anderson, D. J. (1980). *Physiological Processes in Plant Ecology: Towards a Synthesis with Atriplex.* Springer-Verlag, Berlin & New York.

Pons, T. L. (1983). Significance of inhibition of seed germination under the leaf canopy in ash coppice. *Plant, Cell and Environment,* **6,** 385–392.

Stienstra, A. W. & Markusse, M. M. (1978). Growth and mineral composition of *Halimione portulacoides* under different salinity conditions. *Verhandelingen der Koninklijke Nederlandse Akademie van Wetenschappen, Afdeling Natuurkunde, Reeks 2,* **71,** 127–129.

Van der Meyden, R., Weeda, E. J., Adema, F. A. C. B. & De Joncheere, G. J. (1983). *Heukels/Van der Meyden Flora van Nederland,* Wolters-Noordhoff, Groningen.

Waisel, Y. (1972). *Biology of Halophytes.* Academic Press, London.

Yeo, A. R. (1983). Salinity resistance: Physiologies and prices. *Physiologia Plantarum,* **58,** 214–222.

7

Establishment and Peri-establishment Mortality

Robert E. L. Naylor

School of Agriculture
The University
Aberdeen, Scotland

I. INTRODUCTION

One for the rook, one for the crow, one to die and one to grow

This old farming saw embodies an observation often ignored, that only a small proportion of the seeds in the soil progresses to becoming plants. In agriculture the farmer can opt to treat his seed with bird repellent, insecticide, fungicide and establishment promoters in order to minimise the proportion of unsuccessful seeds, but it is rare to observe complete germination even in such highly managed ecosystems as arable crops. The growing of temperate cereals is regarded as a pinnacle of achievement in crop technology and yet in a survey of 468 crops of winter barley sown in autumn 1981, the average establishments from seedbeds which growers described as good, medium or poor were 77, 78 and 57%, respectively (Rance 1982).

In crops in which each seed gives rise to a single unit of product or in which the growth of adjacent single individuals cannot fully compensate for an absent plant, then poor establishment can be an important component of yield. The data of Jaggard *et al.* (1983) on sugar beet establishment and yields suggest that crops sown in early spring produce a loss in sugar yield of about 25–30% that is not attributable to bolting.

The response of the agriculturalist to the likelihood of incomplete establishment is to increase the number of seeds sown. The incorporation of 'field factors' into calculations for seed rate is an empirical response to variable establishment. The seeding rate is compensated by adjusting the result of a germination test (to determine the proportion of viable seeds) by a proportion derived from past experience of performance in the expected seedbed conditions, i.e. the anticipated proportional establishment. It is suggested that in carrots this factor is 0.8 in good conditions but may be only 0.5 in cold soils (McConnell 1968).

STUDIES ON PLANT DEMOGRAPHY:
A FESTSCHRIFT FOR JOHN L. HARPER

Viable seeds contribute to the gene pool of a species. The failure of a seed to germinate restricts the set of individuals which may leave descendants. Concern with the role of the individual has been one characteristic of Harperian ecology in which a study of deaths in a population provides important information. Many papers by John Harper and his co-workers have included special study of plant mortality and its consequences. In some of his earliest work John Harper studied seed and seedling mortality and it is to this topic I wish to turn: what features of the environment govern the success or failure of a seed to progress to becoming an established seedling. I will consider the definition of a safe site for germination, the filling of safe sites with seeds, the survival of safe sites, and the ways in which the agronomist can improve establishment.

The system which we studied was that of directly reseeded grass swards. Recognition that in commercial farming practice, direct reseeding frequently failed to lead to a well-established sward had led others to numerous experiments to test the efficacy of various insecticides and nematicides, and to monitor putative pests during the early growth of the grass. None of these experiments suggested any consistent cause for the failure of establishment of a new sward; nor did they suggest any treatment likely to improve establishment. However, the experiments summarised here were prompted by observations made during the drilling of the seed and before seedling emergence. This study of the germination ecology of ryegrass seed has permitted the identification of common causes of failure of direct reseeds, and has thus indicated possible preventive measures that the grower might take.

II. THE SAFE SITE

The individual seed germinates in response to the precise set of conditions it experiences in its immediate environment. The locality at the scale of size of the seed that provides such conditions has been termed a 'safe site' by Harper (1977). Each safe site provides the precise requirements of an individual seed for dormancy-breaking stimuli, and for the processes of germination to take place. In addition, the resources for germination processes must be available and there must be freedom from hazards such as diseases, predators or toxic substances. The safe site must also survive for the period until the seedling becomes independent of the maternal seed reserves. The requirements of the seed during this time change and so the the definition of what constitutes a safe site must also change.

A. Prerequisites

Seed is often shed in a dormant state or acquires the inability to germinate in response to environmental conditions. These two conditions have been termed innate and induced dormancy (Harper 1957). There is a plethora of literature on the dormancy and germination of seeds. Matthews (1976) has suggested this is partly because of the ease and speed with which experiments can be done. Only infrequently are the results of experiments in carefully controlled conditions in the laboratory appropriate to the interpretation of observations in natural conditions in the field. There are fewer studies of field germination and changes in dormancy status. However, the extensive work of the Baskins on seasonal changes in the germination response of buried seeds of various species illustrates the necessity to study field populations of seeds (see Baskin & Baskin 1985 for a comprehensive review). Harper (1977) has pointed out that enforced and induced dormancy allow an opportunistic response to environmental conditions.

B. Seed Depth

Although seeds are able to germinate from depth, seedlings are generally derived from seed near the surface (Naylor 1970; Zemanek 1972; Maun 1981). Harper (1957) emphasized that the conditions at the soil surface are often those that break dormancy or permit germination. Thus it is environmental conditions near the soil surface that are relevant to the understanding of seed germination and seedling establishment. In particular, exposure to light and to alternating temperatures may be important.

C. Light

Grime *et al.* (1981) have tried to indicate the ecological relevance of germination characteristics in an examination of 403 species of the flora around Sheffield which differed in life form, geographical distribution, ecology, seed shape, weight and colour. Initial germination of freshly-collected seed ranged from below 10 to over 80%. The germination of most species was promoted by light and in 104 of the species a marked reduction was observed when seed was kept in darkness.

D. Alternating Temperatures

Experience of alternating temperatures seems to be a particularly important stimulus to seeds of many but not all species (Naylor 1984). Testing of germination of seeds at constant temperature may give a false impression of their germinability in the field. Two seed populations of *Poa annua* both had 15°C as their optimum constant temperature for germination, though the levels of germination at this temperature were 92 and 3%. However when tested at alternating temperatures of 15/25°C, germination of both seed populations was complete and faster (Table I). Light and temperature may not operate independently in

TABLE I.

Germination Response to Temperature in Two Populations (A and B) of *Poa annua*[a]

	Temperature (°C)							LSD (5%)
	2	5	15	20	25	30	15/25	
Germination (%)								
population A	1	56	92	71	18	2	97	3.8
population B	0	1	3	1	0	0	100	
Mean germination time (days)								
population A	35	28	16	17	23	30	10	1.8
population B	—	35	22	27	—	—	18	

[a] From Naylor and Abdalla (1982) and unpublished data.

providing a stimulus for germination. The germination of four populations of *P. annua* in the dark was lower in constant than in fluctuating temperatures in which it averaged 90% (Naylor & Abdalla 1982). Light increased the germination at constant temperatures but had little extra effect on the already high germination response to fluctuating temperatures (Fig. 1). Soil temperatures fluctuate more near the surface than at depth, and perhaps a truer statement is that fluctuating temperatures can to some extent substitute for the light stimulus to germination.

E. Water Supply

Germination only occurs from imbibed seed. Imbibition may be immediately followed by phases of germination such as root and shoot protrusion from the seed as in arable crop seed, or the imbibed seed may remain ungerminated for a long period, or indeed it may be dehydrated once more. Thus the water-supplying capacity of the substrate is crucial both as a prerequisite for germination and for the success of the processes that constitute germination.

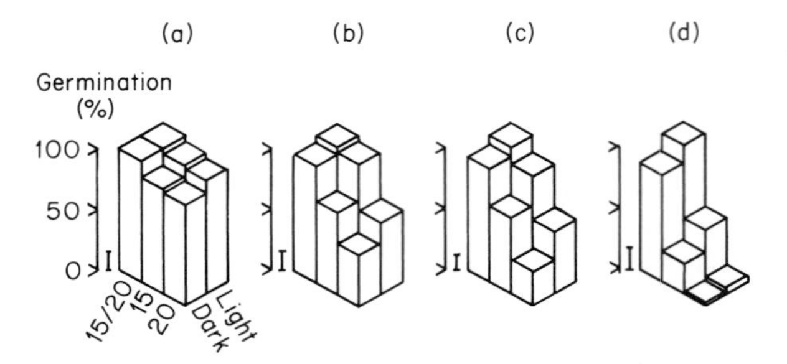

Fig. 1. Germination at 15 or 20°C or at 15/20°C in the light or dark of seed of four populations of *Poa annua*. Bars represent LSD at $p = .05$. [From Naylor and Abdalla (1982).]

Some of the features of seed–substrate interactions have been studied in the context of direct reseeding of grassland. *Direct reseeding* is the technique of re-establishing a grass sward by first killing the old sward with a herbicide and then sowing grass seed directly into drill slits cut into the old sward without creating a seedbed. It is an attractive alternative to conventional cultivation and resowing on soils which are too stony, wet or susceptible to erosion to permit conventional seedbed preparation. However, the use of direct reseeding on sandy soils susceptible to wind erosion has not always proved successful: comparative trials have shown the superior establishment and subsequent higher yields of conventional systems (Naylor, Marshall & Matthews 1983), and glasshouse simulations have confirmed that the lack of an adequate supply of water to the seed is a prime cause of the failure of direct reseeds (Marshall & Naylor 1984b).

When seeds of Italian ryegrass were sown onto the surface of a loamy sand adjusted to different moisture contents, the weight increase of seeds, the proportion that produced a radicle and the proportion that produced a coleoptile differed (Table II). Both radicle and coleoptile protrusion were reduced by lower water tensions and completely prevented below −13 bars. Coleoptile protrusion was more sensitive to reduced water availability than was radicle protrusion. This was also shown in experiments using different concentrations of polyethyleneglycol (PEG) to achieve different water potentials. Both radicle and coleoptile protrusion were prevented in PEG solutions of −11.5 bars although these seeds had taken up as much water as seeds in solutions of −7.3 and −4.5 bars, many of which did germinate. Some ryegrass seeds germinated at −10 bars. However, not all viable seeds exhibited radicle protrusion at −7 bars, and coleoptile protrusion was reduced at −4 bars water potential (Marshall & Naylor 1985). Radicle protrusion is often used as a test criterion for germination, but

TABLE II.

Behaviour of *Lolium multiflorum* on the Surface of a Loamy Sand at Different Moisture Contents[a]

Soil moisture content (% air dry)	Water potential (bar)	Weight gain (%)	Seed (%) with protrusion of	
			Radicle	Coleoptile
15	−2	780	83	83
12	−4	330	77	67
9	−7	35	17	6
7	−13	25	0	0
5	−24	15	0	0
LSD (5%)		150	16	18

[a] From Naylor and Marshall (1983).

coleoptile protrusion is essential for successful seedling establishment. If coleoptile protrusion is taken as the criterion of germination, then a 50% reduction is the response to lowering the water potential to −4 bars.

In the definition of a safe site it is important to consider that different processes or events may be limited by different features of the environment. Thus a different 'environmental sieve' may exist for imbibition, for radicle protrusion, for coleoptile protrusion, and for subsequent growth of these two organs. The information for the impact of water potential on the early germination processes of Italian ryegrasses is summarised in Figure 2.

There is evidence that direct osmotic effects may be implicated in limiting seedling establishment. In 3 years out of 6, the number of sugar beet seedlings established was significantly lower in response to applications of nitrogenous fertilizer (Last *et al.* 1983). In the 3 years when fertilizer did not reduce establishment, sowing was followed by rain. Last *et al.* (1983) observed that the loss of plants in dry years occurred between germination and emergence, and suggested that this loss was due to increased osmotic pressure of the soil solution, preventing water uptake by the germinated seed and seedling.

It is not always poor supply of water that is critical in determining the success of the seed to seedling transition. In vining peas a higher incidence of imbibition damage resulting from rapid water uptake has been suggested as a cause of poor emergence in wet soil conditions (Powell & Matthews 1980). Seed sown into drier soil so that imbibition took place slowly gave better emergence. Similarly, the emergence of poor-quality barley seeds was reduced in wet soil (Perry &

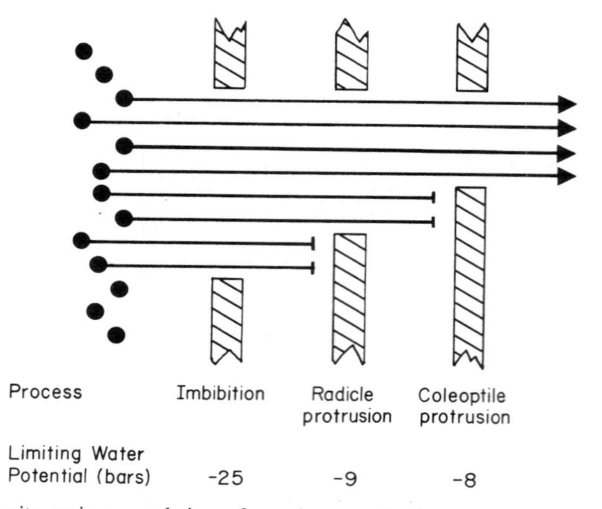

Fig. 2. Microsite environmental sieves for various germination processes as limited by water potential. Values refer to experimental results for Italian ryegrass.

Harrison 1977), and winter barley crops drilled into wet soils were associated with lower percentage establishment (Rance 1982).

Sowing seed onto the surface of the soil or into open drill slits allows the possibility of an effect of seed orientation on water uptake. Seed of Italian ryegrass shows a response to orientation in laboratory experiments. Different seed orientations were obtained on small glass balls in sintered glass funnels maintained with 5-cm water tension. The rate of water uptake did not differ significantly for the first 24 hr (Fig. 3). When the rachilla remained out of contact with the substrate, germination was slower and reached a lower level. The rachilla seems to be an important route for water to enter the caryopsis (Marshall & Naylor 1985). Removal of the lemma and palea also permitted higher levels of germination, which suggests that these accessory parts of the dispersal unit either form an impermeable barrier which directly prevents water's reaching the caryopsis, or that their water potential is lower than that of the caryopsis. Complete soil coverage of the seed is essential to ensure water supply

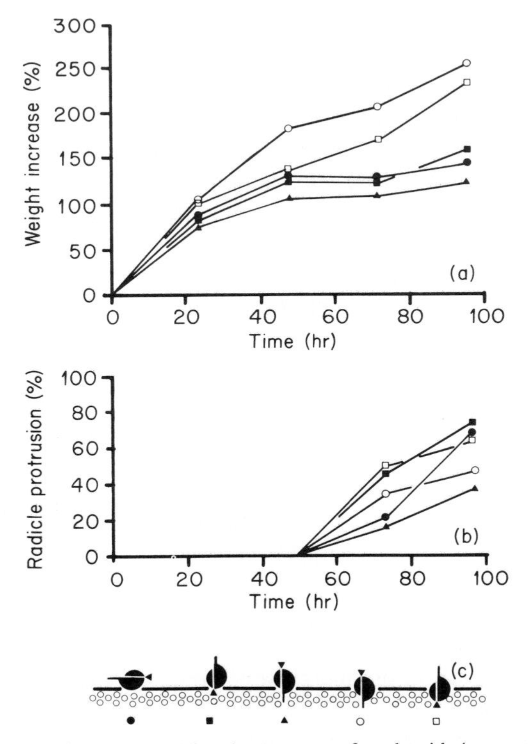

Fig. 3. Percentage weight increase (a) and percentage of seeds with 4-mm radicle protrusion (b) of Italian ryegrass sown in the different orientations illustrated (c). [From Marshall and Naylor (1985).]

to the caryopsis and experimental treatments of adding soil to open drill slits, or of closing the drill slit by foot (size 9) both resulted in improved seedling establishment (Marshall 1982 and unpublished). Ensuring adequate water availability after sowing by irrigation improved the number of emerged seedlings in both laboratory and field studies (Marshall & Naylor 1984b, c).

F. Trash

Grime (1979) has suggested that the accumulation of a layer of persistent litter may restrict the frequency of occurrence of some plant species in vegetation, either by shading or by physical impedance of germination and establishment. In experiments to identify the causes of poor establishment of direct reseeded grass, the presence of surface trash has been shown to be important. Adding trash to conventionally prepared seedbeds lowered ryegrass seedling emergence, and removal of trash improved emergence from direct-drilled plots (Marshall & Naylor 1984b). A large proportion of this effect seems to be purely physical. The direct drilling machine was observed to bury trash and to sow the seed onto desiccated trash rather than into contact with the soil. The trash thus acts as a physical barrier preventing the seed from taking up sufficient moisture to permit radicle or coleoptile protrusion. Additionally, the trash pressed into the drill slit may prevent closure of the slit and cause, consequently, rapid desiccation of the soil adjacent to the seed, which itself may then be uncovered. This has proved important on the loamy sands of our investigations. Reduction of the quantity of grass trash, e.g. by burning or mowing, has resulted in improved seedling emergence in the field (Marshall & Naylor 1984c).

In addition to this physical effect of trash mediated through its effect on water uptake, grass trash may exert a physiological effect through the production of toxins. Compounds toxic to germinating seeds may be released during the decay of crop residues either through the operation of micro-organisms on the trash or as a metabolic by-product of the micro-organisms themselves (Tukey 1969). Inhibition of germination by leachates from plant material has been observed in many crop species including grasses, and decaying plant residues have been shown to contain or to produce substances which are inhibitory to plant growth (Marshall & Naylor 1984a give a brief summary).

G. Freedom from Predators and Diseases

The seeds in the soil are not merely a potential seed bank; they also represent food material for many animals and a potential substrate for colonisation by micro-organisms. The impact of such organisms is sometimes of great economic importance: the principal cause of poor sugar beet establishment on continuous beet plots was feeding by adult pygmy beetles (*Atomaria linearis*). When no insecticides were used, only about 1% of seed produced established seedlings on continuous sugar beet plots, but on plots where sugar beet followed wheat, the establishment was 72% (Thornhill 1984).

One important justification for cultivation of the soil is to reduce the level or risk of attack from predators or diseases, and direct-drilled crops may thus suffer from such attacks (Naylor, Marshall & Matthews 1983). Crop rotation is also justified on similar grounds. Rarely, however, are the effects straightforward. Slug damage to direct-drilled maize can be severe, probably because of the great humidity and soil moisture below the old crop residues (Barry 1969; Cannell 1983), and it has also been noted that open drill slits themselves provide the pest with a convenient pathway from one seed or seedling to the next (Whybrew 1968; Allen 1979).

III. FILLING SAFE SITES WITH SEEDS

The soil manipulations imposed to produce a seedbed can be interpreted as a set of modifications to the habitat aimed at maximising the number of safe sites available. The process of sowing then aims to place each seed into a safe site. Higher seed rates usually result in more seedlings per unit area but often the chance of a seed producing a seedling is lower. Rance (1982) observed this in his survey of 468 winter barley crops. The proximity of one seed to another does appear to decrease the probability of germination and to delay germination.

In natural seed banks the likelihood of the immediate surroundings constituting a safe site is less easy to determine. The large density of buried viable weed seeds and the relative ease with which they can be induced to germinate if soil samples are spread out thinly in warm moist conditions suggest that the likelihood of a seed encountering a safe site in the soil may be low, often around 5%. The data of Froud-Williams, Chancellor and Drennan (1983) on buried weed seed banks in arable cropping systems show that at one site seedlings accounted for only 0.3% of the buried weed seed bank on undistributed plots and 1.9% on ploughed plots. If the transition from seed to seedling is a rare event it is likely that many seeds die in the soil. Jefferies, Davy and Rudmik (1981) observed that most seeds of the salt marsh annual *Salicornia europaea* agg. died without germinating. In natural vegetation as well as in the laboratory, density may influence germination. Inouye (1980) observed that high-density populations of seedlings inhibited subsequent germination in desert annuals.

IV. MICROSITE SURVIVAL

Because it takes time for the seed to proceed through the phases of radicle protrusion, shoot protrusion and growth to the soil surface, and the establishment of photosynthesis in excess of respiration, the safe site must exist over a period of time and continue to provide the conditions and resources for seedling growth. The conditions of the safe site will determine whether or how quickly the seed proceeds to become an established seedling, whether that progress is suspended or indeed whether it is curtailed and the germinating seedling dies. Abrol (1978) has described in detail the many abnormalities cereal seedlings may display in germination tests. Many of the seedling descriptions also apply to seedlings that have germinated in poor conditions in the field: abnormalities of the root and shoot system may occur in response, for example, to waterlogging, capping of the soil, or invasion by pathogens. Thus although the safe site has permitted the onset of germination it has been unable to maintain the quality of environment needed to sustain a young seedling.

In a number of studies mortality of young seedlings has been observed. Sarukhán & Harper (1973) emphasized its occurrence in populations of buttercup species. Similarly, Marks & Prince (1981) observed that in an artificially created population of wild lettuce, *Lactuca serriola,* periods of high mortality followed periods of high recruitment. This death of young seedling also occurs in crops of carrots and onions (Bedford & Mackay 1973). Without detailed information it is not possible to distinguish between mortality of young established seedlings (often an attractive succulent food for herbivores) and failure to establish before seed reserves have become exhausted. This latter phenomenon has been termed 'fatal germination' by Murdoch (1983), who studied its importance in *Avena fatua*. By recovering seed which had been buried at 25-, 75- and 230-mm depths in a loamy sand he was able to show that germination (judged as 2-mm radicle protrusion) was similar at all depths, but that deep burial considerably delayed and reduced emergence. In two experiments in successive years about 40–50% of seeds germinating at 25- and 75-mm depth failed to emerge but 98 and 99% of the deep-sown seed failed to emerge. The data of Last *et al.* (1983) suggested that the osmotic effects of a nitrate-rich soil solution could account for a failure of sugar beet seeds to germinate and of germinated seed to emerge.

Marshall (1982) showed that emergence of Italian ryegrass was significantly lower from direct-drilled plots than from sowings into conventionally cultivated seedbeds. Excavation of the seeds sown showed that in this experiment, particularly on plots in which the seed remained uncovered, some seeds showed radicle protrusion but did not subsequently emerge. It seemed likely that on these soils this fatal germination was due to drying of the soil in the vicinity of the seed. This was studied in a laboratory experiment in which seed was germinated in aerated solutions of water, but batches were transferred at 12-hr intervals to a solution of PEG of -11 bars, and the proportion of seeds with a protruded radicle noted after 150 hr from the start of the experiment. After 192 hr all seeds were transferred back to water to ascertain their capacity to resume growth following 'drought'. Seeds in water continuously showed the first radicle protrusion after 48 hr and ultimately achieved nearly 90% germination. Seed batches which exhibited some radicle protrusion before transfer (i.e. those transferred

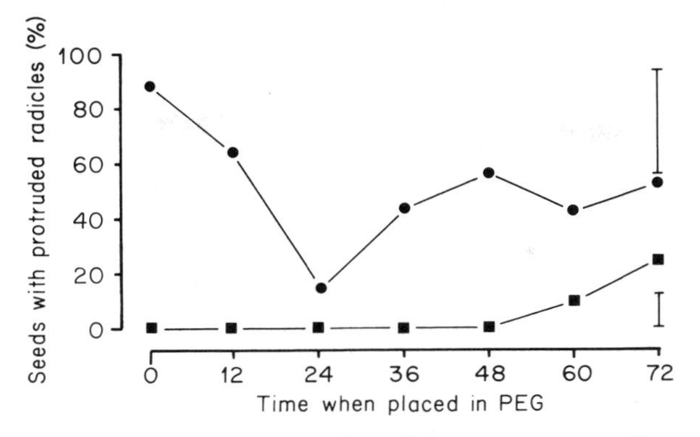

Fig. 4. Radicle protrusion after 150 hr (■——■) and after 300 hr (●——●) of perennial ryegrass transferred from water to a PEG solution of water potential −11.5 bars at various times, and all transferred back to water after 192 hr. Bars represent LSD at $p = .05$. [From data in Marshall and Naylor (1985).]

after 48 hr) showed no further radicle protrusion in PEG, but transfer back to water allowed it to recommence although the final germination was reduced (Fig. 4). Transfer to PEG before 48 hr prevented any radicle emergence until the seeds were transferred back to water and then final germination was generally lower (Marshall & Naylor 1985). On the loamy sand in which we were investigating poor establishment a drop in water potential from −2 to −12 bars is possible over a 2- or 3-year day period in summer when evapotranspiration is high. Such a rapid drop in soil moisture content could leave imbibed or just germinated seed in a no longer safe site, and subsequent environmental changes will determine whether the individual seed is allowed to resume growth.

A sequence of periods of soil wetness and dryness may influence subsequent germination. Seed of *Lolium rigidum* was pretreated by imbibing to 155% of its weight, maintained in the light at 20°C for 2 days and then air-dried for 2 days at 20°C. When such seed was sown, its emergence and establishment were advanced by 2 days (Lush & Groves 1981). A higher proportion (15–20% more) of pretreated seeds produced seedlings. Great practical benefits were seen when such pretreated seed was sown in conditions where there was only a short time to achieve successful establishment before the seedbed dried out: when untreated seed failed to establish, 12% of pretreated seed produced established seedlings when surface-sown, and 34% when sown in the top millimetre of the soil.

V. LOADING THE DICE

In agriculture, variable and unpredictable establishment may have important consequences on yield and profitability. There is thus interest in finding and using treatments that improve establishment in the already severely man-modified habitat of the good seedbed. Irrigation, the incorporation of straw to improve moisture retention, the coating of seed with insecticide or fungicide, and pre-treating the seed (as has been described) are all techniques which seek to improve the number of safe sites. On soils where rain may induce capping sufficiently strongly to reduce emergence, the application of cellulose xanthate can improve seedling emergence by improving the water stability of soil aggregates (Page 1980). This then is a treatment which ensures the continued safety of the germination site for the emerging seedling.

Only recently have hazards at the germination site been demonstrated for grass swards. Clements *et al.* (1982) recorded improved seedling stands or herbage dry matter yields in response to pesticide application on 20 of 45 sites studied. Although some of the reduced establishment and performance they observed was due to post-establishment losses, they suggest that damaged or impaired establishment of grassland may be the norm. A not uncommon cause of poor establishment of winter wheat is a waterlogged soil; waterlogging for 3 days or more delayed the emergence and decreased the final number of plants established on both sandy loam and clay soils (Thomson, Belford & Cannell 1983). Calcium peroxide coatings applied to seeds may have an anti-fungal action, may neutralize phytotoxic acids produced in anaerobic soils, or may slowly release oxygen with benefit to the germinating seed.

Experiments with perennial ryegrass (cv Talbot) seeds showed that coating seeds with calcium peroxide delayed mean germination time from 70 to 80 hr without altering final percentage of germination when the experiments were carried out in petri dishes with 7 ml of water added to the seed test paper. This

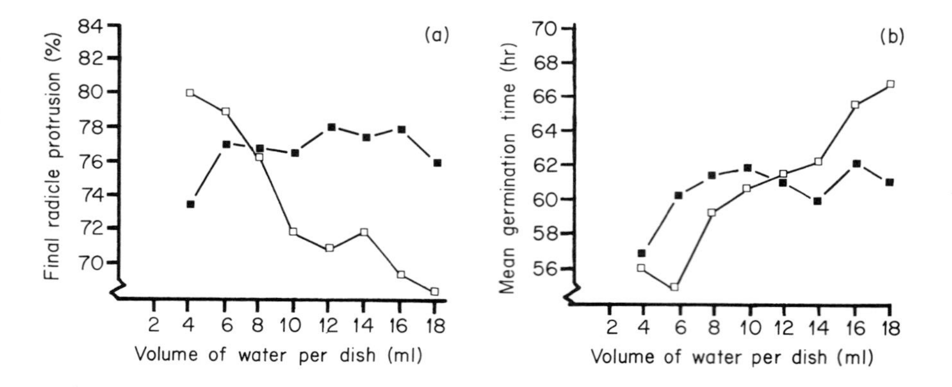

Fig. 5. Final radicle protrusion (a) and mean germination time (b) of seeds of perennial ryegrass either uncoated (□——□) or coated (■——■) with calcium peroxide, in different volumes of water.

volume of water was almost completely retained by the seed test paper. The seeds coated with calcium peroxide showed a consistent final germination and mean germination time over the range of 6–18 ml of water (Fig. 5). A similar response of final germination was seen when seeds were sown in sand. Other experiments suggested that the coating could supply about half of the oxygen requirement of the seed some 3 days after onset of imbibition, by which time some seeds showed radicle protrusion, and about 10% by 6 days, when radicle protrusion was complete.

In field sowings calcium-peroxide-coated seed showed a significant improvement of establishment in dry but not in moist soils (Marshall & Naylor 1984c), and there were indications that seed coating influenced long-term survival of established seedlings.

REFERENCES

Abrol, B. K. (1978). A survey of seedling abnormality in cereals. *Journal of the National Institute of Agricultural Botany (Great Britain)* **14,** 433–457.

Allen, H. P. (1979). Renewing pastures by direct drilling. *Changes in Sward Composition and Productivity* (Ed. by A. H. Charles & R. J. Haggar), pp. 217–222. Occasional Symposium No. 10. British Grassland Society, Hurley.

Barry, B. D. (1969). Evaluation of chemicals for control of slugs on field corn in Ohio. *Journal of Economic Entomology,* **62,** 1277–1279.

Baskin, J. M. & Baskin, C. (1985). Life cycle ecology of annual plant species of cedar glades of southeastern United States. *The Population Structure of Vegetation* (Ed. by J. White), pp. 371–398. Dr. W. Junk, Dordrecht.

Bedford, L. V. & Mackay, D. B. (1973). The value of laboratory germination measurements in forecasting emergence of onion and carrot seed in the field. *Journal of the National Institute of Agricultural Botany (Great Britain),* **13,** 50–62.

Cannell, R. Q. (1983). Crop establishment in relation to soil conditions and cultivations. *The Yield of Cereals,* pp. 45–66. Royal Agricultural Society of England National Agricultural Center, Stoneleigh.

Clements, R. O., French, N., Guile, C. T., Golightly, W. H., Lewis, S. & Savage, M. J. (1982). The effect of pesticides on establishment of grass swards in England and Wales. *Annals of Applied Biology,* **101,** 305–313.

Froud-Williams, R. J., Chancellor, R. J. & Drennan, D. S. H. (1983). Influence of cultivation regime upon buried weed seeds in arable cropping systems. *Journal of Applied Ecology,* **20,** 199–208.

Grime, J. P. (1979). *Plant Strategies and Vegetation Processes.* Wiley, Chichester.

Grime, J. P., Mason, G., Curtis, A. V., Rodman, J., Band, S. R., Mowforth, M. A. G., Neal, A. M. & Shaw, S. (1981). A comparative study of germination characteristics in a local flora. *Journal of Ecology,* **69,** 1017–1059.

Harper, J. L. (1957). The ecological significance of dormancy and its importance in weed control. *Proceedings of the 4th International Conference on Plant Protection, 1957,* pp. 415–420.

Harper, J. L. (1977). *Population Biology of Plants.* Academic Press, London.

Inouye, R. S. (1980). Density-dependent germination response by seeds of desert annuals. *Oecologia,* **46,** 235–238.

Jaggard, K. W., Wickens, R., Webb, D. J. & Scott, R. K. (1983). Effects of sowing date on establishment and bolting and the influence of these factors on yields of sugar beet. *Journal of Agricultural Science,* **101,** 147–161.

Jefferies, R. L., Davy, A. J. & Rudmik, T. (1981). Population biology, of the salt marsh annual *Salicornia europaea* agg. *Journal of Ecology,* **69,** 17–31.

Last, P. J., Draycott, A. P., Messem, A. B. & Webb, D. J. (1983). Effects of nitrogen fertilizer and irrigation on sugar beet at Brooms Barn 1973–8. *Journal of Agricultural Science,* **101,** 185–205.

Lush, W. M. & Groves, R. H. (1981). Germination, emergence and surface establishment of wheat and ryegrass in response to natural and artificial hydration-dehydration treatments. *Australian Journal of Agricultural Research,* **32,** 731–739.

McConnell, P. (1968). *The Agricultural Notebook.* 15th edition. Iliffe Books Ltd., London.

Marks, M. & Prince, S. (1981). Influence of germination date on survival and fecundity in wild lettuce, *Lactuca serriola. Oikos,* **36,** 326–330.

Marshall, A. H. (1982). Establishment of four ryegrass seed lots under different cultivation treatments. *Annals of Applied Biology,* **100,** Supplement, 3, 104–105.

Marshall, A. H. & Naylor, R. E. L. (1984a). The effect of leachates from grass trash on establishment of ryegrass. *Annals of Applied Biology,* **105,** 75–86.

Marshall, A. H. & Naylor, R. E. L. (1984b). Reasons for poor establishment of direct reseeded grassland. *Annals of Applied Biology,* **105,** 87–96.

Marshall, A. H. & Naylor, R. E. L. (1984c). Some factors influencing the establishment of direct-reseeded grass. *Crop Research,* **24,** 23–35.

Marshall, A. H. & Naylor, R. E. L. (1985). Ryegrass seed-water relationships. *Seed Science and Technology* (in press).

Matthews, S. (1976). Seed in relation to ecology. *Advances in Research and Technology of Seeds,* **2,** 86–106.

Maun, M. A. (1981). Seed germination and seedling establishment of *Calamovilfa longifolia* on Lake Huron sand dunes. *Canadian Journal of Botany,* **59,** 460–469.

Murdoch, A. J. (1983). Environmental control of germination and emergence in *Avena fatua. Aspects of Applied Biology,* **4,** 63–69.

Naylor, R. E. L. (1970). The prediction of blackgrass infestations. *Weed Research,* **10,** 296–299.

Naylor, R. E. L. (1984). Seed ecology. *Advances in Research and Technology of Seeds,* **9,** 61–93.

Naylor, R. E. L. & Abdalla, A. F. (1982). Variation in germination behaviour. *Seed Science and Technology,* **10,** 67–76.

Naylor, R. E. L. & Marshall, A. H. (1983). Effect of soil conditions on establishment of direct reseeded grass. *Aspects of Applied Biology,* **4,** 79–85.

Naylor, R. E. L., Marshall, A. H. & Matthews, S. (1983). Seed establishment in directly drilled sowings. *Herbage Abstracts,* **53,** 73–91.

Page, E. R. (1980). Cellulose xanthate as a soil conditioner: Field trials. *Journal of the Science of Food and Agriculture,* **31,** 718–723.

Perry, D. A. & Harrison, J. G. (1977). Effects of seed deterioration and seed bed environment on emergence and yield of spring-sown barley. *Annals of Applied Biology,* **86,** 291–300.

Powell, A. A. & Matthews, S. (1980). The significance of damage during inbibition to the field emergence of pea (*Pisum sativum* L.) seeds. *Journal of Agricultural Science,* **95,** 35–38.

Rance, D. (1982). *Pointers to Profitable Winter Barley.* Technical Services Group, Agriculture Division, Imperial Chemical Industries PLC, Billingham, UK.

Sarukhán, J. & Harper, J. L. (1973). Studies on plant demography: *Ranunculus repens* L., *R. bulbosus* L. and *R. acris* L. I. Population flux and survivorship. *Journal of Ecology,* **61,** 675–716.

Thomson, R. J., Belford, R. K. & Cannell, R. Q. (1983). Effect of calcium peroxide seed coating on the establishment of winter wheat subjected to pre-emergence waterlogging. *Journal of the Science of Food and Agriculture,* **34,** 1159–1162.

Thornhill, W. A. (1984). Plant establishment. *Report, Rothamsted Experimental Station for 1983,* p. 42.

Tukey, H. B. (1969). Implications of allelopathy in agricultural plant science. *Botanical Review,* **35,** 1–16.

Whybrew, J. E. (1968). Experimental Husbandry Farm experience with herbicides and tillage systems for cereal growing. *NAAS Quarterly Review,* **80,** 154–160.

Zemanek, J. (1972). Influence of environmental factors and chemical substances on germination and emergence of the weed *Apera spica-venti* (L.) P. Beauv. *Ochrana Rostlin,* **8,** 29–36.

8

Population Biology and the Conservation of Rare Species

H. J. Harvey

Department of Applied Biology
University of Cambridge
Cambridge, England

I. CONSERVATION AND DEMOGRAPHY

The main thesis to be developed in this essay is that the demographic approach to the study of plant populations pioneered by John Harper may make a valuable contribution to the conservation of wild populations whose numbers are in decline or which are at risk of extinction. Harper (1971) has stressed the importance of actuarial data for the informed management of populations of wild plants, a view since reiterated with special reference to species of interest to conservationists by a number of authors (e.g. Bradshaw & Doody 1978; Zeedyk *et al.* 1978; Davy & Jefferies 1981; Harvey & Meredith 1981). That there is still a need to emphasise the potential value of demographic studies to plant conservation is well demonstrated by the apparent absence of any reference to this topic from a recent collection of essays on conservation (Warren & Goldsmith 1983). The demographic data used to support the proposition developed here will also be examined to see whether generalisations may be made regarding the life history characteristics of species which have populations which are small or at risk.

Harper's overview on the demography of populations of plants was presented in two papers with White (Harper & White 1971, 1974); the model developed in the first of these papers provides the core of his later monograph (Harper 1977 p. 29). Alternative, but related, models for presenting the life cycle of a species

STUDIES ON PLANT DEMOGRAPHY:

have been proposed by, amongst others, Sagar and Mortimer (1976) and Whitson and Massey (1981). The quantification of the components of a detailed, generally accepted, model may have benefits, such as the identification of particular stages of life histories which are critical to the maintenance of populations and the comparison across species of general features in the life cycle. Sagar and Mortimer's model is detailed and includes all stages in the life cycle of plants so far identified as being of interest to the demographer. Its utility has been demonstrated by their review of the life histories of certain species of weed and by its adoption by other demographers (for example in White 1985). It is this model which will be adopted here.

Conservationists (Lucas & Synge 1978; Reveal 1981) have introduced carefully defined terms (e.g. 'endangered', 'rare' and the synonyms 'threatened' and 'vulnerable') to describe the status of species of plant which have small or declining populations. In certain cases these definitions could be improved; for example, Harper (1981) has noted the dangers of using the term 'rare' on a limited geographical basis and has stressed that abundance may vary naturally with time. Lists of the species falling into particular categories have been produced (e.g. Perring & Farrell 1983) and it would have been desirable to illustrate this review by reference to such species. Unfortunately the demography of few of these species has been studied and in order to provide examples it has been necessary to draw upon studies of any species which is of interest to conservationists because of its scarcity in a particular geographical area.

The fact that a species is scarce in an area presents the population biologist with problems. Two of the most prominent of these are whether the population, or the community in which it occurs, can be studied without causing irreparable damage and whether the sample size is large enough to produce reliable results. In addition limits are almost certain to be imposed on experimental manipulation of the ecosystem and on the collection of plant material for detailed laboratory study. These constraints are absent in most demographic studies reported to date, which have generally concentrated on species which are relatively common in their habitats. Another contrast may also be noted. A high proportion of the demographic studies so far completed have been conducted in habitats, such as grazed grassland, arable land, sand dunes, floors of abandoned quarries, or woodland ground layers, where the vegetation is short or sparse or both and where individual plants may be identified and mapped with relative ease. The population biologist concerned with the conservation of rare species (e.g. Bradshaw 1981) cannot choose the species to be studied and must operate in the communities, be they dense fen, scrub or upland grassland, where the species occur. Some of these may be uncongenial to the investigator; others may have a structure which prevents detailed recording of all stages of the life cycle.

It might be argued that demographic studies are unduly time-demanding when simple counts of the number of individuals in plots subject to different treatments (e.g. Godwin 1941; Wells 1971) might indicate the most effective way of maintaining a population. There is some validity in this view, but it may be countered by two arguments. Firstly, as already noted, there will be places where a species is so scarce that experimental management, as opposed to the continuation of the management under which the species has already survived, will not be permitted. In such cases a demographic approach may reveal possible limiting stages in the life cycle and suggest an alternative management.

The second counter argument is that numbers are no indication of important features such as the age structure of, or flux of individuals through, a population; for example, a particular management may allow established plants to survive but prevent recruitment. The work of Rowell (1983) at Wicken Fen illustrates this danger. Rowell has shown that the traditional period for the harvesting of *Cladium mariscus,* on a 4-year cycle, was May and June, in contrast to the recent practice of cutting between July and November. The population of *Cladium* appears to have been maintained under late summer or autumn cutting, but Conway (1942) found no seedlings or recruitment from seed under this regime. In a preliminary study Rowell showed that seed sown on bare peat or into stands of *Cladium* cut in May or in the preceding autumn can produce seedlings in late spring or early summer. Seeds sown into stands of *Cladium* uncut for 4 years did not produce seedlings. The survival and growth of the resulting seedlings, and of seedlings transplanted into the field after initial growth in a glasshouse, were best on bare peat or in plots cut in May. These data suggest that recruitment from seed is unlikely unless cutting at the traditional time is reintroduced.

The identification of those stages in the life history which restrict the size of the population, such as germination and establishment in *Cladium,* is one of the ways in which demography may aid conservation. Such aid depends upon its being possible to remove the restriction by some alteration of management or environment. The removal of one constraint will not automatically result in a larger population, for another phase in the life cycle may become limiting. The identification of the ultimate factor limiting the size of a population may require a perturbation experiment, but such experiments are unlikely to be permitted in those circumstances in which conservation is called for (Greig-Smith & Sagar 1981). Recommendations, derived from demographic studies, for changes in management will therefore have to be confirmed by trial. Another value of demographic studies to conservationists is the insight into the general features of the life history of species which they may provide. There may, therefore, be a variety of ways in which demographic studies of scarce or rare species can provide a logical basis for the planning of conservation management, the design of monitoring programmes and the interpretation of fluctuations in population size.

II. DEMOGRAPHIC STUDIES INVOLVING SMALL OR DECLINING POPULATIONS

Demographic studies have been carried out on several species which, at least in the study area, are rare or uncommon or have declining populations. Some of the results of these studies are summarised in this section. It must be stressed that in some of these accounts a full actuarial programme in which all individuals were identified and their fate recorded was not attempted. In some cases this was because of the nature of the community in which the species occurred.

A. *Peucedanum palustre*

One of the few studies embarked upon with the expressed aim of discovering the reasons for the low numbers of a species at a site, and with the intention of providing guidelines for management which might increase its abundance, is that of Meredith (1978) on *Peucedanum palustre* growing in *Cladium* communities at Wicken Fen, Cambridgeshire, UK (Harvey & Meredith 1981). *Peucedanum palustre* (Umbelliferae) is a semi-rosette hemicryptophyte which generally behaves as a herbaceous perennial. Rosettes are monocarpic, with inflorescences emerging in July or August, but the genet is maintained by the development of basal axillary buds. Data were collected on the production and dispersal of seed, the fate of shed seed, the size and seasonal state of the seed bank, the germination of seed, seedling emergence and survival and the survival of established plants. Observations on seedlings are extremely difficult in uncut stands of *Cladium*. Established plants produced large numbers of seed (about 4500 per plant), except when flowering stems were removed before seeds were ripe (September–October), during the 3- to 4-yearly harvesting of *Cladium*. Dispersal of shed seed (in part at least by water) appeared to be extremely limited, with only a very small proportion of such seeds moving more than a few metres from the parent plant. In closed communities about three-quarters of introduced seed was predated, probably by small mammals, but such predation was much lower (at about 30%) in recently cut areas. Investigation of the natural seed bank revealed few buried viable seed (from 20 to 150 per m^2), but seed which was artifically buried survived well, mostly in a state of enforced dormancy. The majority of seedlings emerged, between April and July, in areas cut the previous autumn and very few were noted under a dense cover of *Cladium*. The density of seedlings in recently cut areas was sometimes high (to 300 per m^2) but survival to the autumn was very low (1–5%), with predation, possibly from slugs or snails, appearing to be an important cause of mortality. The year-to-year survival of established plants was about 70%; many deaths seemed to follow the harvesting of the vegetation. Sufficient plants were recruited from seedlings to maintain the size of the population.

Meredith's study was conducted in *Cladium* communities cut between July and November at a site where the water table and the abundance of *Peucedanum* had fallen since the 1930s (Dempster & Hall 1980; Rowell 1983). Subsequent studies (Rowell 1983) have shown that *Cladium* at Wicken was traditionally harvested in May and June. Changes in the time of harvesting or the level of the water table might remove limitations at certain phases of the life cycle. Cutting in spring rather than in late summer or autumn should prevent the loss of seed through the destruction of flowering shoots and might increase total seed production at the site by one-third or one-quarter, depending on the harvesting interval. Cutting earlier in the year might also increase seedling survival, perhaps through a reduction in competition or an effect on grazing predators. The survival of established plants, which might lose leaves from the basal rosette but not flowering stems, might also be enhanced. It is less clear whether cutting in May and June would increase the number of seedlings emerging. Some effect is at least likely compared with cutting in July, which can result in considerable regrowth of *Cladium* by the following spring. Raising the water table could affect the survival of seedlings by making the site less suitable for molluscs (Brindley 1925) but the major effect might be on the dispersal of seed, which could be increased if winter flooding were more extensive. In the absence of flooding, seed is unlikely to reach naturally those areas of Wicken Fen from which the species is absent. Casual observations during 1984, when a large number of seedlings of *Peucedanum* were found in an area cut in May of that year, support the contention that cutting earlier in the year may increase recruitment.

B. *Fritillaria meleagris*

An elegant, but slightly less complete, study is that of Zhang (1983) on the major Swedish population of the rare bulbous perennial *Fritillaria meleagris* (Liliaceae) growing in tall-grass meadow communities near Uppsala. Established plants were monitored more closely than in Meredith's study and were classified by 'age-states'. The annual production of seed and the number of seed in the soil at different seasons were determined and laboratory studies conducted on the germination of seed. Because of the nature of the vegetation only very limited observations were made on naturally occurring seedlings but the emergence of seedlings derived from sown seed was monitored. The size of the population remained stable from 1981 to 1982 and rose somewhat in 1983, but there was a considerable flux of individuals between the various age-states. Large individuals survived better than small and mortality appeared to decline with increasing age. Recruitment from seed appeared to be low. Predation of both seed and established plants by small mammals was noted; established plants could survive this grazing. The vegetation of the study area was mown in mid-June, when most flower capsules were ripe but were not dry enough to release seeds. The removal of such capsules in the hay crop reduced the number of seeds reaching the soil by about two-thirds. Zhang suggested that delaying the hay cut to late June or early July would increase the number of seeds reaching the soil and might lead to greater recruitment.

C. Rare Plants in Upper Teesdale

Data are available over a period of up to 12 years for 10 species, including national rarities, growing in upland grassland communities in Upper Teesdale, UK (Doody 1975; Bradshaw & Doody 1978; Gibbons 1978; Bradshaw 1981). In addition to numbers at three or more dates in each year, there is supporting information on flowering, seed production and recruitment from seed and by clonal growth. The species differ considerably in their life history characteristics and in the relative importance of longevity, clonal growth and recruitment from seed as means of maintaining population numbers. In several species (*Viola rupestris, Primula farinosa, Carex ericetorum* and *Gentiana verna*) seed production was limited by grazing sheep and rabbits. Grazing by rabbits was sufficiently intense at one period to cause a marked decline in the number of plants of *Polygala amarella*. Gibbons (1978) clearly demonstrated that mortality in *Gentiana, Polygala* and *Viola rupestris* × *riviniana* was independent of age but suggested that this was not the case with *Viola rupestris,* in which mortality seemed to decrease with increasing age. He also noted a marked variation between cohorts, of all species, in the rate at which numbers in a cohort declined. Gibbons also discussed the relevance of the studies to the management of the populations, particularly of *Polygala* and *Gentiana,* in which grazing considerably reduced the number of seeds produced. He noted that grazing might be essential to maintain a habitat suitable for the species, by limiting the growth of more competitive species, but suggested that the exclusion of grazing from the immediate vicinity of flowering plants during the flowering and fruiting stages might result in increased production of seed. In the case of *Polygala,* a species in which all recruits are derived from seed, the consequent benefit would be a larger population of plants. The study populations of *Gentiana* were maintained by clonal growth, and no regeneration from seed was recorded. Here Gibbons suggested that increased production of seed might lead to a wider dispersal of the different colour forms of the species.

D. Orchids in Chalk Grassland

Wells (1981) has reported data collected for up to 16 years on the numbers in the populations of three orchids (*Spiranthes spiralis, Aceras anthropophorum* and *Herminium monorchis*) at two chalk grassland sites in Bedfordshire, UK. Populations were monitored only once each year so that data are available on the survival of individual plants and on flowering but are lacking on other stages of the life cycle. In *Aceras* and *Herminium* both recruitment and mortality varied markedly from year to year, whilst in *Spiranthes* recruitment varied but mortality, as a percentage of the total population, was fairly constant. In *Spiranthes* the chance of survival over time appeared to be constant for a particular cohort but to vary, between 4 and 15%, between cohorts. In *Aceras* annual mortality varied between 1 and 18%, but Wells suggests that the risk of death is more or less constant. Wells notes that the complex life histories of orchids make relating changes in abundance to management activities difficult.

III. SOME GENERALISATIONS

The studies just reviewed were of very different duration, were carried out in a wide range of habitats, involved the collection of different types of data, were based on samples of a wide range of sizes and investigated species with contrasting life histories. Given this diversity it is unlikely that many generalisations about the demography of rare species will be possible. It is, however, interesting to examine some of the data (Table I) in relation to both the components of Sagar and Mortimer's (1976) model of the life cycle and to previous generalisations (e.g. Harper 1977) regarding the life history characteristics of particular categories of plant. The studies also provide an opportunity to assess the impact of predators and pathogens, viewed by Harper (1981) as potentially important factors in determining the numbers of individuals in populations of rare species. In the following remarks only polycarpic species will be considered, as there are too few examples of scarce monocarpic species to justify comparisons.

The year-to-year survival of established plants of species in Table I varies over a wide range, but in half of the cases the annual mortality is greater than 25%. This pattern differs markedly from that for the sample of 17 temperate forest herbs reviewed by Bierzychudek (1982). In this latter group there was only 1 species in which individuals had an average life span of less than 5 years. whilst in 15 species this average was more than 10 years. The sample on which Table I is based is too small, and the other data sets with which it may be compared are too few, to justify any generalisations regarding the relative mortality rates of established individuals in rare and common species. The contrast may, however, be noted and should be examined again when more data are available.

In his review of the natural dynamics of populations of herbaceous perennials Harper (1977) suggested that recruitment from seed generally contributed little to the maintenance of populations, although he recognised *Ranunculus acris* and *R. bulbosus* as exceptions. This view was supported by Cook (1979), but subsequent studies have revealed further exceptions, such as the bulbous geophytes *Allium ursinum* (Ernst 1979) and *Erythronium japonicum* (Kawano, Hiratsuka & Hayashi 1982) (both Lilaceae) in temperate woodlands. Table I shows that, in the habitats in which they were studied, *Peucedanum* and *Polygala* are further

TABLE I.

Features of the Life-Tables of Some Polycarpic Species of Conservation Interest[a]

	Peucedanum palustre[b]	*Fritillaria meleagris*[c]	*Polygala amarella*[d]	*Carex ericetorum*[d]	*Gentiana verna*[d]	*Viola rupestris*[d]	*Viola riviniana*[d]	*Viola rupestris* × *riviniana*[d]
Annual mortality of established individuals, %	30	4–20	31	26	30	14	13	12
Annual seed production, numbers per plant	4500	92	377	26	1183	73	56	0
Mortality of seed in first year, %	30–75	46–59	?	?	?	?	?	—
Size of seed bank at germination period, numbers per m^2	20–150[f]	150[g]	?	?	?	?	?	—
Numbers of seedlings emerging per m^2	0–300	[i]	29[h]	0	0	3[g]	10[h]	0
Mortality of seedlings in first growing season, %	99	?	54[h]	—	—	?	?	—
Recruitment of ramets, number per original plant	?	?	0	0.36	0.37	0.16	0.16	0.18
Mortality risk over time	?	Not constant	Constant[e]	?	Constant[e]	Not constant[e]	?	Constant[e]

[a] Values are means unless a range is given.
[b] Meredith (1978).
[c] Zhang (1983).
[d] Doody (1975).
[e] Gibbons (1978).
[f] Samples to 15 cm.
[g] Samples to 5 cm.
[h] Calculated from available data.
[i] Five percent of sown seed emerged as seedlings under natural vegetation.

species which do not fit Harper's model. The validity of the generalisation must now be in doubt. For the remaining species in Table I recruitment from seed was low in relation to numbers in the population. The orchids studied by Wells (1981) appear to be intermediate between those species in which establishment from seed is virtually an annual event and those in which it rarely occurs. The average annual recruitment from seed of these orchids was low but large numbers of plants become established in certain years. Two factors, low levels of seedling emergence and high levels of seedling mortality, appear to contribute to low recruitment from seed. Data for *Peucedanum* and *Fritillaria* strongly suggest that in these species low levels of germination and seedling emergence are a consequence of the presence of a cover of vegetation. Seedling mortality was particularly heavy in *Peucedanum*.

In some species only small numbers of viable seed may be available to germinate, as a result of either low levels of seed production or high levels of seed predation. Humans, operating through the management they impose, have a major impact on the seed production of certain species. In some cases (e.g. *Platanthera leucophaea,* Bowles 1983) management may prevent flowering; in others (e.g. *Peucedanum, Fritillaria*) flowers may be removed before seed is mature. It is now recognized that in the majority of species the greatest number of individuals die during the seed phase of the life cycle (see reviews by Sagar & Mortimer 1976; Hickman 1979; Cavers 1983).

Continuing studies at Wicken Fen suggest that *Taraxacum palustre,* a dandelion of restricted distribution in the UK, may be another species to show the characteristics of low seed production and poor recruitment from seed. In some years the majority of capitula may be predated by small mammals (Table II), and only a small amount of seed produced. Over 4 years of study only 1 seedling has been found, compared with almost 450 vegetative individuals.

TABLE II.

The Fate of Capitula of *Taraxacum palustre* in 4 Years at Wicken Fen, Cambridgeshire[a]

	Year			
	1981	1982	1984	1985
Total number of rosettes located	209	403	279	163
Total number of capitula emerging	164	590	206	166
Number of capitula predated	144	474	47	97
Percentage of predation	87.8	80.3	22.8	58.4

[a] In 1981 recording started after the peak of emergence of capitula. In 1983 no recording was possible because of flooding of the site. Most predation in 1981 and 1982 was due to small mammals; there was some predation by birds in 1984.

Another of the generalisations regarding herbaceous perennials proposed by Harper (1977) was that the mortality of established plants was likely to be independent of age. Subsequent studies (e.g. Cook 1979; Meagher & Antonovics 1982) have shown that mortality is frequently more closely related to the size than to the age of plants. Three of the species in Table I, and also two of the orchids studied by Wells (1981), show the classic Deevey type 2 survivorship curve, but in *Fritillaria* and *Viola rupestris* the mortality risk falls as plant size increases. Both Gibbons (1978) and Wells (1981) reported differences in survivorship between cohorts, a finding noted by Harper (1977).

The studies reviewed in this paper yield little information on the effects of pathogens but provide clear evidence that predators may have a considerable impact on populations. Predation may occur at any stage in the life cycle. In some cases (e.g. *Fritillaria, Taraxacum*) grazing may remove the majority of leaves on established plants. Such plants generally appear to survive but their reproductive capacity is likely to be reduced. Predation of flowers is a common feature in Teesdale, and also with *Taraxacum* and *Pedicularis* (Menges, Gawler & Waller 1984), and may considerably reduce seed production. In those species for which data on the fate of seed are available such as *Peucedanum, Carlina* (Greig-Smith & Sagar 1981), *Pedicularis, Lesquerella* (Morgan 1983) and possibly also *Fritillaria,* predation seems to be a major cause of losses, with small rodents being identified as important predators in several cases. There is little information on the causes of the death of seedlings, but predation seems to be important in *Peucedanum* and is suspected in *Fritillaria*. The intensity of predation may vary considerably from year to year (e.g. Wells 1967; Bradshaw 1981) (Table II), emphasising the need for long-term studies before the factors affecting the life cycle can be accurately identified.

The ubiquity of predation in the studies reviewed here accords with the general finding (e.g. Cook 1979; Harper 1983) that it is one of the major causes of mortality in plant populations. There is now good experimental evidence (e.g. Parker & Boot 1981; Louda 1982) to support previous suggestions (e.g. Greig-Smith & Sagar 1981; Harper 1981) that predators may be important in determining the abundance of some species of plant. It is still unclear, however, whether rare species are present in small numbers because they are more prone to predation than are more common species. That such differences between species may be significant is suggested by the finding by Landa and Rabinowitz (1983) that scarce species of prairie grass tend to be more palatable to the grasshopper *Arphia sulphuarea* than do more common grasses.

IV. CONCLUSIONS

This review of the limited data available on the life history characteristics of rare species reveals no unique pattern; nor is there any consistent difference between rare and common species. There is perhaps a suggestion that established plants of rare species have a lower survival rate than do those of common species, but there appear to be no differences in the form of the survivorship curve, the role of recruitment from seed in maintaining the population, or the impact of predation.

The examples considered suggest that demographic studies of rare species are almost certain to provide valuable information on general features of the life cycle and may in certain circumstances enable the identification of those stages in the cycle which contribute to populations being small. Where data are limited, as in the orchids studied by Wells (1981), then obtaining indications of the status of the population, for example whether or not recruitment is occurring, may be all that is possible. When, however, sufficient data are available, as in *Peucedanum* and *Fritillaria,* then it should be an easy step from the identification of critical phases in the life cycle to the formulation of management policies which overcome the constraints. Given this potential it is to be hoped that population biologists will follow the lead of conservationists and begin studies of the demography of species in which the results of investigations can be put to some practical use.

ACKNOWLEDGMENTS

I am grateful to Dr. R. B. Gibbons for permission to draw at length from his thesis regarding the possible management of the Teesdale rarities.

REFERENCES

Bierzychudek, P. (1982). Life histories and demography of shade-tolerant temperate forest herbs: A review. *New Phytologist,* **90,** 757–776.

Bowles, M. L. (1983). The tallgrass prairie orchids *Plantanthera leucophaea* and *Cypripedium candidum:* Some aspects of their status, biology and ecology, and implications towards management. *Natural Areas Journal,* **3,** 14–37.

Bradshaw, M. E. (1981). Monitoring grassland plants in Upper Teesdale. *The Biological Aspects of Rare Plant Conservation* (Ed. by H. Synge), pp. 241–251. Wiley, Chichester.

Bradshaw, M. E. & Doody, J. P. (1978). Plant population studies and their relevance to nature conservation. *Biological Conservation,* **14,** 223–242.

Brindley, H. H. (1925). The Mollusca of Wicken Fen. *The Natural History of Wicken Fen* (Ed. by J. S. Gardiner), pp. 154–161. Bowes & Bowes, Cambridge.

Cavers, P. B. (1983). Seed demography. *Canadian Journal of Botany,* **61,** 3578–3590.

Conway, V. M. (1942). *Cladium mariscus* (L.). *Journal of Ecology,* **30,** 211–216.

Cook, R. E. (1979). Patterns of juvenile mortality and recruitment in plants. *Topics in Plant Population Biology* (Ed. by O. T. Solbrig, S. Jain, G. B. Johnson & P. H. Raven), pp. 207–231. Macmillan, London.

Davy, A. J. & Jefferies, R. L. (1981). Approaches to the monitoring of rare plant populations. *The Biological Aspects of Rare Plant Conservation* (Ed. by H. Synge), pp. 219–232. Wiley, Chichester.

Dempster, J. P. & Hall, M. L. (1980). An attempt at re-establishing the swallowtail butterfly at Wicken Fen. *Ecological Entomology,* **5,** 327–334.

Doody, J. P. (1975). *Studies in the population dynamics of some Teesdale plants.* Ph.D. thesis, University of Durham.

Ernst, W. H. O. (1979). Population biology of *Allium ursinum* in northern Germany. *Journal of Ecology,* **67,** 347–362.

Gibbons, R. B. (1978). *Further studies in the population dynamics of some Teesdale plants.* Ph.D. thesis, University of Durham.

Godwin, H. (1941). Studies in the ecology of Wicken Fen. IV. Crop-taking experiments. *Journal of Ecology,* **29,** 83–106.

Greig-Smith, J. & Sagar, G. R. (1981). Biological causes of local rarity in *Carlina vulgaris. The Biological Aspects of Rare Plant Conservation* (Ed. by H. Synge), pp. 389–400. Wiley, Chichester.

Harper, J. L. (1971). Grazing, fertilisers and pesticides in the management of grasslands. *The Scientific Management of Animal and Plant Communities for Conservation,* Symposium of the British Ecological Society, 11. (Ed. by E. Duffey & A. S. Watt), pp. 15–31. Blackwell Scientific Publications, Oxford.

Harper, J. L. (1977). *Population Biology of Plants.* Academic Press, London.

Harper, J. L. (1981). The meanings of rarity. *The Biological Aspects of Rare Plant Conservation* (Ed. by H. Synge), pp. 189–203. Wiley, Chichester.

Harper, J. L. (1983). A Darwinian plant ecology. *Evolution from Molecules to Man* (Ed. by D. S. Bendall), pp. 323–345. Cambridge University Press, London.

Harper, J. L. & White, J. (1971). The dynamics of plant populations. *Proceedings of the Advanced Study Institute on Dynamics of Numbers in Populations* (Ed. by P. J. den Boer & G. R. Gradwell), pp. 41–63. Centre for Agricultural Publishing and Documentation, Wageningen.

Harper, J. L. & White, J. (1974). The demography of plants. *Annual Review of Ecology and Systematics* **5,** 419–463.

Harvey, H. J. & Meredith, T. C. (1981). Ecological studies of *Peucedanum palustre* and their implications for conservation management at Wicken Fen, Cambridgeshire. *The Biological Aspects of Rare Plant Conservation* (Ed. by H. Synge), pp. 365–378. Wiley, Chichester.

Hickman, J. C. (1979). The basic biology of plant numbers. *Topics in Plant Population Biology* (Ed. by O. T. Solbrig, S. Jain, G. B. Johnson & P. H. Raven), pp. 232–269. Macmillan, London.

Kawano, S., Hiratsuka, A. & Hayashi, K. (1982). Life history characteristics and survivorship of *Erythronium japonicum:* The productive and reproductive biology of plants. V. *Oikos,* **38,** 129–149.

Landa, K. & Rabinowitz, D. (1983). Relative preference of *Arphia sulphurea* (Orthoptera: Acrididae) for sparse and common prairie grasses. *Ecology,* **64,** 392–395.

Louda, S. M. (1982). Limitation of the recruitment of the shrub *Haplopappus squarrosus* (Asteraceae) by flower and seed-feeding insects. *Journal of Ecology,* **70,** 43–53.

Lucas, G. & Synge, H. (1978). *The IUCN Plant Red Data Book.* IUCN, Morges, Switzerland.

Meagher, T. R. & Antonovics, J. (1982). The population biology of *Chamaelirium luteum,* a dioecious member of the lily family: Life history studies. *Ecology,* **63,** 1690–1700.

Menges, E., Gawler, S. C. & Waller, D. M. (1984). *Studies into the Population Biology of the Furbish Lousewort: 1983 Technical Report.* Mimeographed Report. Holcomb Research Institute, Butler University, Indianapolis, Indiana.

Meredith, T. C. (1978). *The ecology and conservation of* Peucedanum palustre *at Wicken Fen.* Ph.D. thesis, University of Cambridge.

Morgan, S. (1983). *Lesquerella filiformis:* An endemic mustard. *Natural Areas Journal,* **3,** 59–62.

Parker, M. A. & Boot, R. B. (1981). Insect herbivores limit habitat distribution of a native composite, *Machaeranthera canescens. Ecology,* **62,** 1390–1392.

Perring, F. H. & Farrell, L. (1983). *British Red Data Books: 1. Vascular Plants.* 2nd edition. S.P.N.C., Lincoln.

Reveal, J. L. (1981). The concepts of rarity and population threats in plant communities. *Rare Plants Conservation: Geographical Data Organisation* (Ed. by L. E. Morse & M. S. Henifin), pp. 41–47. New York Botanical Garden, Bronx, New York.

Rowell, T. A. (1983). *History and management of Wicken Fen.* Ph.D. thesis, University of Cambridge.

Sagar, G. R. & Mortimer, A. M. (1976). An approach to the study of the population dynamics of plants with special reference to weeds. *Applied Biology,* **1,** 1–47.

Warren, A. & Goldsmith, F. B. (Eds.) (1983). *Conservation in Perspective.* Wiley, Chichester.

Wells, T. C. E. (1967). Changes in a population of *Spiranthes spiralis* (L.) Chevall. at Knocking Hoe National Nature Reserve, Bedfordshire, 1962–65. *Journal of Ecology,* **55,** 83–99.

Wells, T. C. E. (1971). A comparison of the effects of sheep grazing and mechanical cutting on the structure and botanical composition of chalk grassland. *The Scientific Management of Animal and Plant Communities for Conservation* Symposium of the British Ecological Society 11. (Ed. by E. Duffey & S. A. Watt), pp. 497–515. Blackwell Scientific Publications, Oxford.

Wells, T. C. E. (1981). Population ecology of terrestrial orchids. *The Biological Aspects of Rare Plant Conservation* (Ed. by H. Synge), pp. 281–295. Wiley, Chichester.

White, J. (Ed.) (1985). *The Population Structure of Vegetation.* Dr. W. Junk, Dordrecht.

Whitson, P. D. & Massey, J. R. (1981). Information systems for use in studying the population status of threatened and endangered plants. *Rare Plant Conservation: Geographical Data Organisation* (Ed. by L. E. Morse & M. S. Henifin), pp. 217–236. New York Botanical Garden, Bronx, New York.

Zeedyk, W. D., Farmer, R. E. MacBryde, B. & Baker, G. S. (1978). Endangered plant species and wildlife management. *Journal of Forestry,* **76,** 31–36.

Zhang, L. (1983). Vegetation ecology and population biology of *Fritillaria meleagris* L. at the Kungsangen Nature Reserve, Eastern Sweden. *Acta Phytogeographica Suecica,* **73,** 1–92.

II

Biology of Invasive and Weedy Species

9

Invading Plants: Their Potential Contribution to Population Biology

Richard N. Mack

Department of Botany
Washington State University
Pullman, Washington, USA

I. INTRODUCTION

Few other words shared by ecologists and laymen conjure up as many images as *invasion,* the incursion of organisms into a new area. Furthermore, in general usage there is the implication that the new arrivals will be detrimental to the natives or to the new territory. As invaders, plants have also been termed *aliens, immigrants, exotics, adventives, neophytes* or simply, *introduced species*. These immigrations sometimes result in the permanent establishment of populations in the new range as expressed by the term *naturalized,* which prefaces an ever-increasing list of taxa in local floras. Whatever the outcome, the immediacy of many invasions provides the opportunity to circumvent an ironic dilemma in population biology: many of the events that might explain the differences in organisms from place to place and from time to time (Harper 1977) have already occurred (e.g. microdifferentiation and coevolution, including interspecific competition), usually occur too slowly to measure reliably (e.g. range expansion and contraction), or contain presently undecipherable trends (e.g. annual changes in the sizes of populations).

The importance of these plant invasions to population dynamics and the transformation of landscapes did not escape the attention of Victorian biologists, including Darwin and Wallace. In providing empirical evidence for the 'geometrical ratio of increase' inherent in any species, Darwin (1872) pointed to the rapid spread of alien "cardoon and a tall thistle" in Argentina. Wallace (1905) reported that his correspondents in New Zealand saw *Rumex acetosella* covering "hundreds of acres with a sheet of red". Even then such observations may have

STUDIES ON PLANT DEMOGRAPHY:

sparked theory. Nägeli presented in 1874 the first known mathematical treatment of a fundamental topic in population biology; it dealt with the outcome of a species' introduction into an area already supporting another species (Harper 1974). It is a happy coincidence for my purposes here that in his model Nägeli chose plants.

The neglect of the population biology of invading plants is but one aspect of the general failure of latter-day biologists to build on a distinguished heritage (Harper 1967). Aside from weed control (Dewey 1897) and an early and exhaustive enumeration of the modes of dispersal (Ridley 1930), biologists have been curiously selective in their study of invading plants. Biological control is a profound demonstration of the awareness that biotic constraints eventually curb population growth among invaders. Geneticists have long made use of invaders in examining founder effect, genetic drift, and introgressive hybridization. And although islands have long fascinated biologists, it is the biology of natives and not the consequences of aliens on these islands that has attracted most attention (Carlquist 1974; Bramwell 1979).

Other attributes of plant invasions have lain unutilized, or at least underutilized. For example, in assessing the relative frequency of immigration versus extinction there are no parallels using plants to the revealing experiments with animal addition (Crowell 1973) or fumigation and recolonization (Simberloff & Wilson 1969; Rey 1981). More commonly, island biogeography in general, plus species packing, niche overlap, and the susceptibility of islands to invasion, have been approached through correlations among biotas (Grant & Grant 1982), laboratory experimentation (Robinson & Dickerson 1984), or models (Roughgarden 1974; Case 1981); all are informative but are not complete substitutes for field experimentation. The possibilities offered by invaders to plant population biology are perhaps best illustrated by Chapter 1 in John Harper's ''Population Biology of Plants'' (1977). His introductory remarks on continuous population growth rely on ''very odd and exceptional plants'', i.e. mobile aquatic plants, which are not likely to respond solely to the local limitations of resources. In addition to providing seemingly rapid and measurable examples of population growth, plant invasions allow precise assessment of (a) the circumstances surrounding range expansion, and in turn, population growth; (b) the potential for competition between aliens and natives: and (c) extinctions; these topics are the subject of this chapter. I hope to demonstrate that these themes should not be investigated simply to provide belated parallel examples for subjects built largely upon examples with animals. The ability of some plants such as *Poa bulbosa* to spread via seeds, bulbils, and bulbous culms illustrates just one reason why analogies between plant and animal invaders operate with severe limitations.

Some distinctions among the terms *invader, colonist* and *weed* are necessary. I will use the term *invader* in reference to any taxon entering a territory in which it has never before occurred, regardless of the circumstances. Transoceanic immigrants provide many unequivocal examples, although not the only cases in which the term may be applied (Plummer & Keever 1963). Because the fate of plants, even if they fail to establish at the entry point, will be important to this discussion, all entries will be considered an incursion, i.e. an invasion. Recurring Holocene migrations might not represent invasions by the strict application of the preceding definition, but I consider that distinction unnecessarily restrictive. Information from modern invasions and paleomigrations will prove mutually supportive. In my working definition I will deal with events over large areas, even continents, rather than at a local scale. At the local scale of a community or a habitat all plants must be colonists during succession (*sensu* Harper 1965), but these organisms are not necessarily invaders in my usage. All invaders are colonists, but the converse is not true. Whether a plant is a weed, i.e. unwanted, is mainly a value judgment outside the scope of this paper, but many of the invaders I will illustrate would be considered weeds in their new range.

II. INVADERS AND THE ESTIMATION OF POPULATION PARAMETERS

Predicting the change in a population is deceptively straightforward: if births, deaths, immigration and emigration can be quantified, the fate of the population can be forecast. Among the properties that make invaders intriguing is that (however briefly) immigration, and not the usually overriding flux between births and deaths, plays a major role in affecting subsequent events. Unlike immigration between two adjacent populations, no 'markers' are needed to distinguish immigrants from natives; if an alien can be found, it can be tallied unambiguously. The recurring invasions by many species to the same locality (e.g. contaminants in bales of cotton at a port serving textile mills), provide ready-made treatments through which to examine the importance of differences in the size of founder populations and of the conditions necessary for establishment at the point of entry. Commonly the environmental circumstances lead to local extinction.

As aliens move into a new range they provide rare examples of relatively unrestricted reproductive potential (the births component in population dynamics). In effect these populations may provide the best estimates of the intrinsic rates of population growth outside chemostats, especially if these estimates are compared to actual rates of growth in the home range. Instead of viewing the changing conditions and the various habitats encountered by the invader as complications, I view these as additional treatments allowing estimation of the range in population growth rate under measurable conditions.

Even having left competitors, predators, and parasites behind, aliens eventually reach their ecological and geographical limits in the new range. As they experience these inevitable controls on population growth (an increasing importance of death in the dynamics of the population), invaders may be among our best experimental material for evaluation of the usefulness of the logistic equation as a starting point in modeling population growth. In this regard the circumstances surrounding the decline of an invader through biological control deserve as much attention as events in the initial population build-up. Thus, (a) immigration, (b) a period when births exceed deaths, and then (c) a protracted period in which controls on both range and local abundance cause deaths to approximate births all may be studied stage by stage. The flux among births, deaths, immigration and emigration operates within all populations, but usually the net result is dynamic equilibrium so that detection is impaired, and causation is elusive. Drastic change, whether providing the opportunity for a species to become an invader or exposing natives to an invasion, is equivalent here to rattling the proverbial black box after all has been learned from studying it at rest (Harper 1977). Documentation of the alien's areal spread is a productive first step toward using such change.

III. THE SPREAD OF ALIENS

A. A Qualitative Approach

Regardless of the scale, an invasion takes place across space, and many of its consequences are more easily envisaged by documenting the areal spread of a species. A qualitative yet revealing branch of German phytogeography, *arealkunde* (literally, floristic distribution), has long dealt in part with documenting the first occurrence of aliens across central Europe (Hegi 1929; Meusel 1943; Walter 1954). Such maps may reveal the number of simultaneous entries into the new range, isolated populations, and circumstantial evidence for the pathways of spread. For example, collection records over a 200-year span show both the general westward movement of *Senecio vulgaris* into central Europe and also apparent outlier populations that had established ahead of the overall invasion front (Hegi 1929). With a dense array of sampling sites this qualitative approach can become much more informative by connecting similar dates of first occurrence with isopleths. Using maps prepared in this manner, Elton (1958) was able to detect a pause coinciding with the severe winter of 1917–1918 in the otherwise steady advance of the European starling into North America and to distinguish the breeding population from outlier sightings of the birds. The absence of such maps for plant spread in Elton's seminal work further illustrates my contention that plants have been underutilized in most considerations of the biology (including the population biology) of invasions (but see Mack 1981). Where less information is available maps have recorded only the chronology of first detection among counties or vice-counties, as in the examples compiled by Salisbury (1961) for alien composites in Britain.

A means of estimating the rate of invasion has involved plots over time of the number of first sightings of an alien, often compiled from herbarium specimens. In this manner Lacey (1957) found that *Galinsoga ciliata* had spread much faster in Britain than *Galinsoga parviflora,* especially during World War II when new sites for establishment were created by bomb cratering. Similarly, *Echium plantagineum* was found to be spreading at a faster rate in Australia than two congeners on the basis of reports of infestation in weed districts (Forcella & Harvey 1983). Although such records cannot usually be used to estimate the area occupied, they represent an extensive and largely untapped resource for predicting gross features of an invasion.

B. Spatial Diffusion Models

There have been surprisingly few attempts to formalize predictions of the areal spread of plants (Auld, Menz & Monaghan 1979; Auld & Coote 1980). A remarkable exception to the neglect of this subject was prepared by Skellam (1951); in it he not only modeled areal spread as a random-walk process but predicted the manner of dispersal when the observed rate of spread exceeded hypothetical limits. Modeling such spread through time is much better established, however, for a diverse group of topics ranging from cultural transmission (Cavalli-Sforza & Feldman 1981) to reconstruction of the geographic course of economic change and epidemics (Cliff *et al.* 1981). The conceptual approach in all these subjects has been the same: to consider spread as a spatial diffusion process. In its simplest application spatial diffusion describes the movement of a wave front from a point source, a *focus,* across a two-dimensional field, such as a ripple in water. The purposes of this approach are similar, regardless of the phenomenon being followed. What is (or has been) the rate of spread? What is the potential or the realized range of spread? Can the factors limiting or enhancing the spread be identified? Does the spread occur in a predictable fashion, i.e. what are the pathways? When and why does the spread depart from a wave phenomenon? And most commonly in connection with invaders, could the invasion be checked and the invader eliminated with such information?

The similarity between a plant invasion and the areal spread of disease is obvious (Salisbury 1953) and provides useful analogy for predicting the factors that affect spread. These factors include the following: (a) the number and arrangement of introduction points (foci) at the frontier of or within the new range, (b) the timing of new introductions, (c) the amount of habitat adjacent to the foci suitable for the invader, (d) the environmental heterogeneity and spatial irregularity of the range, closely related to (e) the corridors and barriers between suitable habitat, and (f) the initial size of the immigrant population. Additionally the species' life history characteristics must be quantified.

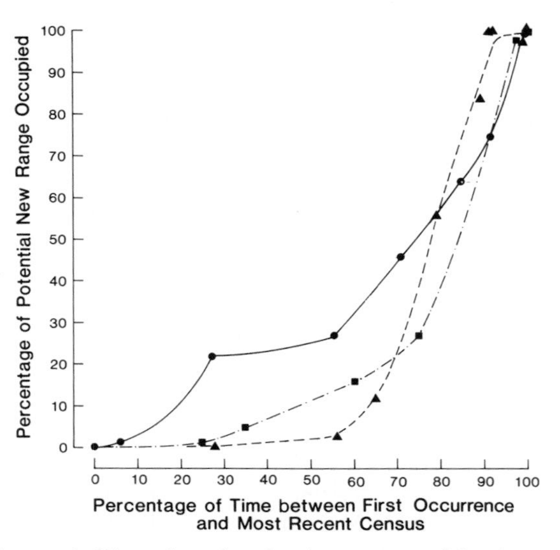

Fig. 1. The spread of three aliens plotted as the percentage of the new range occupied versus the percentage of time between first occurrence and most recent census. Plots are for *Bromus tectorum* (–·■·–) in western North America, 1889–1930 (Mack 1981); *Tamarix pentandra (—●—)* in southwestern United States, 1912–1960 (Robinson 1965); and *Opuntia aurantiaca (—▲—)* in South Africa, 1860–1974 (Moran & Annecke 1979).

The usual pattern of invasion for vascular plants as well as diseases (and even the transmission of cultural innovation) includes slow initial spread in which the alien occurs in a few isolated locales, followed by a phase of rapid range expansion, and a third phase of little or no growth in the areal extent (Salisbury 1953; Mack 1981; Auld, Hosking & McFadyen 1983). Such spread is of course easily comprehended, because it describes a phenomenon with logistic growth; the invaded area initially increases geometrically as a function of the square of the radius of the invading range (Thresh 1983). A linear plot of this range occupation provides a simplistic but nonetheless useful gauge by which to characterize and contrast invasions. For example, in spite of striking differences in their life histories, both *Bromus tectorum* (cheatgrass) and *Opuntia aurantiaca* occupied three-quarters of their new range in only the most recent quarter of their residence time (Fig. 1). Neither species appears to have had density-dependent restriction until just prior to reaching the geographical limits of its new range. Although dissemination of both was facilitated by humans, an annual habit perhaps compensated for the comparatively short time (about 40 years) in which *B. tectorum* occupied its range (Mack 1981) versus the much longer time for spread (over 100 years) taken by the cactus (Moran & Annecke 1979). *Tamarix pentandra* did not have a prolonged lag phase initially but instead displayed a period of slow areal spread (1925–1939) bracketed by periods of log phase

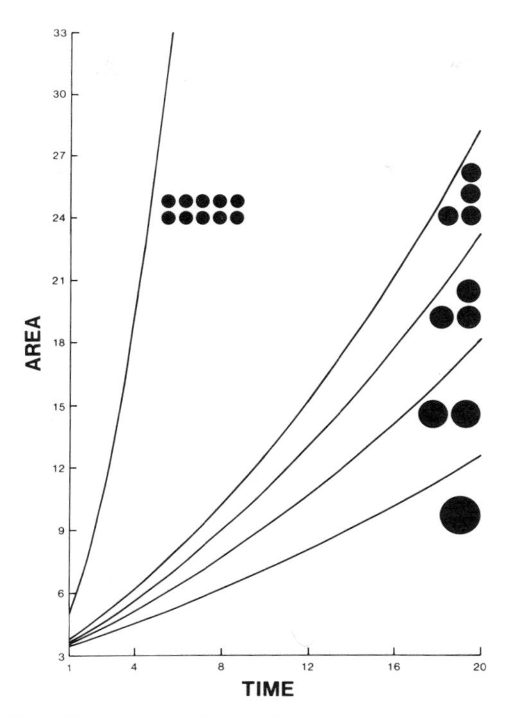

Fig. 2. The addition of foci increases the rate of new range occupation in quadratic fashion even if the total area initially occupied remains constant. In the examples plotted, the area occupied by n = 1, 2, 3, 4 and 10 foci at any time $t(1,2, \ldots ,20)$ is Area $= \pi[r^2 + 2\sqrt{n}\,(rat) + na^2t^2]$, where the rate of growth $a = 0.05$, the initial radius of the large single focus $r = 1$, and the foci in each group are far apart. At $t = 0$ (not shown) each group of foci beside its curve has the same total area.

expansion (Robinson 1965). It is problematical whether this pause in the invasion was induced by the scanty rain in the "dust bowl" years of the 1930s, or by a slackening of reservoir construction along the rivers where the tree is confined or is typical of riverine species that move along ecologically well defined corridors. Only the assembly of many additional examples will establish the importance of the particular circumstances of each invasion versus a species' life history characteristics in determining the shape of the invasion curve.

Each additional focus accelerates the invasion, even if only the number of foci (and not the total area initially occupied) is increased (Fig. 2). The smaller the focus, the greater the fraction of seeds or propagules that will arrive in the uninvaded area. Spread is also enhanced by foci that are far apart, because more area can be covered before population ranges coalesce (Auld, Hosking & McFadyen 1983). The importance of each focus illustrates one reason why the control of aliens is often ineffective: unless eradication is total the few remnant populations simply become the foci for reinvasion. Indeed, the reinvasion would proceed faster than an initial invasion from one focus.

Repeated introductions may be functionally similar to multiple foci in enhancing spread, provided the introductions are not to the same point of entry. In this context, the spread of the weeds of widely planted crops or forage is often much faster than the spread of aliens introduced as contaminants in goods for which there is limited commerce. In the late nineteenth century the parasitic dodders *Cuscuta epithymum* and *Cuscuta pentagona* appeared almost simultaneously in many locations in the Great Basin via commercial seed lots of alfalfa and clover (Hillman 1900). In contrast seeds introduced in the cupules of Turkish oaks used in tanning are known only from the immediate area around one tannery in Cornwall (Grenfell 1983). Multiple undetected introductions complicate any attempt to document natural selection in the taxon's new range, for all the phenotypic variation may be simply a display of the persistence of the phenotypes of different founders in different habitats.

In spite of the difficulty in incorporating anisometric space into spatial diffusion models (Rapoport 1982), identifying the barriers and pathways along the borders of a new range is enormously important to predictions of invasions. Such barriers are increasingly broken through human intervention. Although *Senecio squalidus* had been a resident on stone walls in Oxford since the late eighteenth century, it had only a very limited distribution in Britain until the proliferation of railroads in the nineteenth century (Kent 1964). Similarly, the occupation in 40 years of more than 200,000 km^2 in the intermountain west in North America by *Bromus tectorum* was caused in part by the arrival of the grass just as a comprehensive railroad system was being completed at the end of the nineteenth century (Mack 1981).

Most aliens never display areal expansion simply because too few seeds or propagules immigrate together. Aside from obvious limitations on dioecious species as invaders, other life forms may be restricted. Obligate vegetative propagators would also seem at high risk, since no bet-hedging through recombination and seed dissemination among different microsites is possible. For perennial plants, constraints on establishment caused by the age of first reproduction take on increased importance. In contrast Abbott (1977) found that annuals, especially aliens, have higher extinction rates than perennial immigrants to tropical islands. In this regard it is enigmatic that annuals in general are seemingly underrepresented in island floras (Williamson 1984). The very genotypic limitations within immigrants that potentially lead to genetic drift are much more commonly expressed instead as phenotypes that are unsuited for the introduction site. Even if offspring are produced, inbreeding may restrict the array of phenotypes below that necessary to cope with the new environment. Although chance dissemination can never be discounted, perhaps the failure so far of several widespread tropical weeds throughout a potential new range can be attributed to an insufficient immigrant population [e.g. the absence of *Salvinia molesta* in the Philippines, the absence of *Eichhornia crassipes* in the Niger River (Parker 1977)].

Spatial diffusion models offer a first approximation of the areal spread of invasions, but seldom can the direction of migration be predicted. Salisbury (1933) pointed out that for most recent immigrants into Britain the distribution is not along a continuous front in which spread radiated from a single focus; most patterns are discontinuous. Correlation of the direction of spread with components of the environment is hampered by overwhelming human influence. Although Plummer and Keever (1963) found a slight correlation between the spread of the introduced *Heterotheca latifolia* and the prevailing winds, they also point out that the plant's achenes are readily disseminated along roads and by animals. The invasions of aquatic plants up drainages from a delta would seem to offer the best opportunity to predict the direction of an invasion. The spread of *Elodea canadensis* as it entered northern European rivers in the mid-nineteenth century and advanced inland could be traced and to some extent predicted from collection records (Hegi 1929). Despite its immense potential in predicting the course of invasions, the habitat beyond the current distribution of an alien is seldom characterized as to its susceptibility for invasion. *Senecio vernalis* spread over central Europe on sandy soils and thereby was not regarded as a threat to regions with heavier soils (Meusel 1943).

Climatic change has repeatedly sparked the expansion of a species' range. As plants spread with deglaciation in the late Quaternary their pollen was incorporated into the datable sediments of ponds, lakes and mires along the migration route. In a sense each site became a 'trip wire' from which the earliest date by which a taxon had arrived in the vicinity can be determined. Isopleths of the first occurrence of many taxa in a region have been constructed by using these data (Davis 1983; Huntley & Birks 1983). In very general terms migration rates and migration routes may be estimated, although as yet the overall low density of sampling sites even in North America and Europe makes any calculation of the area occupied very approximate. These records may also reveal the changes in a population locally, assuming that for any taxon there is a direct relation between the number of polleniferous plants and the amount of pollen they produce. Watts (1973) concluded, from his reanalysis of an exceptionally detailed pollen record in varves, that the population of *Pinus strobus* grew logistically over the first 700–1000 years after its arrival. In a similar approach Bennett (1983) found that during migration in Britain in the early Holocene the populations of six arboreal genera doubled locally in less than 200 years; he concluded that the population increases were limited only by the age of first reproduction and not by environment.

These attempts to estimate population size with fossil pollen deal with a serious deficiency in plant population biology. Except the information derived from the aquatic plants referred to earlier, we currently lack data for plants on population change comparable to the results of releasing alien animals on isolated islands (e.g. Scheffer 1951; Klein 1968). In both cases the reindeer populations declined after first displaying exponential growth, a course that perhaps alien plants also show. Yet without the repeated enumeration of plant invaders we are left to speculate on the shape of plots on population *size* over time. Improved predictions on the effect of life history characteristics on population growth also await this information.

IV. INVASION AND DISTURBANCE

The connection between environmental alteration of the new range by humans and the spread of aliens has long been appreciated (Willis 1922) and prompted Harper (1965) to state that "Almost inevitably an invading species becomes established in areas in which some other disturbance has occurred . . .". Drought-tolerant aliens commonly prosper because disturbance often involves removal of a plant canopy and the enhancement of xeric conditions (Baker 1972). But in the absence of directional change in the climate are aliens dependent on such disturbance, and in turn on human intervention? The answer has important consequences for assessing the inevitability of contemporary alien spread and for explaining prehistoric changes in plant communities.

Sufficient examples can now be assembled to indicate that an invasion can proceed without continuing disturbance. Pines from temperate biomes have been repeatedly introduced into tropical and subtropical areas. *Pinus radiata* in Australia was found by Burdon and Chilvers (1977) to invade native eucalypt forest from adjacent plantations, and the spread of *Pinus pinaster* in South Africa has also proceeded without disturbance (Kruger 1977). Alien vines provide further examples. Several species in the composite genus *Mikania* have rapidly spread across arboreal canopies (Holm *et al.* 1977), and *Puereria thunbergiana* (kudzu) is an indiscriminate invader over tree canopies in the southeastern United States (Daubenmire 1978). It is not clear, however, whether these invasions require an initial disturbance to create a focus of establishment. The spread of aquatic vascular plants has undoubtedly been enhanced by the traffic along waterways, but at least the spread downstream would occur without disturbance, e.g. *Eichhornia crassipes* (water hyacinth) and *Salvinia molesta*. As mentioned previously, a potential dividend of examining alien spread will be predictions of the rate of invasion with and without human intervention. For example, did the rate of new range occupation by kudzu or water hyacinth simply track the spread of cultural acceptance (*sensu* Cavalli-Sforza & Feldman 1981) of these species?

V. COMPETITION

It is not surprising that the decline of natives would be explained in part as a consequence of competition with aliens: both Darwin (1872) and Wallace (1902) surmised that such interaction occurred. In the intervening century the anecdotal literature on plant invasions is filled with circumstantial and equivocal references to exotics 'crowding out' or 'overwhelming' natives. In a sense documentation of competition resulting from invasions is only part of a larger controversy in population biology (Schoener 1982; Connell 1983; Strong 1983); demonstrable cases of competition in nature are undoubtedly fewer than once believed.

In light of this legacy of unverified examples of competition, plant invasions display advantages or opportunities as experimental situations. For example, the complicating influence of the "ghost of competition past" (Connell 1980), even if overrated, is not at issue: alien and native have never met. This ability to identify the time when two taxa have first met need not involve only aliens and natives but may also include pairings of alien species. Martin & Harding (1981) found postimmigration selection in response to competition between two alien species of *Erodium* in California. An additional difficulty in measuring competition is that selection in either competitor may soon cause the two populations to interact no longer (Thompson 1982). But new invasions continually occur; the missed opportunity to examine competition in one pairing can be recouped with another example. In addition an invasion often occurs across an exceedingly wide gradient of environments against which competition may be evaluated, ranging from severe chronic disturbance to no disturbance. The opportunities for comparing and contrasting the performances of an invader under diverse conditions could allow partitioning of a response that would be difficult to attribute to anything other than competition. Intriguing situations in which the experiment has in effect already been performed include *Tamarix pentandra* in southwestern United States (Turner 1974), *Casuarina equisetifolia* in Florida (Austin 1978; Morton 1980) and *Rhododendron ponticum* in the British Isles. The range of soil types, including sandstone, muck, glacial till and even limestone pavement (Kelly 1981) on which *R. ponticum* may invade is impressive; each soil type represents a separate treatment for evaluation.

Perhaps overlooked in the course of repeated reference to the role of disturbance in enhancing invasions has been the displacement of native colonizers, i.e. populations that presumably would have increased with increasing disturbance. Such species may provide the least ambiguous examples of competitive displacement. For example, *Festuca microstachys, Festuca octaflora* and *Bromus carinatus* were presettlement colonizers within the intermountain steppe in western North America. Today most of this region has been converted to arable fields or rangelands, yet these species are not even common members of these new ecosystems. Instead aliens, including *Salsola kali, Poa pratensis* and especially *Bromus tectorum,* dominate (Mack 1981). Both field and glasshouse experiments indicate that in the case of *Bromus tectorum,* the precocious alien usurps resources from the natives (R. N. Mack & D. A. Pyke, unpublished data).

VI. EXTINCTION

Darwin (1872) and Wallace (1902) were so impressed by the spread of plant invaders on islands that they suggested that the extinction of natives could be caused by the aliens. To my knowledge, however, we still lack not only an irrefutable example of plant extinction caused in this manner, but even enough demonstrable examples of native–alien interaction to predict the likelihood of extinction through competition as opposed to predation. Despite the astonishing rate of extinction among the native Hawaiian flora [possibly 12% of the taxa since European contact (Fosberg & Herbst 1975)], causes of extinction may be more easily studied through examination of the much more frequent local extinction of aliens.

There are at least two investigative levels to the study of extinctions: the etiology of invasions that fail to result in 'naturalizations' and the test of theoretical predictions on rates of extinction. There must have been many immigrations between biomes (tropical to temperate biomes, etc.) in the last 100 years alone. Yet in the vast majority of cases the alien has failed to persist; the new environment has apparently been beyond the species' ecological amplitude. Nevertheless, intriguing examples in which the explanation is not apparent abound. Even after repeated introductions *Tillandsia usneoides* (Spanish Moss) has not spread beyond gardens in Hawaii (Neal 1965). Although *Casuarina equisetifolia* is spreading in southern Florida, *Casuarina glauca* persists only locally, its spread possibly hampered by an inability to set fruit (Morton 1980). Similarly *Veronica peregrina* has not spread as extensively in Britain as *Veronica filiformis* (Bangerter 1964). Good (1964) hypothesized that the failure of *Viola odorata* and Primulas to establish in New Zealand was caused by lack of appropriate insect pollinators. In the same steppe in which *Bromus tectorum* (cheatgrass) is often dominant (Mack 1981), nine other annual congeners have been introduced, yet none displays the areal distribution and abundance of cheatgrass. The biologies of these other annual bromes differ, but in general these species lack the winter hardiness, and the ability to germinate near 0°C comparable to that of *B. tectorum* (Hulbert 1955). Seldom has the failure of an alien been specifically examined, although in one such investigation Nilsen and Muller (1980a, b) concluded that *Schinus terebinthifolius,* a widespread alien in Florida and Hawaii, is unable to become naturalized in southern California because of the shrub's slow germination before the cessation of seasonal rains.

Estimations of rates of extinction are inextricably tied to island biogeography and more recently to MacArthur and Wilson's (1967) theory of island equilibrium. Few attempts, however, have been made to use plant immigration and extinction on islands to test this theory; rarer still has been the use of aliens (Abbott 1977; Abbott & Black 1980). Currently there is no equivalent for plants to 'experimental zoogeography' (Simberloff & Wilson 1969; Crowell 1973), yet invaders and their recurring local extinctions provide a multitude of experimental scenarios. So diverse are plant invaders and so frequent are their movements that the time to extinction as dictated by the ratio of birth rates to death rates (as well as the influence of repeated immigration), varying population size, life history characteristics, and competitors could all be examined separately (MacArthur 1972).

VII. INVADERS AND FUTURE EMPHASIS IN POPULATION BIOLOGY

I have mentioned only briefly application of the population biology of invading plants to agriculture, yet much of the information assembled here was originally collected as a result of a practical need to control aliens. Not only will agronomic settings continue to provide examples of invasions, but much of plant population biology and the study of aliens will become increasingly linked as all ecosystems become dominated by those taxa that are tolerant of intense human activity. As a result, the ability to identify likely entry points and pathways; anticipate the direction of spread; determine the effectiveness of barriers and the outcome of competition; and predict the extinction rate of populations will be simultaneously useful to both population biology and agronomy. Even if each of the world's biomes is eventually reduced to impoverished cosmopolitan floras, this alteration will be at least slowed by a more thorough examination of the population biology of invaders than has been so far attempted.

ACKNOWLEDGMENTS

As exemplifies the stimulatory influence that John Harper has in population biology, the idea for this paper came in conversations with him in Bangor in autumn 1983. I also thank Jim White for his thoughtful suggestions to me during the writing of this paper. Additionally I thank B. A. Auld, R. A. Black, A. J. Gilmartin, S. S. Higgins, E. A. Kurtz, J. Major, H. Meister, M. Moody, K. Rice, P. Rundel, J. N. Thompson, and O. B. Williams for information and helpful discussions.

REFERENCES

Abbott, I. (1977). Species richness, turnover and equilibrium in insular floras near Perth, Western Australia. *Australian Journal of Botany,* **25,** 193–208.

Abbott, I. & Black, R. (1980). Changes in species composition of floras on islets near Perth, Western Australia. *Journal of Biogeography,* **7,** 399–410.

Auld, B. A. & Coote, B. G. (1980). A model of a spreading plant population. *Oikos,* **34,** 287–292.

Auld, B. A., Menz, K. M. & Monaghan, N. M. (1979). Dynamics of weed spread: Implications for policies of public control. *Protection Ecology,* **1,** 141–148.

Auld, B. A., Hosking, J. & McFadyen, R. E. (1983). Analysis of the spread of tiger pear and parthenium weed in Australia. *Australian Weeds,* **2,** 56–60.

Austin, D. F. (1978). Exotic plants and their effects in southeastern Florida. *Environmental Conservation,* **5,** 25–34.

Baker, H. G. (1972). Migrations of seeds. *Taxonomy, Phytogeography and Evolution* (Ed. by D. H. Valentine), pp. 327–347. Academic Press, London.

Bangerter, E. B. (1964). *Veronica peregrina* L. in the British Isles. *Proceedings of the Botanical Society of the British Isles,* **5,** 303–313.

Bennett, K. D. (1983). Postglacial population expansion of forest trees in Norfolk, UK. *Nature (London),* **303,** 164–167.

Bramwell, D. (1979). Introduction. *Plants and Islands* (Ed. by D. Bramwell), pp. 1–10. Academic Press, London.

Burdon, J. J. & Chilvers, G. A. (1977). Preliminary studies on a native Australian eucalypt forest invaded by exotic pines. *Oecologia,* **31,** 1–12.

Carlquist, S. (1974). *Island Biology.* Columbia University Press, New York.

Case, T. J. (1981). Niche packing and coevolution in competition communities. *Proceedings of the National Academy of Sciences of the U.S.A.,* **78,** 5021–5025.

Cavalli-Sforza, L. L. & Feldman, M. W. (1981). *Cultural Transmission and Evolution: A Quantitative Approach.* Princeton University Press, Princeton, New Jersey.

Cliff, A. D., Haggett, P., Ord, J. K. & Versey, G. R. (1981). *Spatial Diffusion: An Historical Geography of Epidemics in an Island Community.* Cambridge University Press, Cambridge.

Connell, J. H. (1980). Diversity and the coevolution of competitors, or the ghost of competition past. *Oikos,* **35,** 131–138.

Connell, J. H. (1983). On the prevalence and relative importance of interspecific competition: Evidence from field experiments. *American Naturalist,* **122,** 661–696.

Crowell, K. L. (1973). Experimental zoogeography: Introductions of mice to small islands. *American Naturalist,* **107,** 535–558.

Darwin, C. (1872). *On the Origin of Species.* 6th ed. John Murray, London.

Daubenmire, R. (1978). *Plant Geography.* Academic Press, New York.

Davis, M. B. (1983). Holocene vegetational history of the eastern United States. *Late-Quaternary Environments of the United States* (Ed. by H. E. Wright), pp. 166–181. University of Minnesota Press, Minneapolis.

Dewey, L. H. (1897). Migration of weeds. *United States Department of Agriculture Yearbook, 1896,* pp. 263–286.

Elton, C. S. (1958). *The Ecology of Invasions by Animals and Plants.* Chapman & Hall, London.

Forcella, F. & Harvey, S. J. (1983). Relative abundance in an alien weed flora. *Oecologia,* **59,** 292–295.

Fosberg, F. R. & Herbst, D. (1975). Rare and endangered species of Hawaiian vascular plants. *Allertonia,* No. 1.

Good, R. (1964). *The Geography of the Flowering Plants.* 3rd ed. Wiley, New York.

Grant, B. R. & Grant, P. R. (1982). Niche shifts and competition in Darwin's Finches: *Geospiza conirostris* and congeners. *Evolution (Lancaster, Pa.),* **36,** 637–657.

Grenfell, A. L. (1983). More on tan bark aliens. Adventive News 26. *Botanical Society of the British Isles News,* No. 35.

Harper, J. L. (1965). Establishment, aggression, and cohabitation in weedy species. *The Genetics of Colonizing Species* (Ed. by H. G. Baker & G. L. Stebbins), pp. 245–265. Academic Press, New York.

Harper, J. L. (1967). A Darwinian approach to plant ecology. *Journal of Ecology*, **55**, 247–270.

Harper, J. L. (1974). A centenary in population biology. *Nature (London)*, **252**, 526–527.

Harper, J. L. (1977). *Population Biology of Plants*. Academic Press, London.

Hegi, G. (1929). *Illustrierte Flora von Mittel-Europa. Band VI*. Vol. 2. A. Pichler's Witwe & John, Vienna.

Hillman, F. H. (1900). Clover seeds and their impurities. *University of Nevada, Agricultural Experiment Station, Bulletin*, **47.**

Holm, L. G., Plucknett, D. L., Pancho, J. V. & Herberger, J. P. (1977). *The World's Worst Weeds*. East-West Center, University of Hawaii Press, Honolulu.

Hulbert, L. C. (1955). Ecological studies of *Bromus tectorum* and other annual bromegrasses. *Ecological Monographs*, **25**, 181–213.

Huntley, B. & Birks, H. J. B. (1983). *An Atlas of Past and Present Pollen Maps for Europe: 0-13,000 Years*. Cambridge University Press, Cambridge.

Kelly, D. L. (1981). The native forest vegetation of Killarney, south-west Ireland: An ecological account. *Journal of Ecology*, **69**, 437–472.

Kent, D. H. (1964). *Senecio squalidus* L. in the British Isles-4, southern England (1940 →). *Proceedings of the Botanical Society of the British Isles*, **5**, 210–213.

Klein, D. R. (1968). The introduction, increase, and crash of reindeer on St. Matthew Island. *Journal of Wildlife Management*, **32**, 350–367.

Kruger, F. J. (1977). Invasive woody plants in the Cape fynbos with special reference to the biology and control of *Pinus pinaster*. *Proceedings of the Second National Weeds Conference of South Africa*. A. A. Balkema, Cape Town, S. Africa.

Lacey, W. S. (1957). A comparison of the spread of *Galinsoga parviflora* and *G. ciliata* in Britain. *Progress in the Study of the British Flora* (Ed. by J. E. Lousley), pp. 109–115. Botanical Society of the British Isles, London.

MacArthur, R. H. (1972). *Geographical Ecology*. Harper & Row, New York.

MacArthur, R. H. and Wilson, E. O. (1967). *The Theory of Island Biogeography*. Princeton University Press, Princeton, New Jersey.

Mack, R. N. (1981). Invasion of *Bromus tectorum* L. into western North America: An ecological chronicle. *Agro-Ecosystems*, **7**, 145–165.

Martin, M. M. & Harding, J. (1981). Evidence for the evolution of competition between two species of annual plants. *Evolution (Lancaster, Pa.)*, **35**, 975–987.

Meusel, H. (1943). *Vergleichende Arealkunde*. Vol. I. Borntraeger, Berlin.

Moran, V. C. & Annecke, D. P. (1979). Critical reviews of biological pest control in South Africa. 3. The jointed cactus, *Opuntia aurantiaca* Lindley. *Journal of the Entomological Society of South Africa*, **42**, 299–329.

Morton, J. F. (1980). The Australian pine or beefwood (*Casuarina equisetifolia* L.), an invasive "weed" tree in Florida. *Proceedings of the Florida State Horticultural Society*. **93**, 87–95.

Neal, M. C. (1965). In gardens of Hawaii. *Special Publications Bernice P. Bishop Museum*, No. 50, Honolulu, Hawaii.

Nilsen, E. T. & Muller, W. H. (1980a). A comparison of the relative naturalizing ability of two *Schinus* species (Anacardiaceae) in southern California. I. Seed germination. *Bulletin of the Torrey Botanical Club*, **107**, 51–56.

Nilsen, E. T. & Muller, W. H. (1980b). A comparison of the relative naturalizing ability of two *Schinus* species (Anacardiaceae) in southern California. II. Seedling establishment. *Bulletin of the Torrey Botanical Club*, **107**, 232–237.

Parker, C. (1977). Prediction of new weed problems, especially in the developing world. *Origins of Pest, Parasite, Disease and Weed Problems* (Ed. by J. M. Cherrett & G. R. Sagar), pp. 249–264. Blackwell Scientific Publications, Oxford.

Plummer, G. L. & Keever, C. (1963). Autumnal daylight weather and camphor-weed dispersal in the Georgia Piedmont region. *Botanical Gazette (Chicago)*, **124,** 283–289.

Rapoport, E. H. (1982). *Areography*. Pergamon, Oxford.

Rey, J. R. (1981). Ecological biogeography of arthropods on *Spartina* islands in northwest Florida. *Ecological Monographs*, **51,** 237–265.

Ridley, H. N. (1930). *The Dispersal of Plants throughout the World*. L. Reeve, Kent, U.K.

Robinson, J. V. & Dickerson, J. E., Jr. (1984). Testing the invulnerability of laboratory island communities to invasion. *Oecologia*, **61,** 169–174.

Robinson, T. W. (1965). Introduction, spread and areal extent of saltcedar (*Tamarix*) in the western states. *Geological Survey Professional Paper (U.S.)*, **491-A.**

Roughgarden, J. (1974). Species packing and the competition function with illustrations from coral reef fish. *Theoretical Population Biology*, **5,** 163–186.

Salisbury, E. J. (1933). The East Anglian flora: A study in comparative plant geography. *Transactions of the Norfolk and Norwich Naturalist's Society*, **13,** 191–263.

Salisbury, E. J. (1953). A changing flora as shown in the study of weeds of arable lands and waste places. *The Changing Flora of Britain* (Ed. by J. E. Lousley), pp. 130–139. Botanical Society of the British Isles, London.

Salisbury, E. (1961). *Weeds and Aliens*. Collins, London.

Scheffer, V. B. (1951). The rise and fall of a reindeer herd. *Scientific Monthly*, **73,** 356–362.

Schoener, T. W. (1982). The controversy over interspecific competition. *American Scientist*, **70,** 586–595.

Simberloff, D. S. & Wilson, E. O. (1969). Experimental zoogeography of islands: The colonization of empty islands. *Ecology*, **50,** 278–296.

Skellam, J. G. (1951). Random dispersal in theoretical populations. *Biometrika*, **38,** 196–218.

Strong, D. R. (1983). Natural variability and the manifold mechanisms of ecological communities. *American Naturalist*, **122,** 636–660.

Thompson, J. N. (1982). *Interaction and Coevolution*. Wiley, New York.

Thresh, J. M. (1983). Progress curves of plant virus disease. *Advances in Applied Biology*, **8,** 1–85.

Turner, R. M. (1974). Quantitative and historical evidence of vegetation changes along the upper Gila River, Arizona. *Geological Survey Professional Paper (U.S.)*, **655-H.**

Wallace, A. R. (1902). *Island Life*. 3rd ed. Macmillan, New York.

Wallace, A. R. (1905). *Darwinism; an Exposition of the Theory of Natural Selection, with Some of its Applications*. 3rd ed. Macmillan, New York.

Walter, H. (1954). *Grundlagen der Pflanzenverbreitung II. Arealkunde*. Ulmer, Stuttgart.

Watts, W. A. (1973). Rates of change and stability in vegetation in the perspective of long periods of time. *Quaternary Plant Ecology* (Ed. by H. J. B. Birks & R. G. West), pp. 195–206. Blackwell Scientific Publications, Oxford.

Williamson, M. (1984). Sir Joseph Hooker's lecture on insular floras. *Biological Journal of the Linnean Society*, **22,** 55–77.

Willis, J. C. (1922). *Age and Area*. Cambridge University Press, Cambridge.

10

Proso Millet (Panicum miliaceum *L.*): *A Crop and a Weed*

Paul B. Cavers and Marguerite A. Bough

Department of Plant Sciences
The University of Western Ontario
London, Ontario, Canada

I. INTRODUCTION

In his introduction to "The Biology of Weeds", John Harper (1960) mentioned the special problems in taxonomy and evolution that are provided by weedy species. He also noted that human activities "break down geographical barriers and permit a ready flow of potential weed species from one country to another". He further commented that "often a plant species which is relatively innocuous in its country of origin may develop alarming weedy potentialities if it is introduced to new territories". In this paper we will discuss proso millet (*Panicum miliaceum* L.), grown as a crop in Canada since the time of the earliest French settlers (Dekker *et al.* 1981) but now also recognized as a troublesome weed in many parts of the country. As a weed this species is well described by the preceding statements by Harper (1960).

II. PROSO MILLET

A. The Crop

Proso millet was one of the earliest graminaceous species to be domesticated as a cereal crop. Native to Eurasia (Rachie 1975), it has been grown since at least 4000–5000 B.C. in southern Scandinavia, central Europe, China and India (Grabouski 1971; Rachie 1975; Anderson & Martin 1949). It was the chief grain crop

STUDIES ON PLANT DEMOGRAPHY:
A FESTSCHRIFT FOR JOHN L. HARPER

in Europe from about 1200 B.C. to the Roman period (L. Oestry, personal communication). In recent times production has been concentrated in eastern Europe, Russia, China and India (Rachie 1975). It is used primarily for human consumption after cooking, baking and brewing.

In North America, proso millet has been grown primarily for feeding to livestock and as a principal constituent in seed mixtures for wild birds (Strand & Behrens 1979). Many hundreds of varieties are grown around the world, and new ones are regularly licensed in the United States (Robinson 1980; Hinze 1972).

A very popular variety and one of the earliest to be licensed in the United States is Turghai, which was imported from Russia in 1903 by the USDA (Anonymous 1982). Recent varieties have been introduced for specific purposes [e.g. 'Dove' proso millet was introduced from India to attract and feed mourning doves in the southeastern United States (Robinson 1971)]. Much of the proso millet grown in Canada is used for birdseed mixtures and is not of named varieties.

In North America, as in much of the rest of the world, proso millet is often a second-choice crop or 'catch-crop'. It has a comparatively low yield but matures quickly and is thus planted in fields where winter wheat, winter barley or similar crops have been winter-killed or otherwise destroyed. Consequently, proso millet is often a low-value crop and there is a tendency to use domestic seed sources.

B. The Weed

Farmers who have grown proso millet have often noted that volunteer plants arise in succeeding crops. Over the last 20 years we have had several specimens of proso millet brought in for identification from local farms. In addition, we have made a survey of herbarium specimens of proso millet from across Canada, and the labels were often marked "escape from cultivation". Nevertheless, it is only within the past 15 years that this species has risen to prominence as an important weed problem in North America.

The first reports of large-scale infestations came from Minnesota and Wisconsin in 1970 (Strand & Behrens 1979). It spread rapidly, primarily by means of harvesting equipment, and by 1979, proso millet was described as the greatest single threat to continued row crop production in southern Wisconsin and Minnesota (Harvey 1979).

This major weed biotype differs in several respects from the cultivated material. It has a more open panicle and the seeds shatter readily. The most clear-cut difference is in seed (caryopsis) colour, which ranges from a dark olive-brown to almost black at maturity. This biotype seems indistinguishable from a black-seeded biotype that occurs as a weed in Hungary (Terpó-Pomogyi 1976; Dr. A. Terpó, personal communication), Russia (Lysov 1975) and Middle Asia (Hilbig

& Schamsran 1980). Zhukowsky (1964) stated that this biotype constituted a separate species, *Panicum spontaneum* Lyss., but Oestry and de Wet (1981) characterized it as only a subspecies of *P. miliaceum*.

The black-seeded biotype spread to Canada in the 1970s. It now occurs in several major infestations in southern Ontario (Fig. 1), ranging from Huron County in the west to Glengarry County in the east (Dekker *et al.* 1981).

There is also a widespread and dense infestation of proso millet covering thousands of hectares in the corn-growing area of southern Manitoba (Fig. 1). This population differs markedly in appearance from the black-seeded biotype, particularly in its striped, grey-green seed and non-shattering panicle. In our common garden tests this biotype is very similar to an old crop variety named 'Crown', and we are calling this material the crown biotype. Crown is the only variety licenced in Canada, and it has been grown under licence in Manitoba as recently as 1983 (B. Andrews, personal communication). (For further descriptions of Canadian biotypes see Bough, Colosi & Cavers 1986). There are a few populations of crown-type proso millet in southern Ontario and Quebec, but each is restricted to one or two farms.

In addition to the two major weed biotypes, there exist in southern Ontario, Manitoba and Quebec small and scattered populations of proso millet which

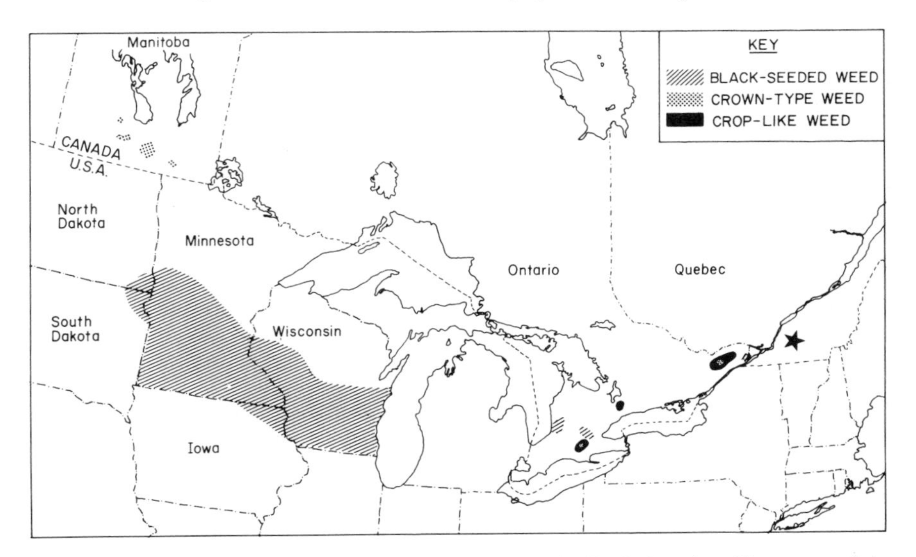

Fig. 1. Some major areas of proso millet infestation in North America. The extent of the Wisconsin–Minnesota infestation is taken from Harvey (1979). More recent information is that black-seeded proso millet has been spreading southward into Illinois and Iowa (R. G. Harvey, personal communication). In Quebec, in the vicinity of Montreal, (★), crown-type weed and other crop-like weed populations (including one with very dark red seeds) occur, but we have yet to determine the full extent of these infestations.

closely resemble major crop types. Although many of these populations are remarkably persistent, most of them do not seem to be spreading to neighbouring farms.

III. CROP AND WEED RACES

Harlan (1965) pointed out that most of the common cultivated plants have companion weed races. There are weed carrots, weed tomatoes, weed sunflowers, weed lettuces, weed radishes, and many others. The grass family provides some of the most important and well-known examples, including weed oats, weed barley, weed wheats, weed rices, weed corns (maizes), weed ryes, weed sorghums and weed sugar canes. De Wet (1975) pointed out some important differences between crop and weed races of cereals; these are summarized in Table I. We will be comparing the crop, crop-like weed and weed biotypes of proso millet for each of these attributes.

A. Seed Size

We have determined seed weights for many different populations of proso millet in southern Ontario. In all comparisons seeds of the black-seeded biotype

TABLE I.

Comparison of Attributes of Crop and Weed Races within a Single Graminaceous Species[a], Indicating Where Various Proso Millet Weed and Crop Races Fit into the Scheme[b]

Crop races	Weed races
Larger seeds (C, C-W, Crown)	Smaller seeds (B)
Greater seedling vigour (C, Crown, some C-W)	Lesser seedling vigour (B, some C-W)
Rapid, complete germination, little dormancy (C, C-W, Crown)	Delayed, intermittent germination, stronger dormancy (B, dark-red C-W, possibly some Crown and C-W)
Non-fragile inflorescences (C, C-W, Crown)	Fragile inflorescences (fertile florets easily disarticulated) (B)
Uniform population maturity (C, some C-W, some Crown)	Non-synchronous maturity; non-synchronous tillering (B, orange-red-seeded C-W, some other C-W, some Crown)
Apical dominance (fewer but larger inflorescences) (C, C-W, Crown)	No apical dominance (smaller, more numerous inflorescences) (orange-red-seeded C-W, B had less than others, as did some Crown)
No natural seed dispersal mechanisms (C, C-W, Crown, B)	Effective seed dispersal mechanisms

[a] Adapted from de Wet (1975).
[b] C = crop races, B = black-seeded weed, C-W = crop-like weeds except Crown.

were significantly lighter than those of the crop-like weeds or the crops. For example, Moore and Cavers (1985) recorded mean weights of 3.69, 3.75 and 3.72 mg/seed for mid-season collections from widely separated populations of the black-seeded biotype in Huron, Glengarry and Waterloo counties. Another experiment (M. A. Bough & P. B. Cavers, unpublished data) has shown that mean seed weight for the black-seeded biotype declines throughout the season from about 4.3 mg in the second week of production to 3.0 mg for the last seeds produced. By contrast three crop-like weed populations had mean weights of 4.90–5.99 mg/seed and samples from three crops grown for bird seed had mean weights ranging from 5.72 to 6.23 mg/seed.

We have received samples of seeds of proso millet from Dr. A. Terpó in Hungary. The black-seeded population, which when grown was similar to our black-seeded weed, had the lightest seeds (4.74 mg/seed), whereas the other populations, with seeds of a variety of colours, had mean weights per seed ranging from 5.09 to 6.04 mg.

Our results for the various 'crown' populations are quite different (Table II). The weed population from Manitoba had significantly heavier seeds than any of the other crop-like weed populations of this type. In fact, they are in the same range as the weights of the crop seeds mentioned previously and in Table II. Conversely, the least weedy of the five weed populations, the Cadman population from Oxford County, Ontario, had the lightest seeds of the five crown populations. However, the weights for all crown populations were well above the values recorded for seeds of any populations of the black-seeded biotype.

In summary, the overall pattern of seed weight in proso millet follows the prediction in Table I. The heaviest seeds, in general, are those of the crop material; the lightest are those of the black-seeded weed.

TABLE II.

Mean Weight per 100 Seeds for Six Widely-variable and Widely-separated 'Crown' Populations of Proso Millet (All Collected in 1983)

Population	Mean weight per 100 seeds (mg) ± S.D.
Manitoba, Carman area: weed population	624.8 ± 4.0
Manitoba, Carman area: weed allowed to grow as a crop (not totally mature when plants cut)	501.5 ± 48.8
Manitoba, Plum Coulee: crop grown under licence	613.5 ± 4.9
Ontario, Carleton County: shatters relatively easily; many weedy characters	553.7 ± 4.1
Ontario, Dundas County: non-shattering, open panicle	541.3 ± 7.4
Ontario, Oxford County: non-shattering open panicle, less effective weed	468.7 ± 3.7

B. Seedling Vigour

From Table I we can predict that crop races would have greater seedling vigour than weed races. Moore and Cavers (1985) compared three black-seeded weed populations, three crop-like weed populations and three crop populations of proso millet for seedling vigour over the first 21 days of growth. They found that seedling size at each of three harvests was related to original seed size. Seedlings from the heaviest (crop) seeds were the largest on each occasion, and seedlings from the lightest seeds (of the black-seeded weed) were the lightest. There was very little difference between any two populations in the *rate* of growth.

A recent experiment by C. A. Vis (unpublished) confirms the preceding findings. Three biotypes were used: the black-seeded weed, the 'crown' weed from Manitoba and a golden-seeded crop-like weed from Ontario (Table III). After 17 days of growth under similar conditions seedlings of the black-seeded weed were much lighter than those of the other two biotypes. However, the total root lengths at the same harvest were very different (Table III). Seedlings of the black-seeded weed tended to produce long, slender roots, whereas those of the crown biotype were shorter and thicker. Presumably the long slender roots would be more efficient in enabling the plant to utilize a larger volume of soil.

C. Germination Patterns and Dormancy

De Wet (1975) suggested that crop races show rapid complete germination, whereas weed races exhibit delayed intermittent germination.

In a series of germination tests in growth cabinets over the past 3 years, all seed samples of the crop biotypes and those of the crop-like weeds (including the crown population from Manitoba) have shown rapid complete germination over a range of light and temperature conditions. Germination was finished within 5

TABLE III.

Seedling Development of Three Biotypes of Proso Millet 17 Days after Emergence (Experiment Run in Greenhouse, November 1983)

Parameter measured	Black-seeded weed (Huron Co., Ontario)	'Crown' type crop-like weed (Carman, Manitoba)	Golden-seeded crop-like weed (Durham Co., Ontario)
Total dry weight per plant (mean of 20 plants, in grams)	0.0277	0.0758	0.0829
Mean total root length (mean of 10 plants, in millimetres)	151	119	168

days in almost every case. In contrast, germination patterns for all black-seeded populations in the same tests were extremely intermittent and continued for several weeks to several months. Moore and Cavers (1985) found that with black-seeded weed samples there is usually a small flush of germination followed by irregular germinations over a long period.

O'Toole (1982) discovered that field germination of the black-seeded populations was even more intermittent than we found in the laboratory. In general, a new flush of seedlings appeared after each cultivation during the growing season. Because cultivation patterns differ greatly among different crops it is not surprising that the germination patterns of the black-seeded weed were very different in neighbouring fields containing corn, white bean and barley crops.

The lack of any apparent specialized dormancy mechanisms in seeds of crop-like weeds makes it difficult to understand how they can persist as weeds. In several experiments we noted that seeds of crop-like weed biotypes buried at depths to 20 cm (plough depth) throughout the winter showed poor survival. By the following spring the majority had rotted, although there were variations between biotypes and depths. In the event of a mild autumn these crop-like weed biotypes will germinate readily. These seedlings are killed once the temperature falls. Persistence of such populations must depend on a very high initial seed input.

In contrast, the black-seeded weeds have much greater dormancy. Very few fresh seeds will germinate, and survival in the soil over winter is high; up to 90% remain viable and dormant by the following spring under some conditions.

D. Fragility of Inflorescences

Dr. Linda Oestry (personal communication) has spent several years studying variability in proso millet and found that there is a strong correlation between a tendency of ripe spikelets to shatter and weedy material. Apparently, in panicles of weedy populations a layer of callus forms strongly at the base of the spikelet. Characteristically a hollow cup is left after the spikelet has broken away.

The Canadian populations of proso millet fit into two more or less clear-cut groups in terms of their tendency to shatter. All populations of the black-seeded weed drop their seeds shortly after they ripen. All other populations tend to retain their seeds. Even the 'crown' population from Manitoba retains its seeds very well. One Manitoba farmer with a massive population of this biotype in his soil simply allowed the millet to grow in 1983 and harvested a bumper crop.

Once again our findings for proso millet are in general agreement with the predictions of Table I.

E. Synchrony of Tillering and Maturity

From Table I we would expect crop races to have very uniform maturity across a population, which would be effected through synchronous tillering. Crops of proso millet usually exhibit such uniformity because the seeds are all planted at the same time, the plants produce no tillers or very few, and all plants, selected to have little genetic variation, tend to mature at the same time. Some crop-like weed populations behave similarly to crops, but others, such as the crown population from Manitoba, have numerous tillers formed at different times and thus ripen seeds over a period of several weeks. An orange-red-seeded crop-like weed found near Ottawa, Ontario, possesses extreme variation in ripening. Plants of this population form numerous late tillers. Many seeds on these tillers do not ripen before the plant is killed by fall frosts.

Since seeds of the black-seeded biotype exhibit intermittent germination in the field, one usually finds plants at many stages of development in a single population. In addition, plants with sufficient space to form tillers [in dense stands few tillers are formed and the population is less variable (Harper 1965)] usually have several panicles at different stages of ripeness. Thus, it is not surprising that seeds of the black-seeded biotype are shed over periods of 3 months or more in some crops (O'Toole & Cavers 1983). In general, many of the weed and crop-like weed populations have the asynchronous tillering and seed maturity that are described as typical of graminaceous weed races (Table I).

F. Apical Dominance

The characteristic of apical dominance is closely tied in with tillering and plant maturity. Crop populations tend to produce a single panicle from the apical bud even when the plants are widely spaced. Black-seeded weed populations and the orange-red-seeded crop-like weed have the greatest tendency to produce inflorescences on tillers arising from lateral bud development. This pattern parallels closely the one predicted for crop and weed races in Table I.

G. Dispersal Mechanisms

In Table I, we suggested that crop races generally have no natural seed dispersal mechanisms, and it appears that all populations of proso millet can be placed into this category. O'Toole & Cavers (1983) made a detailed study of seed dispersal from plants of the black-seeded weed in three crops: barley, corn and white beans. In all cases a vast majority of seeds were found on the soil within 1 m of the parent plant (Fig. 2). Wind had little effect in the dispersal of these seeds, which are heavy compared to those of most other weeds.

S. McCanny & P. B. Cavers (unpublished data) have shown that large numbers of proso millet seeds can be dispersed both between and within fields, through being picked up and deposited by farm machinery. Combines in particular were found to distribute seeds effectively over distances up to 50 m from a source within a field. Combines also carried up to 13,000 seeds at a time out of a

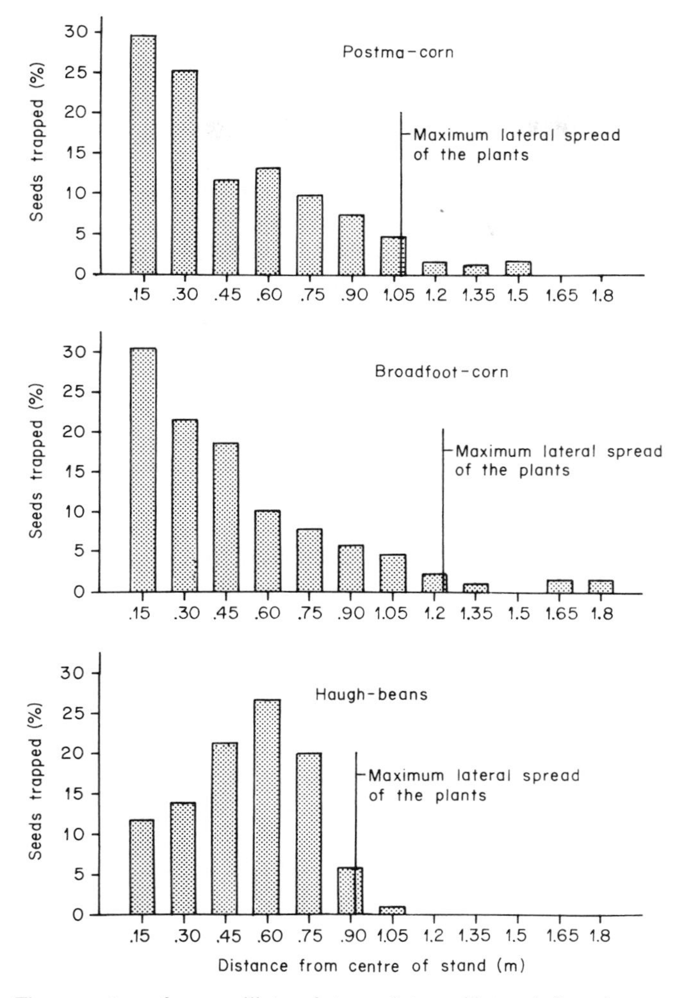

Fig. 2. The percentage of proso millet seeds trapped at equal intervals from the centre of a 50 × 50-cm^2 stand of this weed. The values for each field are the averages of three different sampling sites. [From O'Toole & Cavers (1983).]

field. A comparison between black-seeded and golden-seeded weed populations revealed that the black seeds were taken up in greater numbers by corn (maize) combines. In addition, all members of our research group can attest to the fact that black and Manitoba crown seeds of proso millet stick tenaciously to perspiring human beings. Proso millet seeds are also transported in mud that sticks to shoes, boots, and vehicle wheels.

Since proso millet is a prime ingredient in many birdseed mixtures, it is not surprising that many farmers believe that wild birds are important agents of dispersal for the weed populations. An unpublished study by Stephen McCanny does not support this hypothesis. McCanny used ring-necked doves (*Streptopelia rosegrisea*) in a laboratory feeding experiment. The doves were presented with 13 g (c. 2000 seeds) of one sample of proso millet seeds after they had been without food for 24 hr. Uneaten and regurgitated seeds were removed and weighed after a 2-hr feeding period. McCanny found that the crop seeds were eaten readily, the crop-like weed seeds to a lesser extent, those of the crown-type weeds even less and seeds of the black-seeded weed to a negligible extent. Indeed, the birds vigorously regurgitated the black seeds whenever possible.

Predation probably causes the destruction of vast numbers of proso millet seeds. In another study, Stephen McCanny followed the fates of seeds of different colours which were left on the soil surface. Seeds from all samples were eaten in large quantities, but he has not as yet been able to determine the identity of these seed predators.

In summary, none of our populations of proso millet possesses reliable innate means of seed dispersal. However, the actions of birds and mammals (especially man) can distribute the seeds over wide areas.

IV. DISCUSSION

How do the various Canadian populations of proso millet fit into the scheme outlined in Table I? A reconsideration of the scheme put forward in Table I indicates that the black-seeded weed matches the description of weed races quite well, whereas the crops and all of the crop-like weeds fit more closely with the characteristics of crop races. We shall discuss the ecological implications of this crop–weed complex in more detail.

Large seed size is desirable in improving the value of a crop biotype (de Wet 1975). In addition, the larger seed size exhibited by the crop populations provides more carbohydrates and other resources to allow a seedling to make rapid early growth (Harper, Lovell & Moore 1970) and also (in proso millet) to survive hazards such as drought (D. R. J. Moore & P. B. Cavers, unpublished). Large seed size may be important in enabling proso millet to do well as a crop of dry areas (Anderson & Martin 1949). The large size of the crown-weed seeds from Manitoba may be essential for early survival in that relatively dry climate.

Small size may help to render seeds of the black-seeded weed less attractive to birds (and other predators) than crop or crop-like weed seeds. Also, seeds of the black-seeded weed usually germinate after soil disturbance and in open microsites. Large seed size and great seedling vigour would be less important in such situations. Of greater value would be the production of large numbers of seeds, and a plant that has smaller seeds can produce more of them from the same amount of resources (Harper 1957). In addition, black-seeded proso millet allocates proportionally fewer resources to leaf, stem and root production and more to seeds (M. A. Bough, unpublished).

Most seeds of the crop and crop-like weed populations of proso millet exhibit enforced dormancy (Harper 1959) and germinate as soon as moisture and temperature conditions are favourable. The lack of innate dormancy (Harper 1959) in the crop-like weed populations is probably the single most important defect that they possess in their developing role as weeds. Many of these crop-like weeds can be eliminated from a field if they are prevented from ripening seeds for one or two seasons. Their seed banks (Harper 1977) are soon exhausted through germination, predation, rotting, etc., of seeds in the soil. In contrast, the prolonged innate dormancy and intermittent germination patterns of the black-seeded populations ensure that new crops of seedlings will continue to be produced, even though several successive populations of growing plants are eradicated by herbicides (Harper 1957).

The amount of shattering exhibited by inflorescences determines much of the pattern of seed loss from the parent plant. On the one hand, plants of the crop biotypes and the crop-like weeds tend to retain their seeds until the end of the growing season. The vast majority of these seeds reaches the soil within a brief period, and it is difficult for a soil-surface predator to consume all of them before a protecting cover of snow or ice covers them. This phenomenon is a variant of the predator satiation hypothesis discussed in more detail by Cavers (1983). Disadvantages of this pattern of seed loss are (a) that birds can detect and consume ripe seeds while they remain attached to the panicle, and (b) there is less chance of some seeds being buried beneath the surface before winter. Seeds of the black-seeded weed are released over periods of weeks and months. In this way only a few seeds are available to soil-surface predators at a single point in time, and the seeds may be buried or otherwise protected before predators can find them.

If seeds produced early in the season are killed by herbicides or mowing, new tillers can still produce viable seeds. Also, crown or black-seeded weed populations can continue to ripen seeds in longer growing seasons, whereas crop and some crop-weed biotypes have a determinate manner of growth, and their seed number is fixed at a relatively early point in the season. The disadvantage to the crown and black-seeded weed populations is that many unripened seeds on late tillers are destroyed by frost. Also, a late-season climate may be unsuitable for the formation of viable seeds since M. A. Bough (unpublished) found that many of the later inflorescences bore empty glumes.

In summary, proso millet offers a fascinating complex of biotypes for study. Crop varieties display all of the characters that have been selected for during the process of domestication. The black-seeded biotype has undoubtedly existed as a weed for many centuries in Eurasia, and it possesses a suite of characters associated with successful weeds. Finally, there is an array of crop-like weeds which have experienced mixed success in their new role as weeds. For those biotypes that continue to survive in this role there will be further intensive selection in the next few years, the outcome of which we await with interest.

ACKNOWLEDGMENTS

We wish to thank our colleagues Joe Colosi, Dwayne Moore, Jim O'Toole, Stephen McCanny and Cathy Vis for many helpful suggestions and for allowing us to cite unpublished information from their work. We appreciate the opportunity to work with proso millet on farms in Ontario and Manitoba and to discuss this weed problem with local people in each area. We thank Doris Rowley for technical assistance and Donna Irwin and Stefani Tichbourne for typing the manuscript.

Financial assistance from the Natural Sciences and Engineering Research Council of Canada, in the form of an operating and a strategic grant, is gratefully acknowledged.

REFERENCES

Anderson, E. & Martin, J. H. (1949). World production and consumption of millet and sorghum. *Economic Botany,* **3,** 265–288.

Anonymous (1982). Variety trials of farm crops. *Miscellaneous Report—Minnesota, Agricultural Experiment Station,* **24.**

Bough, M. A., Colosi, J. C. & Cavers, P. B. (1986). The major weedy biotypes of proso millet (*Panicum miliaceum* L.) in Canada. *Canadian Journal of Botany* (in press).

Cavers, P. B. (1983). Seed demography. *Canadian Journal of Botany,* **61,** 3578–3590.

Dekker, J. H., McLaren, R. D., O'Toole, J. J. & Colosi, J. C. (1981). *Proso Millet.* Factsheet 81-067. Ontario Ministry of Agriculture and Food, Toronto.

de Wet, J. M. J. (1975). Evolutionary dynamics of cereal domestication. *Bulletin of the Torrey Botanical Club,* **102,** 307–312.

Grabouski, P. H. (1971). Selective control of weeds in proso millet with herbicides. *Weed Science,* **19,** 207–209.

Harlan, J. R. (1965). The possible role of weed races in the evolution of cultivated plants. *Euphytica,* **14,** 173–176.

Harper, J. L. (1957). Ecological aspects of weed control. *Outlook on Agriculture.* **1**(5), 197–205.

Harper, J. L. (1959). The ecological significance of dormancy and its importance in weed control. *Proceedings of the 4th International Congress of Crop Protection, 1957,* Vol. I, pp. 415–420.

Harper, J. L. (1960). Introduction. *The Biology of Weeds* (Ed. by J. L. Harper), pp. xi–xv. Blackwell Scientific Publications, Oxford.

Harper, J. L. (1965). The nature and consequence of interference amongst plants. *Genetics Today, Proceedings of the 11th International Congress of Genetics, 1963,* (Ed. by S. J. Geerts), Vol 2, pp. 465–482. Pergamon Press, Oxford.

Harper, J. L. (1977). *Population Biology of Plants.* Academic Press, London.

Harper, J. L., Lovell, P. H. & Moore, K. G. (1970). The shapes and sizes of seeds. *Annual Review of Ecology and Systematics,* **1,** 327–356.

Harvey, R. G. (1979). Serious new weed threat; Wild Proso. *Crops and Soils,* **31,** 10–13.

Hilbig, W. & Schamsran, Z. (1980). Zweiter Beitrag zur Kenntnis der Flora des westlichen Teiles der Mongolischen Volksrepublik. *Feddes Repertorium,* **91,** 25–44.

Hinze, G. (1972). Millets in Colorado. *Colorado, Agricultural Experiment Station, Bulletin, 553S.*

Lysov, V. N. (1975). Millet—*Panicum* L. *Flora of Cultivated Plants. III. Groat Crops (Buckwheat, Millet, Rice)* (Ed. by A. S. Krotov), pp. 119–236. Kolos, Leningrad.

Moore, D. R. J. & Cavers, P. B. (1985). A comparison of seedling vigour in crop and weed biotypes of proso millet (*Panicum miliaceum* L.). *Canadian Journal of Botany,* **63,** (in press).

Oestry, L. & de Wet, J. M. J. (1981). Seed proteins and systematics of cultivated, weed and wild forms of *Panicum miliaceum* L. *American Journal of Botany, Miscellaneous Series, Publication,* **160,** 76.

O'Toole, J. J. (1982). *Seed banks of* Panicum miliaceum *L. in three crops.* M.Sc. thesis, University of Western Ontario, London, Canada.

O'Toole, J. J. & Cavers, P. B. (1983). Input to seed banks of proso millet (*Panicum miliaceum*) in southern Ontario. *Canadian Journal of Plant Science,* **63,** 1023–1030.

Rachie, K. O. (1975). *The Millets. Importance, Utilization and Outlook.* International Crops Research Institute for the Semi-arid Tropics, Hyderabad, India.

Robinson, L. H. (1971). "Dove" proso millet - new mourning dove food? *Proceedings of the 25th Annual Conference, Southeastern Association of Game and Fish Commissioners, Charleston, South Carolina, U.S.A.,* pp. 137–140.

Robinson, R. G. (1980). Registration of Minsum proso millet. *Crop Science,* **20,** 550.

Strand, O. E. & Behrens, R. (1979). *Identification and Control of Wild Proso Millet.* Agronomy Fact Sheet No. 35. University of Minnesota Extension Service, St. Paul.

Terpó-Pomogyi, M. (1976). Some monocotyledonous weeds spreading in Hungary. *Agrartudomanyi Kerteszeti Egyetem Kozlemenyei,* **40,** 515–527.

Zhukowsky, P. M. (1964). *Kulturnije Rasztyenija i ich szorodicsi (Cultivated plants and their wild relatives),* pp. 228–230. Izd. Kolosz, Leningrad.

11

Population Dynamics of a Few Exotic Weeds in North-east India

R. S. Tripathi

Department of Botany
School of Life Sciences
North-Eastern Hill University
Shillong, India

I. INTRODUCTION

Weeds were once regarded as "slightly improper material for biological studies" (Harper 1960), and research on their biology, except that directly related to their control, remained more or less neglected. However, the publication of "The Biology of Weeds" (Harper 1960) generated more active interest among plant scientists in biological studies of weeds (Tripathi 1977). During the last 25 years or so, these so-called undesirable plants, which are often prolific and persistent and grow 'out of place', have been intensively studied. Although the economic importance and nuisance value of weeds in agriculture have resulted in numerous studies, the realization that they are also excellent material for addressing basic evolutionary and ecological issues has stimulated further interest during recent years.

II. EXOTIC WEEDS IN NORTH-EAST INDIA

North-eastern India (22°–29°30′ N, 80°40′–97°25′ E) covers an area of about 25.5 million hectares (Borthakur 1984). The region is characterised by high rainfall (usually above 2000 mm) and humidity (relative humidity 80%). Large variations in altitude (50–7089 m), topography and soil have resulted in varied ecological conditions which favour growth and multiplication of very many plant species, making north-east India a rich floristic region. The slash and burn agriculture (locally called 'jhum') prevalent in the region, shortening of the jhum

STUDIES ON PLANT DEMOGRAPHY:

cycle (intervening fallow period after which the vegetated land is again cultivated) due to increased human population, and deforestation have created habitats suitable for successful colonisation by arriving exotic species, including weeds. Several exotic weeds have undergone tremendous range expansion in north-eastern India. Many of them are now naturalised and have become important pests of agriculture and plantation crops. Besides having economic implications as pests, some of these weeds have successfully colonised the newly created and disturbed areas in this region. The ever-increasing population and luxuriant growth on the wastelands and jhum fallows are now posing a serious threat to the native flora of the region.

Of these exotic weeds the species of *Eupatorium* and *Galinsoga* (Asteraceae) are of special interest because they have become dominant components of the weed communities in north-east India by virtue of the rapid increase in their population.

III. POPULATION DYNAMICS OF *EUPATORIUM* SPP.

The three species of *Eupatorium* that are noxious weeds in the north-eastern region are *E. odoratum, E. adenophorum* and *E. riparium*. They are native to Mexico and are perennial, producing enormous numbers of seeds annually, which are dispersed long distances by wind.

A. *Eupatorium odoratum*

Eupatorium odoratum is a serious weed in 23 countries of the world, including India (Holm *et al.* 1977). It occurs abundantly on the wastelands and cultivated fields, and along the road-sides and fence rows at lower altitudes (up to 1000 m) in Assam, Meghalaya and adjoining states in north-eastern India. It also infests tea, teak and rubber plantations. It is a successful coloniser in abandoned fields and forest clearings, where it has become a dominant species in early successional communities. *Eupatorium odoratum* is an apomictic allohexaploid (chromosome number 60) (Khonglam & Singh 1980). Sometimes in open moist habitats the plants may grow 3 m high. Seed germination occurs in May–June, and plants show active vegetative growth during July–October, after which flowering starts attaining peak in December. Fruiting is at a maximum in January. The cypsellas are dispersed soon after maturation in February–March. Seeds require light for germination (Auld & Martin 1975). Edwards (1974, 1977) has reported climatic and edaphic ecotypes in this species.

1. Population Flux

Out of a population of 27 adult plants per square metre in June 1977, on a 5-year-old fallow, only about 63% survived until June 1978. Nineteen individuals, a small fraction of the 1358 recruited over a 1-year period, were added to the original population (Fig. 1). There was a net gain of nine plants to the weed population that existed in June 1977, despite a very high seed production (c. 87,900 per square metre), and seed input to the soil (11,240 seeds per square metre). This shows that the populations of seeds and plants decreased with time because of various regulatory forces operating at different stages of the life cycle of the weed. The factors that contribute to the population regulation of *E. odoratum* (as summarised in Fig. 1) have been discussed by Yadav and Tripathi (1981). The effects of the associated vegetation on the reproductive behaviour of this weed are worth examining.

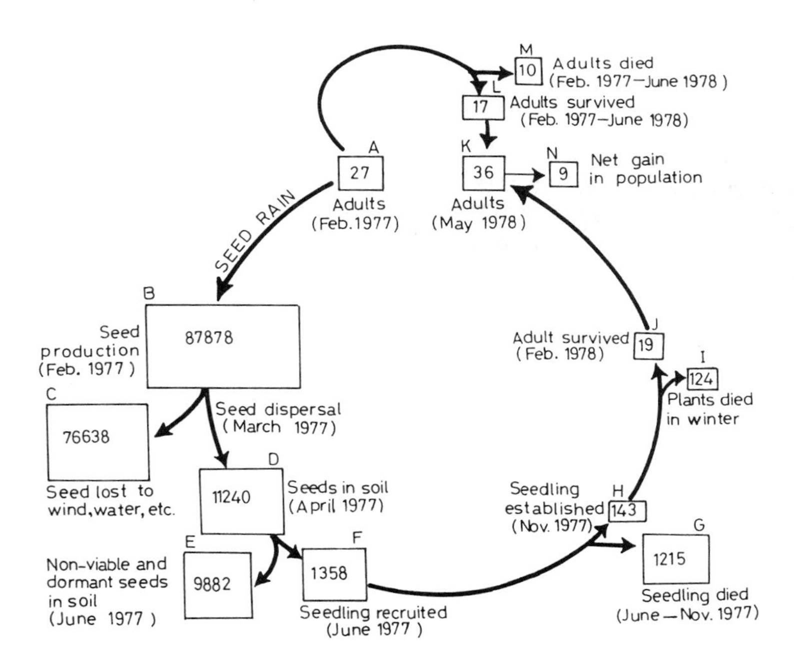

Fig. 1. Schematic summary of the population flux of *E. odoratum* in a 5-year-old fallow. Data in boxes refer to the population of individuals (seeds or plants) per square metre. Box A, Adult population in February 1977 that remained unchanged until June 1977. Box J, Population of adults that survived in February 1978 remained unchanged until June 1978. Other boxes are explained in the figure. [Redrawn from Yadav & Tripathi (1981).]

TABLE I.

Effect of Associated Vegetation on the Reproductive Success of *E. odoratum*[a]

Parameter	*E. odoratum* grown with associates	*E. odoratum* grown without associates	Calculated value of t
Percentage of fertile shoots per plant	28.2	63.8	8.761[b]
Number of capitula per plant	975.3	4958.4	4.774[b]
Seed output per plant	27,699.0	153710.0	4.248[b]

[a] After Yadav and Tripathi (1981).
[b] Differences significant at $p = .05$.

2. *Effect of Associates*

The associated vegetation exercised strong regulatory influence on this weed in nature by causing great reduction in its reproductive growth (Table I). This is also confirmed by more severe mortality of the seedlings of *E. odoratum* in older fallows than in newer fallows and by the reduced reproductive success of the weed on the older fallows (Kushwaha, Ramakrishnan & Tripathi 1981). A net gain of 9 plants to the original population of 27 plants per square metre over a period of 1 year (Fig. 1) implies a high rate of population growth. And this increase occurred when *E. odoratum* was growing in competition with the associated vegetation. Yadav and Tripathi (1981) argued that if the *E. odoratum* population continues to grow at this speed, within a few years it may suppress many useful elements of the original flora of the north-eastern region of India unless some other regulatory forces become operative. A much smaller seedling population in June 1978 compared to that in June 1977 indicates that some effective regulatory forces certainly operated in nature, and light was suspected to be one such factor by Yadav and Tripathi (1981). Kushwaha, Ramakrishnan and Tripathi (1981) also reported that the vigour of *E. odoratum* is drastically reduced in 5-year-old fallows where the vegetation could offer shade to the weed.

B. *Eupatorium adenophorum* and *Eupatorium riparium*

Both *E. adenophorum* and *E. riparium* are noxious weeds growing luxuriantly on roadsides and in abandoned fields, wastelands and sparse pine stands in the hilly region of north-eastern India. *Eupatorium adenophorum* is distributed at c. 550–2000 m altitude, whereas *E. riparium* is limited to about 100–1700 m. They co-exist in the areas with overlapping altitudinal range. They are autotriploid (chromosome number 51) and produce apomictic seeds in large number (Khonglam & Singh 1980). *Eupatorium adenophorum* is an erect, 0.8- to 1.5-m high, many-stemmed weed growing abundantly in open moist places. Germination commences in April–May, and plants grow vegetatively until November–December, after which flowering starts and seeds mature by March–April. The plants senesce after producing seeds. They re-sprout in May–June (wet months in this region) every year and occasionally, one or two shoots emerge from the shoot base. *Eupatorium riparium* is a scrambling herb, growing 30–70 cm high, in moist shady habitats. The new shoots arise from the base of old plants in May–June. The parts of the shoots that come into contact with the soil produce new shoots from the nodes, which in turn give rise to adventitious roots from their base and become partially independent. The plant grows vegetatively until November and flowers in December, and cypsellas are dispersed in March–April. Seed germination commences with the onset of rains in April–May.

1. *Effects of Burning*

The development and composition of the weed communities in the region are greatly influenced by the prevailing slash and burn agriculture. Field observations (Singh 1980) indicate that both weeds grow more abundantly and vigorously on the burnt sites, although they are not fire-resistant. The effect of burning was, therefore, studied on natural populations of *E. adenophorum* and *E. riparium* by burning a small plot in an old stand of *Pinus kesiya* where the understorey vegetation was dominated by these species. The fire was of mild intensity, in which pine needles and herbaceous vegetation were burnt, leaving behind the leafless shoots of *Eupatorium* spp. with viable shoot bases. The population changes were monitored over a 2-year period (April 1977–April 1979).

TABLE II.

Population Flux of Adult Plants of *E. adenophorum* and *E. riparium* in Relation to Burning Over a 2-Year Period (April 1977 to March 1979)[a]

Parameters	*E. adenophorum*		*E. riparium*	
	Burnt	Unburnt	Burnt	Unburnt
No. of plants per square metre in April 1977 (*a*)	18.7	50	66.7	111.7
No. of plants per square metre in March 1979 (*b*)	19.7[b]	26	143[b]	156
Net change (*b* − *a*)	1.0	−24	76.3	44.3
Rate of increase (*b*/*a*)	1.05	0.52	2.14	1.4
Percentage of survival of plants present in April 1977	75	52	71.5	67

[a] After R. S. Tripathi and A. S. Yadav (unpublished).

[b] The individuals recruited through genets are also included. In the unburnt area, however, no seedlings survived.

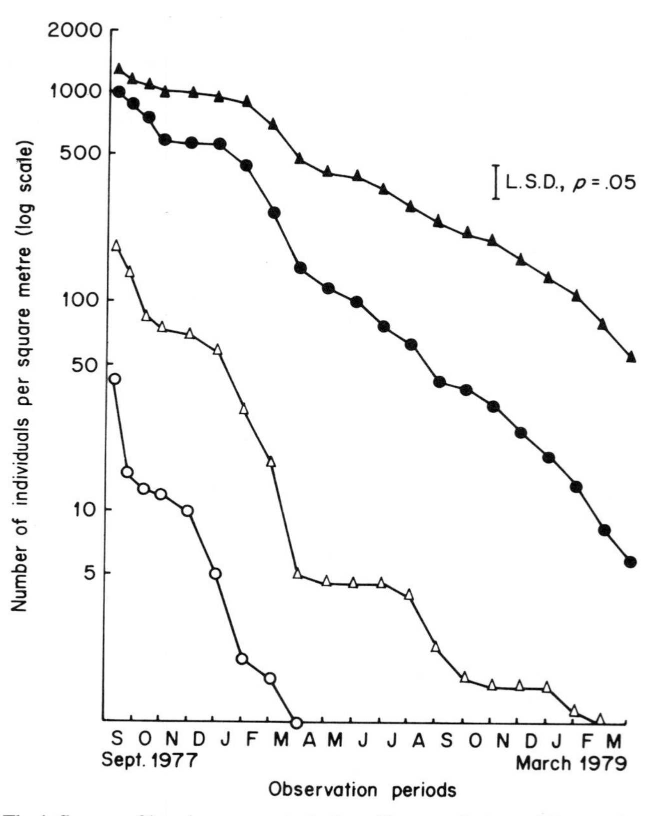

Fig. 2. The influence of burning on survival of seedling populations of *Eupatorium* spp. emerging shortly after fire. Seedling populations of *E. adenophorum* in the burnt (●——●) and unburnt plots (○——○) and of *E. riparium* in the burnt (▲——▲) and unburnt (△——△) plots.

Both weeds showed greater rates of population increase in the burnt plot as compared to the unburnt one (Table II). Seedling survival was greater in *E. riparium* than in *E. adenophorum* in the burnt plot (Fig. 2), where only 23.9% seedlings died in the first 2 months in contrast to 91.8% in the unburnt site. At the end of the study period, none of the seedlings survived on the unburnt plot. A thick layer of pine litter on the ground in the unburnt plot prevented the seedlings from establishing contact with soil, thus causing greater mortality on this site. Besides, more severe competition offered by luxuriantly growing associated herbs on the unburnt site further reduced seedling survival.

C. Seed Population Dynamics of *Eupatorium* Spp.

The three species of *Eupatorium* reproduce mainly by seeds, which are produced in large numbers. The seed population of each of the three *Eupatorium* spp. in soil is quite high (Yadav & Tripathi 1982). Most seeds were present in the top 2-cm soil layer, although some seeds of *E. adenophorum* and *E. riparium* were also present in soil to a depth of 10 cm. In the seed population recovered from soil the proportion of germinable seeds was larger in *E. adenophorum* and *E. riparium* than in *E. odoratum,* although the fraction of non-viable seed population was greater in *E. odoratum.* Viability and germination of the seeds recovered from soil was much lower compared to the freshly collected seeds. Survival of buried seeds of *Eupatorium* spp. declined exponentially with time, and a large proportion of viable seeds acquired enforced and induced dormancy during burial (Yadav & Tripathi 1982).

IV. POPULATION DYNAMICS OF *GALINSOGA* SPP.

The genus *Galinsoga,* though a native of tropical America, has now spread throughout the world. *Galinsoga parviflora* alone grows as a weed of 32 crops in 38 countries between 54° N and 40° S (Holm *et al.* 1977). *Galinsoga ciliata* and *G. parviflora* are now reported from India as ruderal and cropland weeds. In the north-eastern hill region, they grow as noxious weeds of maize, potato and other horticultural crops. Both species grow abundantly on moist habitats and jhum fallows with *E. riparium, E. adenophorum, Drymaria cordata, Panicum brevifolium, Osbeckia crinata, Hypochaeris radicata,* etc., as their common associ-

TABLE III.

Density and Half-life (±S.E.) of Seedling Cohorts of *Galinsoga* spp. Emerging at Different Times in Wasteland and Cropland (Both Species Considered Together)[a]

Date cohort was first observed		Density/0.25 m²	Half-life in weeks
Wasteland			
26 April	1980	437.6 ± 38.7	3.1 ± 0
8 May	1980	39.3 ± 2.6	3.1 ± .3
20 August	1980	128.6 ± 17.9	3.5 ± .3
10 March	1981	405.3 ± 18.8	2.9 ± .3
20 April	1981	155.0 ± 7.8	2.2 ± 0
12 July	1981	110.6 ± 0.3	1.6 ± 0
Cropland			
28 April	1980	500.3 ± 16.8	7.9 ± .6
19 May	1980	137.6 ± 31.8	3.0 ± 0
22 August	1980	201.0 ± 27.4	3.0 ± .6
18 March	1981	335.0 ± 21.6	2.0 ± 0
20 April	1981	106.0 ± 15.9	2.0 ± .3
27 July	1981	81.3 ± 7.5	1.4 ± .3

[a] After Rai and Tripathi (1984).

ates. Their enormous seed production and short life span of about 2 months have enabled them to invade disturbed habitats and open areas created by deforestation and slash and burn agriculture.

The seeds of *Galinsoga* spp. germinate in spring (March–April) when the temperature rises and the first rains are received. They lack dormancy and can germinate immediately after dispersal under favourable conditions. Both species normally have three seedling cohorts, which appear during March–April, April–May and July–August (Table III). The emergence of seedlings at different times reduces intra- and inter-specific competition through displacement of their temporal niches. It also ensures successful completion of the life cycle of the weeds even if seedling emergence at a given time is followed by adverse growth conditions.

A. Population Flux

A study of population flux on two contrasting habitats, wastelands and croplands, over a period of about 2 years revealed that the seedling cohort

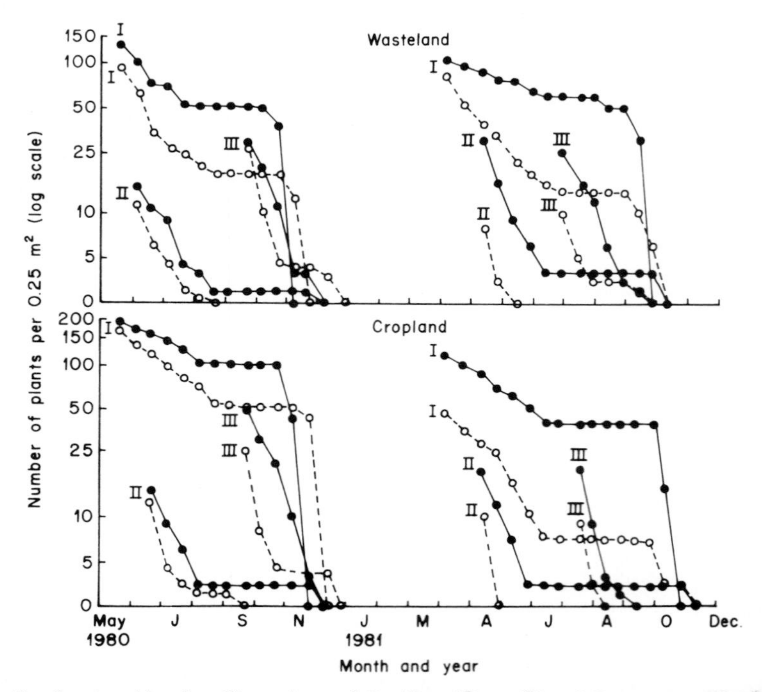

Fig. 3. Survivorship of seedling cohorts of *G. ciliata* (●——●) and *G. parviflora* (○- - -○) on wasteland and cropland sites during 1980 and 1981. I, II, III, cohorts appearing in March–April, April–May, and July–August, respectively.

appearing in March–April was larger in size than the two subsequent emerging cohorts, as a result of availability of more seeds (Fig. 3). It had longer half-life (Table III) and greater survival and produced more seeds (Rai & Tripathi 1984). The maximum standing population of the weeds was observed in April 1980 and March 1981 and subsequently declined as a result of seedling mortality (Rai 1982). Cohort III, appearing in July–August 1980, had larger population than cohort II, presumably due to germination of seeds contributed by the earlier cohorts. In 1981, however, population size of these two cohorts did not differ significantly. The plants senesced during November–December, because of prevailing low temperature and soil moisture depletion. These conditions also hampered germination of the seeds which entered into an enforced dormancy. Thus the population was reduced to nil in winter. In general, both weeds showed almost similar survival pattern. The three cohorts of a given species differed, however, in their survivorship (Fig. 3).

The increasing seed input to the soil does not proportionately increase the seedling populations of the *Galinsoga* spp. (Rai & Tripathi 1983), because the emergence depends upon the interaction between the available 'safe sites' (*sensu* Harper, Williams & Sagar 1965) and soil seed bank (Harper 1977; Yadav & Tripathi 1981). The surviving individuals reacted to density by reduced flowering (Fig. 4). Further, the density increase also adversely influenced the other components of fecundity (Rai & Tripathi 1983). Under moisture-deficient conditions the seed output–input ratios of *G. ciliata* and *G. parviflora* were less than 1 at a sowing density of 270 seeds per pot (equivalent to 8100 seeds per square metre), irrespective of soil texture (Table IV). Thus seed output from a given

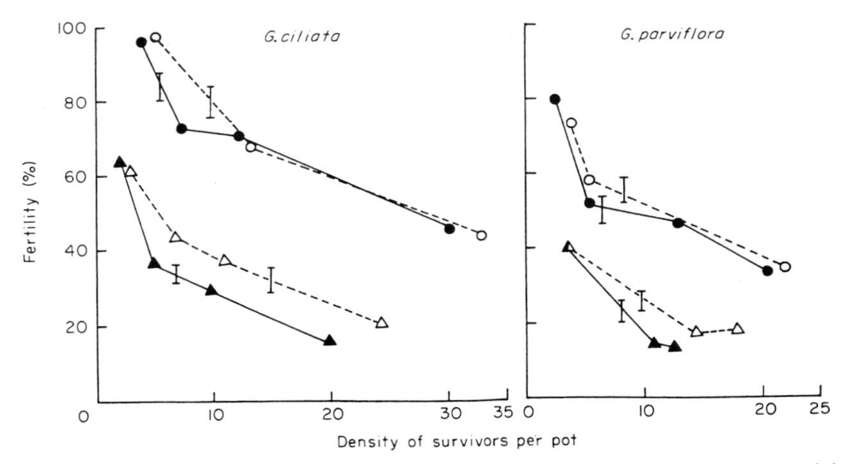

Fig. 4. Percentage of fertile plants of *G. ciliata* and *G. parviflora* in relation to surviving density: ——, clay loam soil; - - - -, sandy loam soil; ○ and ●, high moisture level; △ and ▲, low moisture level. Vertical bars represent LSD at 5% level. [After Rai & Tripathi (1983).]

TABLE IV.

Seed Output–Input Ratio of *G. ciliata* and *G. parviflora* in Relation to Sowing Density at Two Soil Moisture Regimes and Textures[a]

Moisture regime	Clay loam soil: Sowing densities per pot 10	30	90	270	Clay loam soil: Mean ± S.E.	Sandy loam soil: Sowing densities per pot 10	30	90	270	Sandy loam soil: Mean ± S.E.
G. ciliata										
High	54.8	21.2	9.6	2.9	22.1 ± 11.5	71.7	23.6	9.6	2.6	26.8 ± 15.5
Low	6.6	2.5	1.0	0.3	2.6 ± 1.4	9.9	4.1	1.4	0.4	3.9 ± 2.1
G. parviflora										
High	35.4	13.5	4.2	1.0	13.5 ± 7.7	68.9	27.6	6.5	1.7	26.2 ± 15.3
Low	4.4	1.8	0.6	0.2	1.8 ± 0.9	5.7	3.2	1.2	0.3	2.6 ± 1.2
Mean ± S.E.	25.3	9.8	3.9	1.1		39.0	14.6	4.7	1.3	
	±12.1	±4.6	±2.1	±0.6		±18.1	±6.4	±2.0	±0.6	

[a] After Rai and Tripathi (1983).

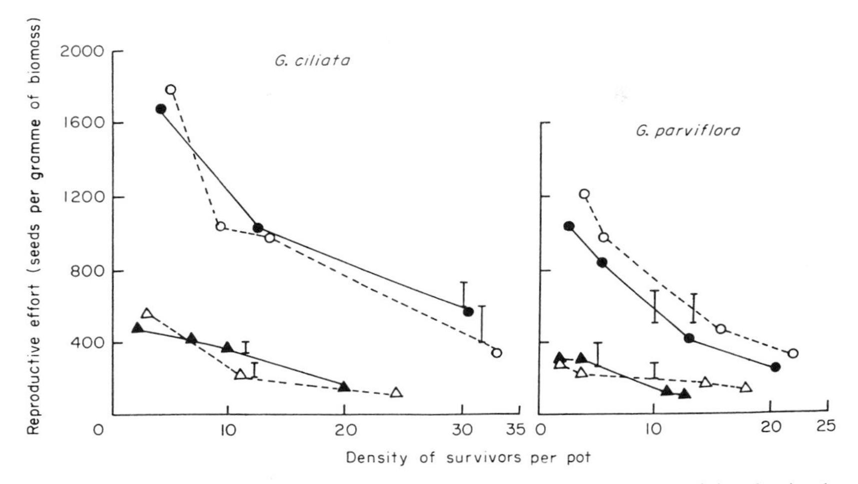

Fig. 5. Reproductive effort of *G. ciliata* and *G. parviflora* in relation to surviving density; key as in Fig. 4. [After Rai & Tripathi (1983).]

population depends on seed input and environmental conditions affecting seed viability, germination and other processes contributing to seeding in the plants. Rai & Tripathi (1983) argued that wastage of resources by non-reproducing individuals may be critical, particularly in situations in which the production of seeds is less than the number of seeds sown.

The reproductive effort (*sensu* Bazzaz & Carlson 1979) of both species decreased with density (Fig. 5). The lower moisture supply caused a significant reduction ($p < .01$) in reproductive effort irrespective of soil texture. *Galinsoga parviflora,* however, showed greater reproductive effort in sandy loam soil than in clay loam under high soil moisture regime, which signifies the role of soil texture–soil moisture interaction in population regulation of these weeds. High soil nitrogen promoted growth and seeding in both species and minimised the density effect (J. P. N. Rai & R. S. Tripathi, unpublished). This indicates that apart from density, the texture, moisture and nitrogen content of soil also play important roles in regulation of the weed populations.

B. Effect of Associates

High density of associated vegetation in cropland, as observed during the second year of study, regulated the populations of *Galinsoga* spp. by reducing seedling recruitment and enhancing mortality (Rai & Tripathi 1984). The controlling influence of the associated vegetation has been further demonstrated by the reduced germination of *Galinsoga* seed introduced to the established vegetation (Rai 1982). The deleterious effect of associated vegetation on these weeds was largely accomplished through reducing available light. This was confirmed

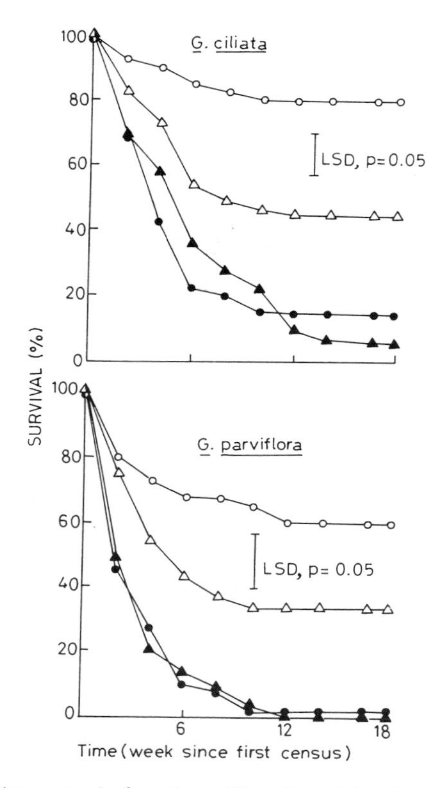

Fig. 6. Survival (percentage) of *in situ* seedlings (triangle) and transplants (circle) of *G. ciliata* and *G. parviflora* as affected by established vegetation. Solid symbols represent uncleared plots (with associates) and open symbols cleared plots (without associates).

by an experiment in which low light intensity reduced their survival and growth (Rai 1982).

The effect of established vegetation on survival and growth of the *in situ* emerged seedlings of *G. parviflora* was so great that none of them survived (Fig. 6). However, in *G. ciliata* a few seedlings managed to survive, but they showed much suppressed growth and failed to reproduce. Thus the two species of *Galinsoga* responded differently to the presence of associates. Transplanted seedlings (5-day-old, two-leaved) grew better than the *in situ* emerged seedlings, signifying that competition effect is reduced with age of the introduced individuals. This is in agreement with Cavers and Harper (1967), who observed better survival of phytometers than the seed-borne individuals of *Rumex* spp. Tripathi and Harper (1973) also reported that the plants of *Agropyron repens* and *A. caninum* established from tillers proved more aggressive than their counterparts produced from seeds.

V. CONCLUSION

The studies on population dynamics and growth of these exotic weeds in relation to burning, age of the jhum fallows, associated vegetation and varied density, and light and soil conditions indicate that they are particularly successful on disturbed habitats. Being aggressive, they are spreading fast and are set to pose serious threat to certain useful elements of the native flora. Of course, their populations may become stabilised in course of time, but before this happens, there is a risk that many useful native species could be suppressed or even eliminated. Further, the disturbed habitats are being created at a much faster pace than ever before in the region. This brings the impending problems connected with rapid increase in populations of these weeds into sharp focus. Evidently, there is an urgent need to undertake intensive studies on exotic weeds with particular reference to their population dynamics, analysis of factors contributing to their remarkable success in the region, niche divergence, and possible impact on the native flora. A comparative study of the population behaviour, individual fitness and reproductive strategies of important exotic weeds in the country of their origin and newly invaded areas may be revealing and rewarding.

ACKNOWLEDGMENTS

I thank Dr. H. N. Pandey and Dr. J. P. N. Rai for their valuable suggestions and comments.

REFERENCES

Auld, B. A. & Martin, P. M. (1975). The autecology of *Eupatorium adenophorum* Spreng. in Australia. *Weed Research,* **15,** 27–31.

Bazzaz, F. A. & Carlson, R. W. (1979). Photosynthetic contribution of flowers and seeds to reproductive effort of an annual colonizer. *New Phytologist,* **82,** 223–232.

Borthakur, D. N. (1984). Potentials of agriculture in the northeastern region *Resource Potential of North East India. Vol. II. Living Resources* (Ed. by R. S. Tripathi), pp. 1–8. Meghalaya Science Society, Shillong.

Cavers, P. B. & Harper, J. L. (1967). Studies in the dynamics of plant population. I. The fate of seeds and transplants introduced into various habitats. *Journal of Ecology,* **55,** 59–71.

Edwards, A. W. A. (1974). The ecology of *Eupatorium odoratum* L. V. Effect of moisture on growth of *E. odoratum. International Journal of Ecology and Environmental Sciences,* **1,** 61–69.

Edwards, A. W. A. (1977). The ecology of *Eupatorium odoratum.* L. VI. Effect of habitats on growth. *International Journal of Ecology and Environmental Sciences,* **3,** 17–22.

Harper, J. L. (Ed.) (1960). *The Biology of Weeds.* Blackwell Scientific Publications, Oxford.

Harper, J. L. (1977). *Population Biology of Plants.* Academic Press, London.

Harper, J. L., Williams, J. T. & Sagar, G. R. (1965). The behaviour of seeds in soil. I. The heterogeneity of soil surfaces and its role in determining the establishment of plants from seeds. *Journal of Ecology,* **53,** 273–286.

Holm, L. G., Plucknett, D. L., Pancho, J. V. & Herberger, J. P. (1977). *The World's Worst Weeds: Distribution and Biology.* Hawaii University Press, Honolulu.

Khonglam, A. & Singh, A. (1980). Cytogenetic studies on the weed species of *Eupatorium* found in Meghalaya, India. *Proceedings—Indian Academy of Sciences, Section B,* **89B,** 237–241.

Kushwaha, S. P. S., Ramakrishnan, P. S. & Tripathi, R. S. (1981). Population dynamics of *Eupatorium odoratum* in successional environments following slash and burn agriculture. *Journal of Applied Ecology,* **18,** 529–536.

Rai, J. P. N. (1982). *Studies on population regulation of two annual weeds,* Galinsoga ciliata *(Raf.) Blake and* G. parviflora *Cav.* Ph.D. thesis, North-Eastern Hill University, Shillong, India.

Rai, J. P. N. & Tripathi, R. S. (1983). Population regulation of *Galinsoga ciliata* and *G. parviflora:* Effect of sowing pattern, population density and soil moisture and texture. *Weed Research,* **23,** 151–163.

Rai, J. P. N. & Tripathi, R. S. (1984). Population dynamics of different seedling cohorts of two co-existing annual weeds, *Galinsoga ciliata* and *G. parviflora,* on two contrasting sites. *Acta Oecologica [Series]: Oecologia Plantarum,* **5,** 357–368.

Singh, A. (1980). *Studies on population dynamics of* Eupatorium odoratum L., E. adenophorum *Spreng. and* E. riparium *Regel.* Ph.D. thesis, North-Eastern Hill University, Shillong, India.

Tripathi, R. S. (1977). Weed problem—an ecological perspective. *Tropical Ecology,* **18,** 138–148.

Tripathi, R. S. & Harper, J. L. (1973). The comparative biology of *Agropyron repens* (L.) Beauv. and *A. caninum* (L.) Beauv. I. The growth of mixed populations established from tillers and from seeds. *Journal of Ecology,* **61,** 353–368.

Yadav, A. S. & Tripathi, R. S. (1981). Population dynamics of the ruderal weed *Eupatorium odoratum* and its natural regulation. *Oikos,* **36,** 355–361.

Yadav, A. S. & Tripathi, R. S. (1982). A study on seed population dynamics of three weedy species of *Eupatorium. Weed Research,* **22,** 69–76.

12

Weeds and Agriculture: A Question of Balance

B. R. Trenbath[1]

Centre for Environmental Technology
Imperial College
London, England
and
Elm Farm Research Centre
Hamstead Marshall
Near Newbury Berkshire, England

I. INTRODUCTION

Agriculture is aptly described as "man's earliest and most persistent effort to mould his environment by ecological manipulation" (Taylor 1984). Because of its history, agriculture has accumulated a massive literature which, inevitably, provides rich pickings for ecologists in search of data bearing on ecological principles (Harper 1971). The mass of detailed results from controlled experiments done by agricultural scientists and the outcome of huge-scale 'natural' experiments carried out by farmers offer ecologists many opportunities to test hypotheses. However, as scientists who claim to study the fundamentals of interaction within biological communities, ecologists are in turn challenged to respond with guidance on how to ensure that agricultural systems behave satisfactorily. It seems that ecologists have far to go to repay the debt to agriculture for all that they have learned from it.

Among agricultural activities, the management of weeds would seem to be an area in which ecology should be able to make a particularly significant contribution. I suggest this because two of ecology's central concerns are with population dynamics and the ways in which plant species interact both with each other and with the environment. There is, in addition, a pressing reason why ecology

[1]Present address: 12 New Road, Reading, Berkshire RG1 5JD, England.

should mobilise its potential in this direction: in an energy-hungry world, more energy is expended on the weeding of man's crops than on any other single human task (Audus 1976). Indeed, in the least developed countries, about 40% of the total labour time devoted to crop production is spent in weeding (L. J. Matthews, unpublished). Where the area planted to crops is limited by the availability of labour at weeding (e.g. Hinton 1978), improving the effectiveness of weeding or reducing the need for it must have high priority.

An aspect of weeds which is interesting to ecologists is that they combine both harmful and useful qualities, sometimes within a single species (Andres *et al.* 1976). For instance, the alternative Australian names for *Echium plantagineum,* 'Paterson's Curse' or 'Salvation Jane', show the ambivalent attitude of graziers to one species which under some conditions is a 'noxious weed' (Burdon, Marshall & Brown 1983) and under other conditions provides vital emergency feed (Delfosse & Cullen 1981). Sometimes it is through their links with species not so obviously part of the agroecosystem that weeds become socially useful. Thus, because it produces abundant nectar, the same *Echium* species was recently the subject of outcry from Australian apiarists when the Commonwealth Scientific and Industrial Research Organisation declared its intention to control it biologically (Delfosse & Cullen 1981).

Where, like this, weeds have both beneficial and deleterious effects, it becomes rational to manage them towards an 'optimum' level rather than to aim only at their extermination. To show how ecological principles can contribute to management of this special kind of weed species, this essay considers three case studies in which farming systems are vitally dependent on a continued presence of the weeds.

II. FOREST SPECIES IN A SWIDDEN SYSTEM

The first case study concerns the dynamics of the biomass of forest species on a plot used periodically for crop production within a system of rotational cultivation ('swiddening', Kundstadter 1978). When swiddening is practised in the tropical uplands, the forest cover on a plot is usually felled and burned, a sequence of crops is grown there, and then the forest is allowed to regrow until the next crop phase. This cycle of alternating crop and fallow phases has been used in areas such as north Thailand and Papua New Guinea for hundreds or thousands of years and seems to provide a potentially sustainable agricultural system (Sabhasri 1978; Thiagalingam & Famy 1981).

When run according to traditional methods, the system owes its sustainability to the way in which, during the fallow phase, the regenerating forest restores soil fertility lost during the crop phase. The taking of crops leads principally to depletion of organic matter and nutrients, loss of soil (by erosion), soil compaction and a colonisation of the plot by freely-dispersing herbaceous weeds. Pest and disease organisms may immigrate and increase. The biomass of forest species, the tree 'weeds', is also reduced, at first abruptly, by cutting and burning of the above-ground material, but later more gradually, by the regular cutting back of resprouting suckers as part of the weeding procedures of the crop phase. Such woody weeds were estimated as comprising 70% of the weed biomass present in a first rice crop (Nakano 1978).

When cropping stops and the fallow phase commences, the herbaceous weeds and the regenerating shoots from the cut tree stumps form a cover over the plot. The shading of the soil and the fall of leaf litter lead to accumulations of organic matter and nutrients, some brought up from deep in the profile. Protected from the force of the rain and no longer cultivated, the soil recovers its structure. Since the shoots of the evergreen forest trees do not die down in the dry season, these usually outgrow the herbaceous weeds and suppress them. Without hosts, the crop pest and disease organisms disappear. When, after perhaps 10–20 years, the forest cover is again cut and burned, the soil receives a large input of nutrients from the ash. With the aid of this, soil fertility is restored to that at the start of the previous crop phase (e.g. Sabhasri 1978; Hinton 1978; Greenland & Okigbo 1983). The system is thus in equilibrium.

Recent increases of population pressure in swiddening areas have, however, led to types of intensification of the system which are threatening to make it collapse. Although a temporary increase in production may be achieved, under extreme pressure a point is reached where the forest fails to regenerate, and, consequently, crop production may become impossible (UNESCO/UNEP/FAO 1978; Greenland & Okigbo 1983). In the intact swidden system, the forest trees can therefore be viewed in two ways: first as weeds in the crop phase and second as vital restorers of soil fertility in the fallow phase. The successful operation of the swidden system clearly depends on achieving the right level of control of the tree weeds.

In an effort to understand better the working of a swidden system, I have recently built a simulation model of the interactions among its three main components, the forest species ('trees'), the herbaceous weeds ('grass'), and soil fertility (Trenbath, Conway & Craig, 1985; Trenbath 1984). Differential equations were written for each component to express the rate of change of either biomass (for 'trees' and 'grass') or soil fertility (measured in terms of potential yield of a rain-fed rice crop). Different sets of equations were used according to whether the crop or the fallow phase was being simulated. Since the details are being published elsewhere, the approach is described only in outline.

In the crop phase, tree biomass and soil fertility fall exponentially with time while weed biomass rises sigmoidally to reach an eventual plateau. In the fallow phase, tree and grass biomasses rise according to modified Lotka–Volterra competition equations. In these, the competition coefficients (which express the depressive effect of a unit weight of the one on the growth rate of the other) vary linearly in opposite directions as tree biomass increases. This variation in values attempts to represent a diminishing influence of a unit of grass on tree growth (and a corresponding increasing influence in the reverse direction) as tree biomass increases towards its maximum level. The basis for this relationship is threefold. First, it can be inferred that as tree biomass grows, the competitive ability of light-demanding grass almost certainly suffers (Ivens 1983); second, greater tree biomass implies a lowered probability of fire, and this further tips the balance in favour of tree dominance (Nakano 1978). Third, this approach is also in line with any allelopathic influence of the grass (Eussen 1978). The modification to the Lotka–Volterra equations is the addition of factors which give an advantage in potential growth rate to grass at low fertility level and to trees at high fertility level. The basis for this is again in non-quantitative statements such as "[grass] tends to invade as fertility declines" (Ivens 1983) and "low soil fertility slows forest regrowth more than weed growth" (Barker 1984).

Simulations carried out by using the model have given results in which the calculated curves of tree regrowth in various lengths of crop phase have been well matched by the few sets of observations available. In particular, as the crop phase lengthens, the rate of regrowth is slowed ("a widely recognised effect", Nakano 1978). Most significantly, a length of crop phase is eventually reached, at which tree regrowth suddenly fails completely, and, under a regime of seasonal burning, the plot is permanently dominated by grass. This exactly matches experience (Greenland & Okigo 1983; UNESCO/UNEP/FAO 1978; K. F. Jackson, personal communication).

Further simulations with different starting points have shown the system's state space to be divided into two stability domains by a curved plane ('separatrix') rising from the soil fertility axis (Fig. 1). If the system's state is such that its locus in state space is beyond or above the separatrix (as viewed in the figure), then, during a fallow phase, the locus will move towards the equilibrium point marked 'Mature forest'. On the other hand, if the system's locus is in front of the separatrix, it will move towards the other equilibrium point, marked 'Mature grassland'. The separatrix thus constitutes the boundary of the sustainable swidden system. Any schedule of cropping and fallow which allows the locus to cross it has thereby destroyed the system.

This view of the swidden system leads to a reformulation of the agricultural problem in ecological terms: how can the swidden system be managed so as to be maximally productive while keeping its locus safely within the stability domain of the mature forest? Although estimation of the productivity aspect depends on separate calculation, Fig. 1 immediately suggests that the position of the separatrix represents the level of tree–grass balance. The double curvature of the separatrix, however, shows that allowable tree–grass ratios depend on both the absolute values of tree and grass biomasses and the level of soil fertility.

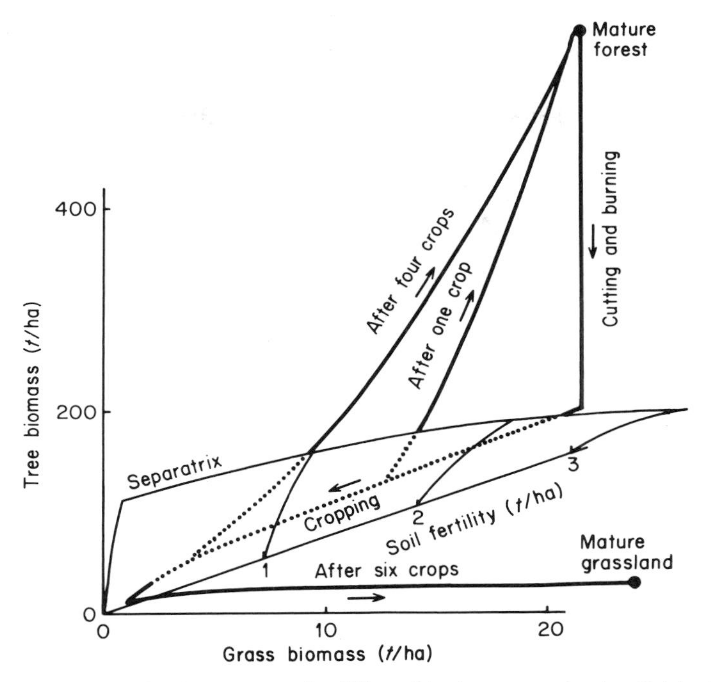

Fig. 1. Three-dimensional state space of swidden cultivation system showing division into two domains each with its equilibrium point. System trajectories calculated by the model show the way that excessive intensification causes a shift, through the separatrix, into the grassland domain. [Redrawn after Trenbath, Conway & Craig (1985).]

Such a conclusion opens the way, at least theoretically, for the development of sets of tables to help farmers and advisory personnel estimate how far any given plot is from the 'dangerous' separatrix. If the plot's locus is still in the forest domain but more distant than a certain safety factor, then the plot could be cropped again before abandonment. If the plot is already in the grassland domain, estimates could be made of, say, how much fertilizer needs to be applied or how much weed biomass needs to be killed to bring the plot back into the forest domain.

III. TREE SPECIES IN A GRAZING SYSTEM

The second case study concerns a problem of severe die-back of *Eucalyptus* species on 1 million ha of grazing land around Armidale (NSW), Australia. The die-back seems to have been caused by continued clearing of eucalypt woodland and individual relict trees to provide more and better pasture for sheep. As this process has continued, several factors have worked together to cause populations of beetles, especially of scarabs, to explode. The three main factors favouring the beetles seem to have been better survival of larvae in plant roots in the greater area of improved pasture, a greater availability and a greater nutritiousness of the leaves of trees growing in and near improved pasture, and reduced predation on the beetles by native bird species as nesting sites are lost. The high populations of adult beetles graze the reduced number of trees so intensively that the trees die (Lehane 1979; Old, Kile & Ohmart 1981).

As in the first case study, the grown trees can be viewed by farmers in two ways: either negatively as weeds, which diminish pasture quality, limit its extent and produce seeds for further infestation; or positively, as sources of shade and shelter for sheep, and of nesting sites for birds preying on beetle pests of pasture. Further positive points, not so strictly agricultural, are the landscape, amenity and conservation value of the trees. For the farmers of the area, the loss of all the native eucalypts would be near-disastrous (Lehane 1979).

The present level of clearing in the area varies around 80% but in the last few years further loss of trees on farm properties has been rapid and largely unplanned. This apparently sudden loss of control over tree and pasture balance suggested a situation rather similar to that in the swidden system. In order to understand better the management options available, a simple model of the system was constructed (Trenbath & Smith 1981) to study the likely interactions among four components, 'trees', 'pasture', 'scarabs' and 'birds'. Since pasture is taken as occupying any space without trees, it is not an independent variable; hence, the model has three state variables: biomasses of trees, scarabs and birds. The trees are the native eucalypts (not any planted exotics), the scarabs considered are only those with obligate dependence of the adults on the native eucalypts, and the birds are the species which effectively depend on the same scarab populations. The other simplifying assumptions have been detailed elsewhere (Trenbath & Smith 1981; Trenbath, Conway & Craig 1985).

In outline, the model consists of three differential equations giving the rate of change of the three state variables. All three have the basic form of sigmoid, logistic curves but grazing and predation terms are added to the equations for tree and scarab biomasses. Maximum biomass ('carrying capacity') of scarabs is given as a scaled product of the amounts of adult and larval food; an effect of pasture improvement is included in the calculation of larval food. Similarly for birds, the maximum biomass is a scaled product of the availability of potential nesting sites (including those in exotic trees) and the quantity of scarabs.

Although several of the model's parameter values are speculative, the use of the model itself in this study is more sophisticated than in the previous one. Here, based on the methods of Ludwig, Jones & Holling (1978) and Walker *et al.* (1980), calculations have given the positions of the lines in state space where both tree biomass and scarab biomass will reach an equilibrium with bird bio-

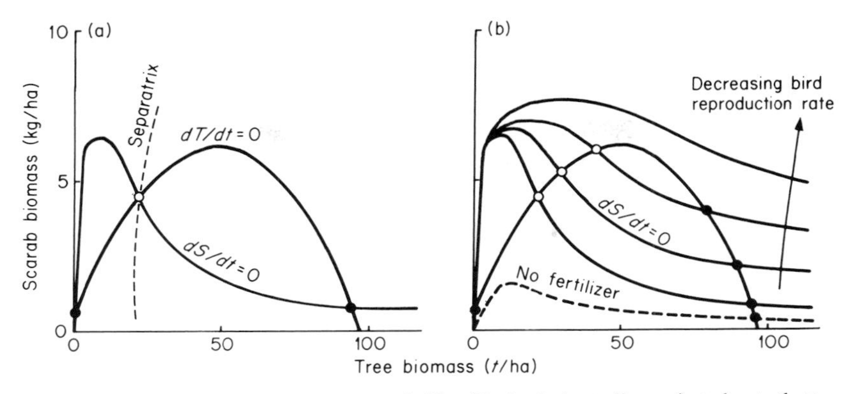

Fig. 2. State space of tree–pasture system in New England, Australia, projected onto the tree–scarab plane. Lines are equilibrium lines for tree biomass ($dT/dt = 0$) and scarab biomass ($dS/dt = 0$). Intersections mark either stable equilibria (filled points) or unstable equilibria (unfilled points). a, Lines calculated from the model by using 'standard' parameter values; the position of the separatrix plane through the unstable point is approximated by the dashed line. b, Effect of calculating scarab equilibrium lines either with decreasing values of bird reproduction rate (continuous lines) or with pastures assumed unimproved ('no fertilizer'). Remaining single equilibria are either the 'high tree' point or the 'low tree' point, respectively. [Redrawn after Trenbath & Smith (1981).]

mass. Where these lines intersect, all three components will be in equilibrium. The results are shown in Fig. 2. To facilitate their representation, the positions of the two equilibrium lines have been projected onto the tree–scarab plane of the three-dimensional state space.

In Fig. 2a are shown the courses of the equilibrium lines when the pastures are assumed to be either heavily fertilized or to have built up considerable soil nitrogen due to fixation by sown clovers. The lines intersect in three places, but simulations with the model have shown that only two of the intersections represent stable equilibrium points. At one of these stable points, the 'high tree' equilibrium, the tree biomass will be near its maximum possible value; at the other, the 'low tree' equilibrium, tree biomass will be almost negligible. In simulations started from a range of different points, the courses taken by the system locus ('system trajectories') have shown that a separatrix plane passes through the unstable equilibrium.

Although the uncertainty of several of the parameter values must be emphasized, the situation represented by Fig. 2a seems to agree with experience. A sudden onset of tree death due to beetle attack is just what would be expected according to the model if tree clearing were to force the system locus to cross the separatrix. This separatrix crosses state space at about 70–80% woodland clearance. Simulations using the model suggest that, without further intervention, the system will move towards a virtually tree-less state, the low tree equilibrium. This agrees with Lehane's (1979) description of parts of the area: "vast tracts are virtual gum tree graveyards—every tree in sight is either dead or clearly dying''. In line with those same simulations, survey has shown that as clearance exceeds 80–90%, biomass of tree-dependent scarabs indeed falls very sharply (Roberts *et al.* 1982).

Given that the model is a good enough simulator of reality to provide farmers with useful ideas on how to manage the 'weed' trees, the problem can be reformulated in ecological terms. As in the swidden system, the problem for most farmers seems likely to be how to keep the system locus within the stability domain of the woodland although staying, for profitability's sake, as close as safely possible to the separatrix? Experience with the model's behaviour again suggests that, in principle, monitoring of areas to determine their positions relative to the local, long-term separatrix would be helpful. Safety factors might be calculated on the basis of the risk of annual variation of climate moving the separatrix by a given distance. Permission for further clearance of a site might only be granted if its locus were sufficiently distant from the average position of the separatrix.

Rehabilitation of areas where the native eucalypts are all ''dead or clearly dying'' is likely to be so costly that it is essential that government policy makers receive advice from ecologists in attempting an overall analysis. I suggest that even a preliminary model such as already described may help structure discussion and clarify options. For instance, to policy makers, the most obvious approach to propose to farmers might be a massive replanting of the lost eucalypts (or of resistant exotics), insecticide aerial spraying to kill the beetles, and destocking of some pastures to favour tree regeneration. However, experience with the model will prompt the ecologist to ask for clarification on the following key issues: the possibility of maintaining the required level of beetle kill over the years needed to establish enough tree biomass to carry the system past the separatrix; the availability of exotic tree species or habitat treatments that will allow replanted trees to serve as nesting sites for predatory birds; the possibility of toxic side effects of insecticide on birds; and the probabilities that give extents and durations of destocking will be sufficient to be effective.

Experience with the model will act not only as a prompt to help frame critical questions, but as a guide to the implications of management interventions. For instance, Fig. 2b shows that if, say, accumulation of insecticide residues causes bird reproduction rate to fall, the separatrix will change its position until the topology of state space suddenly changes. The high tree equilibrium may disappear, leaving a state space with only one equilibrium, the low tree one. This would be a 'catastrophe' in both the usual and Thom (Zeeman 1977) senses. On the other hand, the deterioration of pasture related to destocking could, if severe enough, lead to an advantageous 'catastrophe' (Fig. 2b): the low tree equilibrium could disappear, leaving the state space, this time, with only the high tree equilibrium.

If further research could refine and validate the model, an economic submodel could calculate likely cash flows of various rehabilitation scenarios. Comparisons of cash flows are regularly made in assessing alternatives in physical engineering projects and, I believe, could also be useful here in judging the likely financial viability of proposed major works in which the farmers are to be involved.

IV. ANNUAL GRASSES AND ALTERNATE HUSBANDRY

The third case study concerns the management of annual grasses in a Western Australian farming system where, on a given part of the farm area, periods of wheat cultivation alternate with periods of sheep rearing on pasture. As in the swiddening system, the cropping tends to exhaust the soil while utilization as pasture tends to regenerate it.

The grasses can be viewed either as serious weeds of the wheat or as desirable pasture species. The grasses' prolific seed production and variable seed dormancy allow them to establish at high densities in many wheat crops, where they cause large yield reductions. However, the yield reductions are partly compensated for by the good contribution of grasses to early forage production in pasture and the protection they afford against soil erosion in dry years. Nevertheless, their presence in pasture reduces clover production and nitrogen fixation and lowers the profitability of subsequent wheat crops (Perry *et al.* 1980; Rerkasem, Stern & Goodchild 1980).

In view of the complex mixture of positive and negative effects of the grasses, farmers find it hard to know the best level of grass abundance to seek. For instance, in the past decade, grassy pastures and applications of nitrogenous fertilizer to the wheat crop seemed a viable combination, but, with rising fertilizer prices and difficulties in the herbicidal control of grasses in wheat, there are now suggestions that a total elimination of grasses might be preferable (Perry *et al.* 1980). Other difficult decisions for the farmer concern the ratio of wheat to pasture years in the rotation (primarily governed by relative profitabilities) and the lengths of the crop and pasture phases (Poole 1980; Ewing 1980). Since the best level of grasses in the system will depend on these latter decisions, the management problems facing farmers are formidable.

Hoping to provide a simulator with which to explore some of the consequences of farm-management decisions, a provisional model of key parts of the system has been assembled (B. R. Trenbath & W. R. Stern, unpublished). This model and some preliminary results will be briefly outlined.

The three state variables considered are soil nitrogen and the soil seed pools of ryegrass (the main annual grass) and of subterranean clover. The dynamics of these variables are calculated for a representative part of a farm property. In this first effort, a fixed pattern of climate is assumed for all years. To provide the performance indicators which interest farmers, levels of profit from the wheat and sheep enterprises are calculated as functions of the changing soil nitrogen level, and plant species composition and production in the wheat fields and pasture. The breeding herd of ewes is managed for the production of wool, fat lamb and wethers; a standard herd structure is assumed.

The management options that can be specified include wheat–pasture ratio and duration of phases, level of nitrogen fertilizer application to wheat, level of herbicide applications to control ryegrass in both wheat and pasture, and seed rate of clover sown either at the start of the pasture phase or as an undersown crop in the last season of the wheat phase. The 'environmental' variables specified correspond to those of the Central Wheat Belt of Western Australia and include cost and price structure (fertilizer, herbicides, clover seed, field operations, wheat, wool, sheep carcasses, marketing costs, etc.) and maximum wheat, ryegrass and clover productivities given the c. 400-mm rainfall of the area.

The model is organised around difference equations with 1-year time steps for the three state variables. In the crop phase, the change of soil nitrogen content is found by a full nitrogen budget; the changes in seed pools of the ryegrass and clover are calculated, taking account of germination rates, competitive effects of wheat on biomass production, mortality corresponding to level of herbicide application, and reproductive effort of the plants. Wheat yield is calculated as a function of total weed density and soil nitrogen level. In the pasture phase, change of soil nitrogen follows an empirical curve established for the region but adjusted for clover content of the sward. The competition between ryegrass and clover is based on a de Wit (1960) approach, which takes account of germination rates, plant densities and variation in relative crowding coefficients with soil nitrogen level. Inputs into seed pools are calculated from the seasonal production of dry matter. Seasonal dry matter production also determines stocking rate and hence the amount of wool and carcasses sold off the farm.

Analysis of the first results suggests that from the biological point of view the system of alternate husbandry differs from the cases of swiddening and eucalypt die-back in being globally stable. Here, under a given management regime, state space apparently contains only one equilibrium point to which the system locus moves whatever the starting point. The position of this equilibrium point around which, during the rotation, the system locus oscillates is determined mainly by the relative lengths of the cropping and pasture phases, and by the amounts of fertilizer and herbicide used.

As examples of some preliminary output from the model, the levels of the state variables are graphed for three versions of a rotation where 4 years of wheat alternates with 4 years of sheep pasture (Fig. 3). In the first version, fertilizer nitrogen is added at a rate sufficient to ensure a constant 'satisfactory' wheat yield, but no clover is sown (Fig. 3a). Clover disappears and soil nitrogen losses are compensated for by fertilizer. The income at equilibrium is calculated as

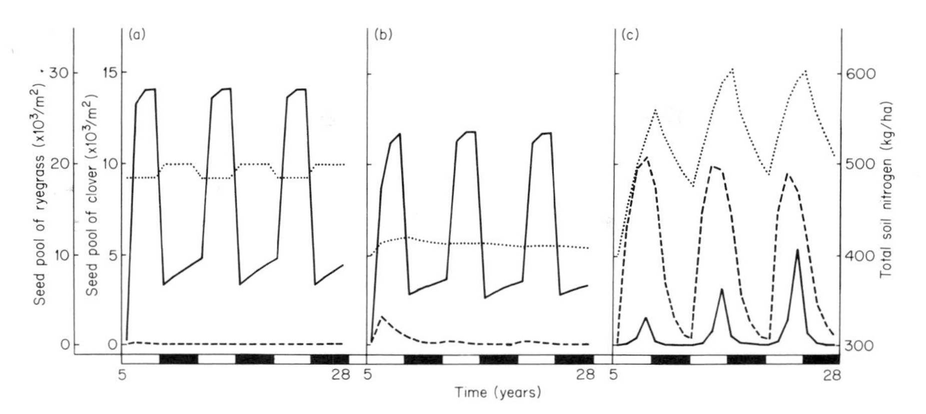

Fig. 3. Simulated time courses of annual ryegrass (——), subterranean clover (- - - -) and total soil nitrogen (····), near equilibrium, in an alternate husbandry system in Western Australia. [From B. R. Trenbath & W. R. Stern (unpublished).] The rotation is 4 years of pasture (unfilled rectangles) and 4 years of wheat (filled rectangles) (the first wheat phase is omitted). a, Fertilizer applied, no clover sown, standard herbicide. b, No fertilizer, clover sown, standard herbicide. c, No fertilizer, clover sown, herbicide application doubled.

$A21.60 per hectare per year. The second version represents a first attempt by a farmer to replace the nitrogen fertilizer with biologically-fixed nitrogen (Fig. 3b). However, in spite of clover's being sown at the start of each pasture phase, the ryegrass population is so high that it still dominates the pasture. Because little nitrogen is fixed, the level of soil nitrogen remains at its very low initial value with almost none 'available'. This system is calculated to lose $A1.90 per hectare per year. The third version is a modification of the second in which the farmer doubles his investment in herbicide (Fig. 3c). Better control of the ryegrass allows clover to dominate the pasture and freely fix nitrogen. The good yields of wheat bring the calculated income up to $A29 per hectare per year.

Whether or not this model is developed to a point at which it can guide quantitative management decisions, it seems already to provide a simulator with enough realism built in to make it a tool useful to trainee farm managers and advisors. In a more frivolous vein, it could immediately provide an after-dinner diversion for visiting farmers who might like to compete on the host's 'micro' to gain the highest theoretical income. Whatever the venue, however, ideas are likely to be sparked by the use of the model.

V. CONCLUSIONS

The three studies considered here have concerned plant species whose status as 'weed' is not altogether clear. In their management a balance must clearly be struck. Such species seem to provide the cases in which a particularly comprehensive and nuanced view of the plant is needed and in which ecology has one of its best opportunities to repay to agriculture part of an old debt of gratitude.

REFERENCES

Andres, L. A., Davis, C. J., Harris, P. & Wapshere, A. J. (1976). Biological control of weeds. *Theory and Practice of Biological Control* (Ed. by C. B. Huffaker & P. S. Messenger), pp. 481–499. Academic Press, New York.

Audus, L. J. (Ed.) (1976). *Herbicides: Physiology, Biochemistry, Ecology*. Vol. I. Academic Press, New York.

Barker, T. C. (1984). Shifting cultivation among the Ikalahans. *Working paper, Series 1, Program on Environmental Science and Management, University of the Philippines at Los Baños, Philippines.*

Burdon, J. J., Marshall, D. R. & Brown, A. H. D. (1983). Demographic and genetic changes in populations of *Echium plantagineum. Journal of Ecology,* **71,** 667–679.

Delfosse, E. S. & Cullen, J. M. (1981). New activities in biological control of weeds in Australia. II. *Echium plantagineum:* Curse or salvation? *Proceedings of the Fifth International Symposium on the Biological Control of Weeds, Brisbane, Australia* (Ed. by E. S. Delfosse), pp. 563–574. Commonwealth Scientific and Industrial Research Organisation, Australia.

de Wit, C. T. (1960). On competition. *Verslagen van Landbouwkundige Onderzoekingen,* **66.8,** 1–82.

Eussen, J. H. H. (1978). *Studies on the tropical weed* Imperata cylindrica *(L.) Beauv. var.* major. Ph.D. Thesis, University of Utrecht.

Ewing, M. A. (1980). The cropping revolution—Legumes are here to stay. *Journal of Agriculture (Western Australian Department of Agriculture),* **3,** 80–81.

Greenland, D. J. & Okigbo, B. N. (1983). Crop production under shifting cultivation, and the maintenance of soil fertility. *Symposium on Potential Productivity of Field Crops Under Different Environments, 1980,* pp. 505–524. International Rice Research Institutes Los Baños, Philippines.

Harper, J. L. (1971). Grazing, fertilizers and pesticides in the management of grasslands. *The Scientific Management of Animal and Plant Communities for Conservation* (Ed. by E. Duffey & A. S. Watt), pp. 15–31. Blackwell Scientific Publications, Oxford.

Hinton, P. (1978). Declining production among sedentary swidden cultivators: The case of Pwo Karen. *Farmers in the Forest* (Ed. by P. Kundstadter, E. C. Chapman & S. Sabhasri), pp. 185–198. University Press of Hawaii, Honolulu.

Ivens, G. W. (1983). *Imperata cylindrica,* its weak points and possibilities of exploitation for control purposes. *Mountain Research and Development,* **3,** 372–377.

Kunstadter, P. (1978). Subsistence agricultural economies of Luá and Karen hill farmers, Mae Sariang District, Northwestern Thailand. *Farmers in the Forest* (Ed. by P. Kunstadter, E. C. Chapman & S. Sabhasri), pp. 74–130. University Press of Hawaii, Honolulu.

Lehane, R. (1979). Requiem for the rural gum tree? *Ecos,* **19,** 10–15.

Ludwig, D., Jones, D. D. & Holling, C. S. (1978). Qualitative analysis of insect outbreak systems: The spruce budworm and forest. *Journal of Animal Ecology,* **47,** 315–332.

Nakano, K. (1978). An ecological study of swidden agriculture at a village in Northern Thailand. *South East Asian Studies,* **16,** 411–446.

Old, K. M., Kile, G. A. & Ohmart, C. P. (1981). *Eucalypt Dieback in Forests and Woodlands.* Commonwealth Scientific and Industrial Research Organisation, Australia.

Perry, M. W., Thorn, C. W., Rowland, I. C., MacNish, G. C. & Toms, W. J. (1980). Pastures without grasses . . . a speculative look at farming in the 80's. *Journal of Agriculture (Western Australian Department of Agriculture),* **4,** 103–109,

Poole, M. L. (1980). The cropping revolution . . . changes in cropping methods. *Journal of Agriculture (Western Australian Department of Agriculture),* **3,** 73–77.

Rerkasem, K., Stern, W. R. & Goodchild, N. A. 1980). Associated growth of wheat and annual ryegrass. 1. Effect of varying total density and proportion in mixtures of wheat and annual ryegrass. *Australian Journal of Agricultural Research,* **31,** 649–658.

Roberts, R. J., Campbell, A. J., Porter, M. R. & Sawtell, N. L. (1982). The distribution and abundance of pasture scarabs in relation to *Eucalyptus* trees. *Proceedings of the Third Australasian Conference on Grassland Invertebrate Ecology, Adelaide, 1981* (Ed. by K. E. Lee), pp. 207–214. South Australian Government Printer, Adelaide.

Sabhasri, S. (1978). Effects of forest fallow cultivation on forest production and soil. *Farmers in the Forest* (Ed. by P. Kunstadter, E. C. Chapman & S. Sabhasri), pp. 160–184. University of Hawaii Press. Honolulu.

Taylor, L. R. (1984). Presidential viewpoint. *British Ecological Society Bulletin,* **15,** 126–130.

Thiagalingam, K. & Famy, F. N. (1981). The role of *Casuarina* under shifting cultivation—a preliminary study. *Nitrogen Cycling in South-East Asian Wet Monsoonal Ecosystems* (Ed. by R. Wetselaar, J. R. Simpson & T. Rosswall), pp. 154–155. Australian Academy of Science, Canberra.

Trenbath, B. R. (1984). Decline of soil fertility and the collapse of shifting cultivation systems under intensification. *The Tropical Rain Forest: the Leeds Symposium* (Ed. by A. C. Chadwick & S. L. Sutton), pp. 279–292. Leeds Philosophical and Literary Society, Leeds.

Trenbath, B. R. & Smith, A. D. M. (1981). Basic concepts for a systems analysis of eucalypt dieback in New England. *Eucalypt Dieback in Forests and Woodlands* (Ed. by K. M. Old, G. A. Kile & C. P. Ohmart), pp. 234–343. Commonwealth Scientific and Industrial Research Organisation, Australia.

Trenbath, B. R., Conway, G. R. & Craig, I. A. (1985). Threats to sustainability in intensified agricultural systems: Analysis and implications for management. *Research Approaches in Agricultural Ecology* (Ed. by S. R. Gliessman). Springer-Verlag, New York (in press).

UNESCO/UNEP/FAO (1978). *Tropical Forest Ecosystems: a State-of-Knowledge Report.* Natural Resources Research XIV. UNESCO, Paris.

Walker, B. H., Ludwig, D., Holling, C. S. & Peterman, R. M. (1980). *Stability of Semi-arid Savanna Grazing Systems.* CRE-80-1. Centre for Resource Ecology, University of Witwatersrand (reprint).

Zeeman, E. C. (1977). *Catastrophe Theory, Selected Papers 1972–1977.* Addison-Wesley, Reading, Massachusetts.

III

The Demographic Interpretation of Plant Form: Application to Plant Competition and Production

13

On the Astogeny of Six-cornered Clones: An Aspect of Modular Construction

A. D. Bell

School of Plant Biology
University College of North Wales
Bangor, Gwynedd, Wales

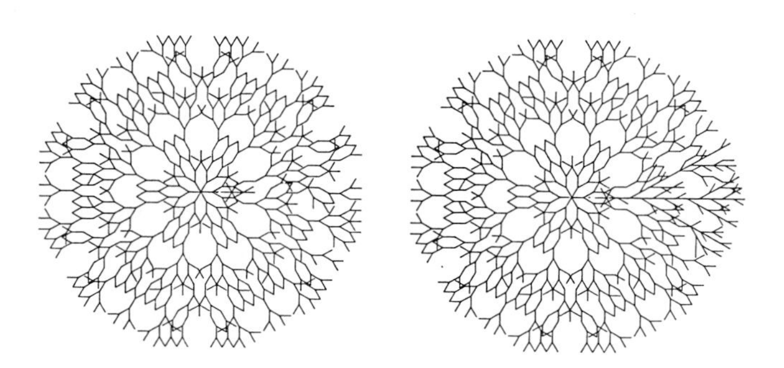

The principles needed to extricate the causes of a snowflake's shape do not arise from Nothing

Kepler (1611)

Constructional pattern and symmetry are tantalizingly elusive features exhibited by many plants. The human inquirer into things natural is excited by precision and regularity whenever they occur. If a natural object or organism demonstrates consistency of form, then it is comfortingly assumed that such symmetry is the consequence of Something rather than Nothing.

The format of this essay owes much to Johann Kepler, mathematician and astronomer, who became obsessed with the search for order in life: his feelings were exemplified by "Die Harmonie Mundi" (1619). In 1611, Kepler presented

STUDIES ON PLANT DEMOGRAPHY

his patron, the Baron Wackher von Wackhenfels, with a lighthearted essay, ''A New Year's Gift, or On the Six-cornered Snowflake'' (Schneer 1960). In his gift, Kepler made great play of presenting Nothing: ''I am well aware how fond you are of Nothing, not so much for its low price as for the sport, as delightful as it is witty, that it affords your pert sparrow; and I can readily guess that the closer a gift comes to Nothing the more welcome and acceptable it will be to you'' (Kepler 1611).

Historians are baffled by the reference to the sparrow, whilst 'Nothing' is presumed to be a private joke, unless it is a pun between the German *nichts* and the Latin *nix* (''nothing'' and ''snow'', respectively). Alternatively, the emphasis on Nothing rather than nothing may be meant to imply the converse, i.e. Everything: the underlying principles. Thus, later in his contemplation of the hexagonal snowflake we find ''But this is folly, to be so carried away. Why, my endeavour to give almost Nothing almost came to Nothing''. ''From this almost Nothing I have almost formed the all embracing Universe itself!'' (Kepler, 1611).

And so, nothing being too much to contemplate, now as then, herewith an attempt at a little something.

I. MODULAR CONSTRUCTION

The quest for a means of interpreting plant shape and form is as old as botany and originates with Aristotle or more specifically with Aristotle's younger contemporary Theophrastus. The development of this contemplative study of plants has been outlined by Arber (1950). Successive interpretations of form have had their heyday and declension, only to be reappraised at a later date (Cusset 1982). Behind them all is either a deliberate or an unwitting search for a basic unit of plant construction. The procession of phytomers, metamers, caulomers and the like has been reviewed comprehensively by White (1979, 1984). Each category has its uses, but one of the most persistent and, in contemporary botany, perhaps the most useful, is that of the apical meristem and the shoot unit into which it might develop. It is the activity of apical meristems, their location, growth and death, which manifests the form of plants. The shoot is the underlying constructional unit of the plant. The dynamic morphology of a plant can be described in terms of a developmental sequence of accumulation, cessation of growth, and loss of this type of module. In addition, each shoot module can be considered in more detail and itself described in metameric terms, perhaps that of node plus axillary meristem and its subtending leaf. Care must be exercised, however. Frequently the vasculature of a leaf links most directly not to its subtended bud or shoot, but to a more proximal (lower) bud or shoot. This is the case, for example, in white clover (*Trifolium repens*) (Williams 1984).

Having arrived at some system for describing plant form in terms of the accumulation of modules of whatever nature, then a study of the dynamics of modular growth becomes possible (e.g. Harper 1978, 1980, 1981, 1985; Harper & Bell 1979). "In (modular) organisms the zygote develops by repeated iteration of structure—a basic unit of construction (of architecture) forms a module that is continually reexpressed. . . . A consequence of looking at plants and modular animals in this way is that demographic methods may appropriately be applied to the process of growth of the product of a single zygote" (Harper 1980). The plant morphologist should be (and is) delighted by this revival of interest in plant form. Morphology is the poor relation in botanical fields and has always been in need of periodic prompting (e.g. Lang 1916; Hamshaw-Thomas 1932; Corner 1946; Sattler 1966; Hallé & Oldeman 1970, as emphasized by Tomlinson 1983). One likes to think that this latest impetus is perhaps the most significant yet. "Form is a sadly neglected aspect of plant growth, particularly in ecology. . . . The demographic approach to plant form gives a new meaning to much classical plant morphology. In particular, it begins to make it possible to quantify the ways in which an individual plant enters into the environment of a neighbour" (Harper 1980). Form is important; different forms and therefore different underlying rules govern the spatial and temporal juxtaposition of modules within their respective genets. The modular development of an organism, its dynamic morphology, must influence subtly the way in which an individual enters the demesne of a neighbour. Are there inevitable consequences when patterns with different architectures meet? There is so much 'background noise' in a natural situation that it is difficult to monitor. Where does one start? Why not eliminate chance and study the growth of patterns pure and simple? One hopes that this will give some insight into the dynamic morphology of 'live' pattern in modular organisms: *astogeny*.

II. ASTOGENY AND PATTERNS OF GROWTH

Astogeny is a most appropriate word to borrow from that sector of the zoological world which busies itself with modular animals. Astogeny has been defined as 'the development of a colony by budding' (Greek *astos*, citizen; *genos*, descent; Holmes 1979). There is a happy parallel in the etymologies of *morphology* and *astogeny*. Morphology has its origins in the works of von Goethe (1790), in which the concept of change in form of leaves is the essential principle (e.g. metamorphosis). Astogeny has been applied to the change in form of progressively produced zooids in developing colonies (q.v. Webster 1971). This is precisely the way a plant is progressively constructed; the embryo derived from the zygote develops into a colony of modules by the process of 'budding', that is, by the production of apical meristems. The earlier in the sequence of astogeny in which any one such meristem participates, then the more crucial the fate of that meristem is likely to be to the establishment and to the form of the individual. Several questions arise from this supposition. Will small differences in the behaviour of meristems (their locations and potentials) be correspondingly more significant for plant construction during establishment astogeny than the behavior of similar meristems which may become active later in the development of the organism? Do subtle differences of form, resulting from differing geometric details of astogeny, affect the relative growth performance of interacting genets? Is the outcome of such interference altered by the timing of the confrontation, in relation to the ages of the participants? Can a depauperate pattern or plant (measured in terms of the numbers of metamers or modules in its construction) withstand the interference of a more robust architecture if it has a longer period to establish itself? All these possibilities presuppose that one pattern of growth in, for example, a stoloniferous plant can actually be 'better' than any other. However, as noted by Harper (1981), it is not wise to assume that "all is for the best in the best of all possible worlds". In this respect the use of computer-generated patterns has its conceptual advantages: there is no 'environmental noise'; the astogeny of any one given pattern can have only one outcome. Interference between two dissimilar patterns during their development will have only one outcome: the performance of each (measured in terms of number of modules born) will be depressed by finite but different amounts in relation to solitary control patterns, and one type of geometry may be found to be more successful than another.

Kepler's philosophy in 1611 was to enjoy for a time the fun of speculation, which fitted in well with the festive nature of his essay, before risking the disappointments that might result from the unsuccessful testing of his hypotheses. "I shall push this notion [of the origin of the six-cornered snowflake] as far as it will take me, and only afterwards shall I test its truth, for fear that the ill-timed detection of a groundless assumption may perhaps prevent me from fulfilling my engagement to discourse about a thing of Naught". In this Keplerian spirit, a little more speculation is thus in order.

To interpret developmental patterns of growth, that is, the behavior of 'citizens' in the population of shoot modules, it is ideally necessary to know why a particular module does what it does. What are the most influential internal and external stimuli, and what will be the response of the apical meristem of the module to various combinations of these stimuli? Only when it is known how a module would have behaved in a given environment is it reasonable to manipulate its experience and draw conclusions from its response. Certain knowledge of what would have happened without interference represents a control for any given situation.

What is needed therefore to begin a rigorous analysis of real botanical pattern is an obligingly conservative plant, the dynamic morphology of which is predictable at the module level. Some plants are conveniently suitable for this, e.g. *Theobroma cacao* (Greathouse & Laetsh 1969), but others are more or less confusingly plastic. Having identified a predictable response pattern of meristems to a standardized environment, the recording of the responses of similar meristems to well-defined variations in the environment could be systematically explored. In real plants the activity of one module in a particular, chosen environment might affect the behaviour of neighbouring modules, setting up a chain reaction of response.

The simplest 'standard' module would appear to be the germinating epicotyl. Computer patterns can be generated in which the subsequent modules develop independently of each other and in which the 'environment' is absent. Presumably their behaviour would be much easier to interpret than the dynamics of real plant morphologies. Kepler (1611) would have enjoyed an interlude in which to duck the weeds and play with a computer lest the plants present that "ill-timed detection of a groundless assumption". So on this occasion why not indulge in the elimination of the environment and all the stochastic problems of plants and generate a series of totally determinate patterns? Then the interference between various modular branching processes can begin to be investigated. The exercises which follow utilize a graphics program (Bell, Roberts & Smith 1979) modified to allow the fate of potential modules to be influenced if confronted by existing modules of the same, or a different, genet (Bell 1984). Taking once again the inspiration from a snowflake, all the patterns shown here commence in a regular six-cornered form. Unfortunately, even confining the initial architectures to the hexagonal format, countless numbers of astogenies are possible; there is a grave danger of just producing so many wallpaper designs (J. L. Harper, personal communication). Therefore, the simulations that are illustrated here are based on an extremely simple morphology.

To begin, every module is of standard length (10 arbitrary units, which may be thought of as metamers) and is able to bear just two daughter modules at its distal end, each daughter being identical to its parent in all respects. The recursive simulation proceeds in cycles, the daughters born in cycle t becoming the parents for cycle $t + 1$. Even with this simple dichotomizing system, the permutations of pattern are endless; 10 variations are presented here, differing solely in the angle of daughter module to parent module. Parameters such as module length, rate of growth, location on parent are all constant. The basic patterns (A–J) are shown:

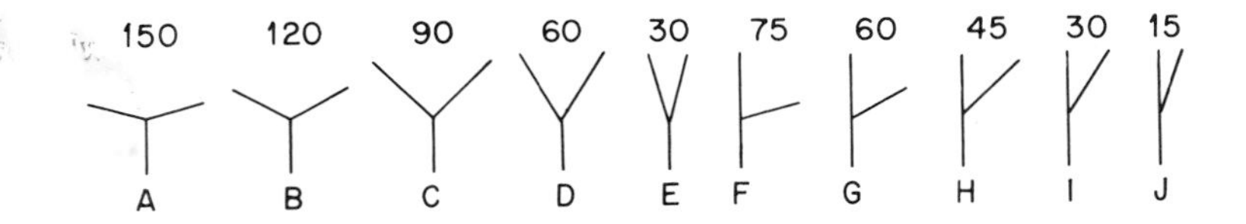

Patterns A–E will form sympodial branchings of a decurrent form with no leaders. Patterns F–J will form linear systems with distinct leaders and, although strictly sympodial in terms of modular construction, will be referred to here as monopodial for convenience. Such a linear sympodial pattern occurs in *Carex arenaria* (Harper & Bell 1979). Both types occur in nature. But uncontrolled proliferation leads to morphological chaos, so a rule to restrict this is necessary. Development of a module is prevented by an excess of neighbours. Such a rule may take the following form: if the growth of a module results in its mid-point being within 14 units of length from the mid-points of five existing modules, it is aborted. At this stage the parameters of interference are totally arbitrary, and a potential module will respond indiscriminately to existing modules of its own, or of another, clone. A change from a threshold of five modules at 14 units' distance would produce a completely new set of patterns. The architectures and rules outlined here are maintained throughout the simulations to be described.

All plants have to become established. Establishment astogeny of clonal plants (but not of colonial animals) is a largely neglected field. Nevertheless, this is the crucial stage in the dynamic morphology of a genet. The younger the clone, the fewer the number of modules in its construction, and therefore the more vulnerable it is to the loss of individual modules. Also, a newly establishing clone should normally develop in a radial manner; at least this seems to be intuitively obvious, despite the many exceptions to be found among monocotyledonous plants in which establishment growth is not radial although it does often differ from adult morphology. The details of pattern in the very early stages of astogeny are therefore likely to differ from later architecture when growth becomes sectorial. Certainly, if fragmentation of the clone takes place, the mature portions, those most often described, must be radial sectors of the overall genet and will have a bias of astogeny in one direction. Old ramets have polarity. Out of deference to Kepler, the initial development of the artificial clonal patterns illustrated here is restricted in every case to six modules departing symmetrically from the central 'germination' site. Possible alternatives would be limitless. As a simulation proceeds, new modules are added at each growth cycle to the distal ends of the developing pattern unless overcrowding prevents this. Once a module has failed to develop it is lost forever: there is no provision in this simulation game for reiteration from dormant module meristems. Comparison of the growth of these different patterns, either developing singly or with mutual interference, is available from listings of the numbers of modules born in each cycle, up to an arbitary 15 cycles. Simple hypotheses can now be tested. For example, in solitary patterns, to what extent is the accumulation of modules affected by branching angle? Do the two types of pattern, monopodial and sympodial, differ in response to change in angle? Are initially successful patterns consistent over time in this respect? Wherever two patterns develop as near neighbours, does the initial relative orientation of the pair have any bearing on the outcome? If the patterns are different is their resistance to interference commensurate with their productivity when developing alone? Does the success of a relatively unproductive pattern improve if it has more time to become established before meeting a neighbour? Some experiments addressed to these questions are recorded here, and results are presented in the spirit of Kepler's gift as graphical snowflakes. In addition, total module number is shown against time; in the case of interclonal interference, module development is compared with the corresponding control suffering only intraclonal interference.

III. SOLITARY PATTERNS: THE EFFECT OF ANGLE

Figure 1 shows the accumulation of modules with time for a range of five daughter-to-parent angles (15°, 30°, 45°, 60°, 75°). The corresponding patterns produced at cycle 15 are shown in Fig. 2. Qualitatively, the sympodial patterns and monopodial patterns differ from one another markedly at higher angles but become more and more similar at lower angles. Productivity, however, is not consistent with change in angle. The numbers in brackets in Fig. 2 indicate the order of the patterns in a hierarchy based on accumulated module total at cycle 15. This order is practically unpredictable before the event, but causes can be guessed at by comparing the diagrams with the step-by-step module accumulation (Fig. 1a,b). Thus, the interclonal interference in the sympodial system with a 75° branching angle is severe, whereas the very narrow angle of 15° allows more modules to be born without approaching neighbours. But it is not possible to be sure of the likely success of a pattern from its early development. For example, at cycle seven, the sympodial pattern 45° is more productive than sympodial pattern 30°, but this position is reversed by cycle 15. Kinks in the growth curves are related to catastrophies in particular cycles. An annulus of holes appears in the sympodial 30° pattern at cycle six, the graph being depressed in this region (Fig. 2). Generally, narrow angles do better than wide angles, and monopodial patterns are more productive than sympodial patterns. In the context of these rigid growth forms, diverging centrifugal lines lessen the chance of sister-module collision. This intrinsic feature of monopodial astogeny, which is typical of stoloniferous plants, is apparent in Fig. 2. The poorest monopodial performer, 60°, is handicapped by its unavoidable commitment to parallel lines.

Two further points may be noted. Firstly, sympodial patterns illustrated here are truly symmetrical, whereas the monopodial patterns occur as either sinistral or dextral forms, depending upon the location of the second order branches. In Fig. 2, the monopodial patterns of 15°, 45° and 60° are sinistral. This effect is most apparent at 75°. Secondly, the narrowest angled monopodial pattern looks very similar to the corresponding sympodial pattern, and has in fact become obligatorily sympodial. Daughters produced at narrow angles crowd the mono-

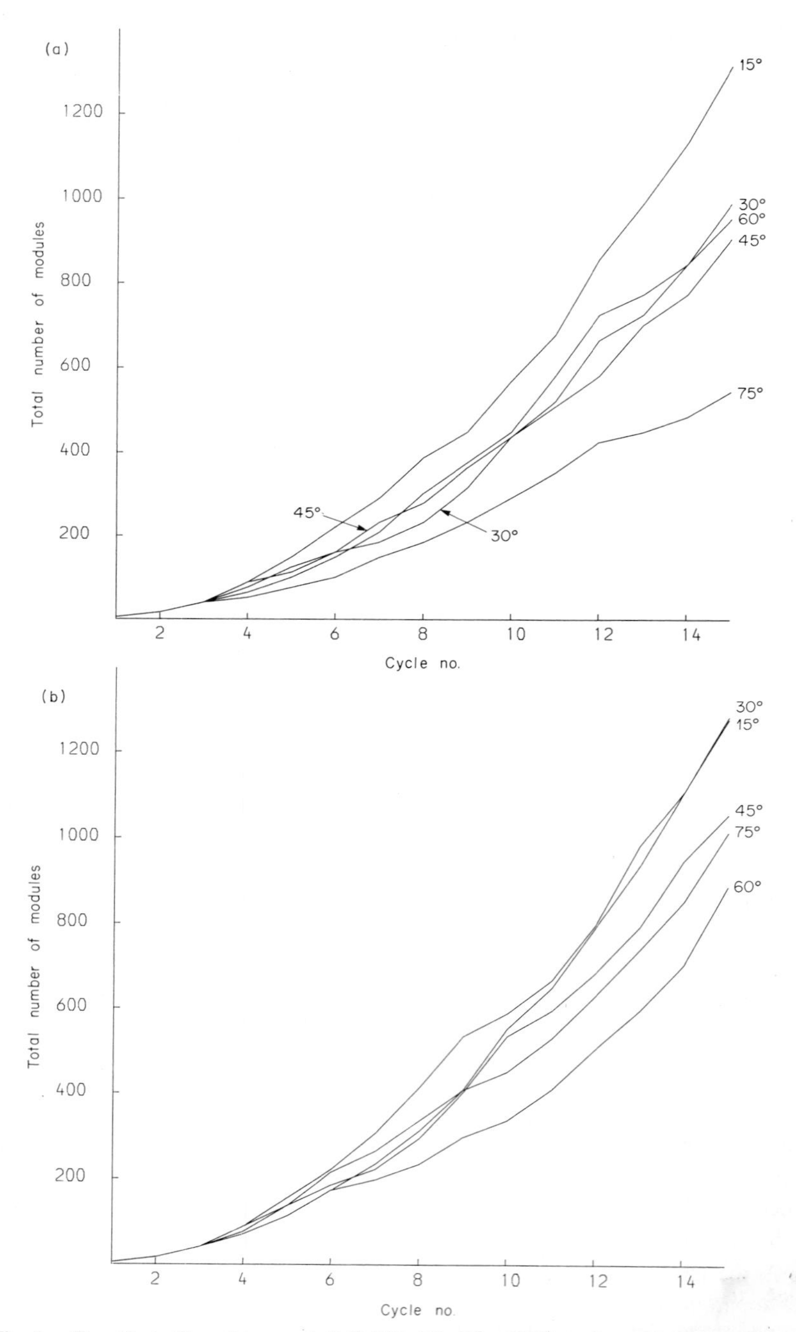

Fig. 1. The effect of branching angle (15°, 30°, 45°, 60° and 75°) on the total number of modules produced. Durations of development, 15 cycles. Rule of growth governing response to intraclonal interference unchanged throughout. a, Sympodial patterns. b, Monopodial patterns.

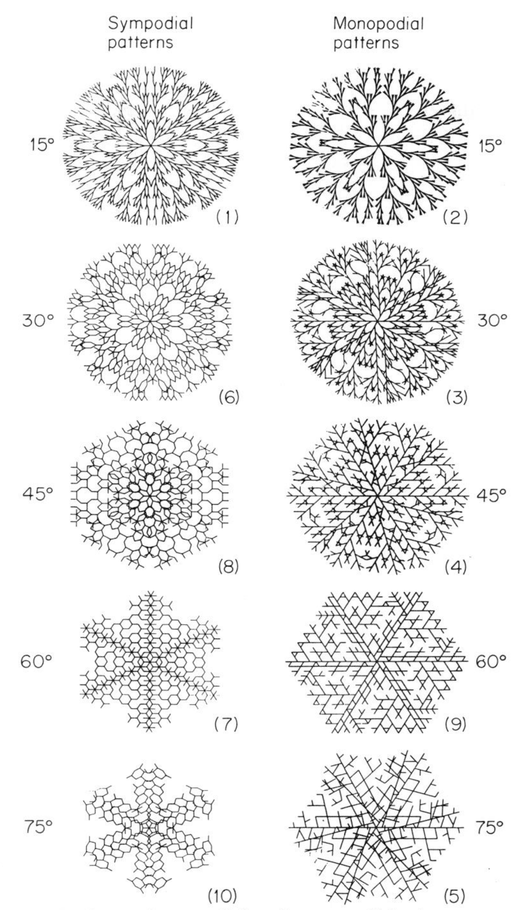

Fig. 2. The change in form of sympodial and monopodial patterns as governed by change in branching angle. Length of developmental sequence, 15 cycles. The figures in parentheses indicate the ranking order of these patterns at cycle 15 based on the total number of modules born. Identical rules of growth apply to each pattern, including rules governing response to intraclonal interference; only branching angle is varied.

podial leader and cause its abortion. By progressively reducing the angle of daughter to parent from 30°, the threshold for this transition in astogeny is found to occur at 23° (Fig. 3). Visually there is a discordancy of radial symmetry at this angle (Fig. 3c). Patterns obtained by altering the angle by 1° intervals above and below 23° appear to be increasingly organised and doily-like (Fig. 3a,e).

Small changes in angle alone can make large changes in both form and productivity (Fig. 1). Are such phenomena predictable in a mathematical manner (Ulam 1966)? What are the underlying principles in the astogeny of simple interactive patterns?

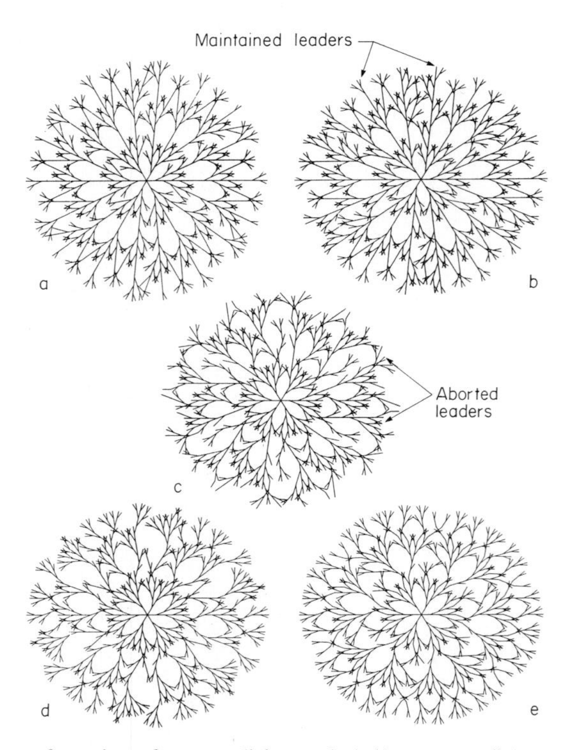

Fig. 3. The transformation of monopodial growth (a,b) to sympodial growth (c,d, and e) brought about by a slight change in branching angle. Length of developmental sequence, 15 cycles. a, 25°; b, 24°; c, 23°; d, 22°; e, 21°. Rules of growth including those governing response to intraclonal interferences as in Fig. 2.

IV. NEIGHBOURS

When two plants grow into each other there will be some degree of interference between them. Details of the mechanisms behind such events are not clear but act at the module population level. Typically existing apical meristems die together with the shoot that bears them, and modules that would have developed in a freestanding situation fail to do so. This being so, then the earlier in the astogeny of the plant that the fate of modules is influenced by neighbours, the greater the impact on the fate of that plant's form. The younger a plant, the more closely it resembles other members of its species in its architecture: the fewer the modules, the smaller the variation in form. Young forms of stereotyped plants or patterns are likely to be more sensitive to subleties of neighbourly interaction.

For example, the consequence of interference between very young plants may be influenced by their respective orientation or angle of presentation. The first two stolons developing from a seedling usually grow in opposed directions. Two such plants germinating close together could develop such that they will either produce their initial stolons oriented in parallel and therefore continually interfering, or at right angles to each other, meeting once and once only.

A. Effect of Orientation on Interference between Simple Patterns

Figure 4a, b, and c shows the outcome at cycle 15 of interference between pairs of sympodial patterns having a branching angle of 30° (see Fig. 2 for the free-growing control). Three orientations of presentation of the initial six-cornered development are tested (see p. 198):

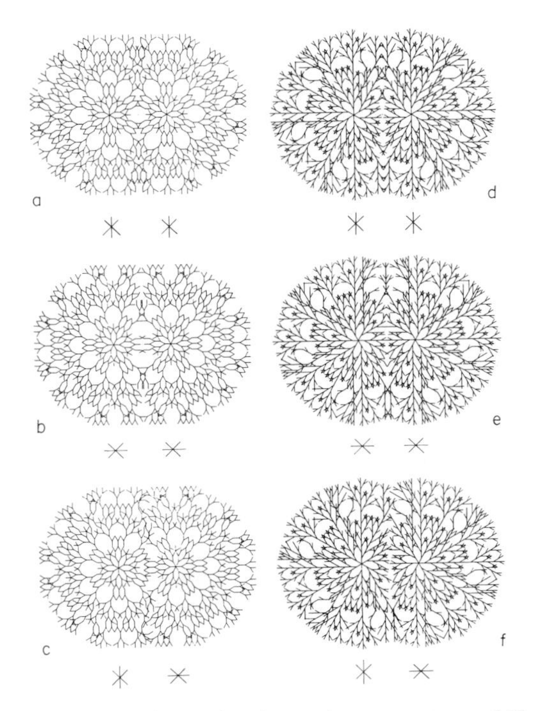

Fig. 4. Effect of initial orientation on interference between patterns of like kind at cycle 15. Distance apart at commencement of growth, 100 units. Length of one module 10 units. All branching angles 30°. a, b, c, Sympodial patterns. d, e, f, Monopodial patterns. Initial orientations as indicated. Rules of growth as in Fig. 2. Those governing response to interclonal interference identical to those governing intraclonal interference.

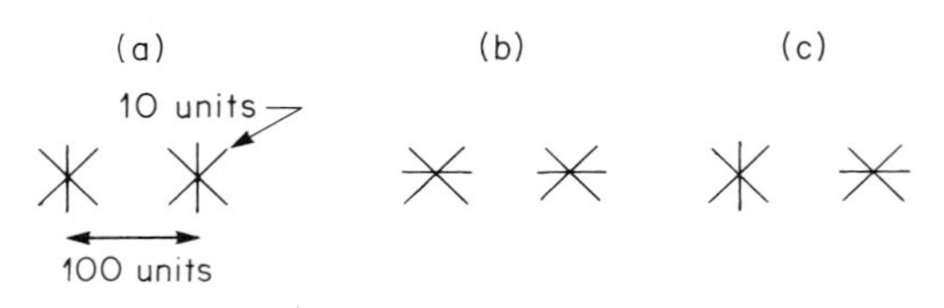

The patterns begin to intermingle at cycle six and are modified in different ways depending upon orientation. The effect on the performance is shown in Fig. 5a, where accumulative productivity at each cycle is shown as a proportion of the control of that cycle. Overall, productivity compared to the control declines steadily until cycle 11 and then temporarily recovers. Study of the patterns in Fig. 4 shows that by cycle 11 all branches heading more or less directly towards each other have ceased to grow, and subsequently colliding modules are doing so in a glancing fashion with consequently fewer mortalities. Also, as the converging clones expand, the zone of interference represents a progressively smaller proportion of the circumference in relation to the growing periphery as a whole:

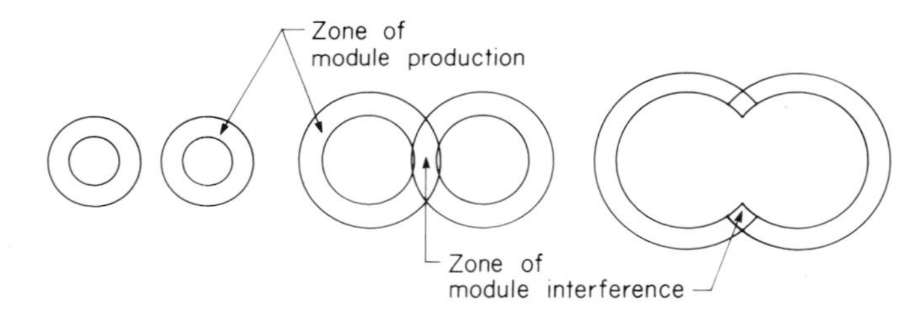

Thus the increase in the proportion of modules lost in relation to the control slows down. The effect of initial orientation on the productivity of patterns can be judged from Fig. 5a. There is an approximately 2% difference in numbers of modules born, depending upon angle of presentation in any given cycle. The monopodial patterns (Fig. 4d,e,f and Fig. 5a) are more conservative in this respect. The opening gambits are more numerous as each pair of orientations has variations depending upon handedness,which itself may affect productivity. This facet is ignored in the patterns shown in Fig. 4. The slopes of productivity compared to controls (Fig. 5a) lack the irregularity at cycle 11 shown for sympodial pattern. Also there is less difference, about 0.5%, which is due to angle of presentation. Orientation therefore has less impact on interference in monopodial

Fig. 5. The effect of different initial orientations on interclonal interference, in terms of total numbers of modules produced. Module number is expressed as a proportion of the appropriate control developing without interference. Rules for intraclonal and interclonal interference are identical and remain unchanged throughout. The graphs resulting from four different initial pairs of orientations (< >, < –, – >, – –) are shown for each of the following patterns: (a) ▩, Sympodial pattern (competing with sympodial pattern); ▧, monopodial pattern (competing with monopodial pattern). (b) ▩, Sympodial pattern (competing with monopodial pattern); ▧, monopodial pattern (competing with sympodial pattern).

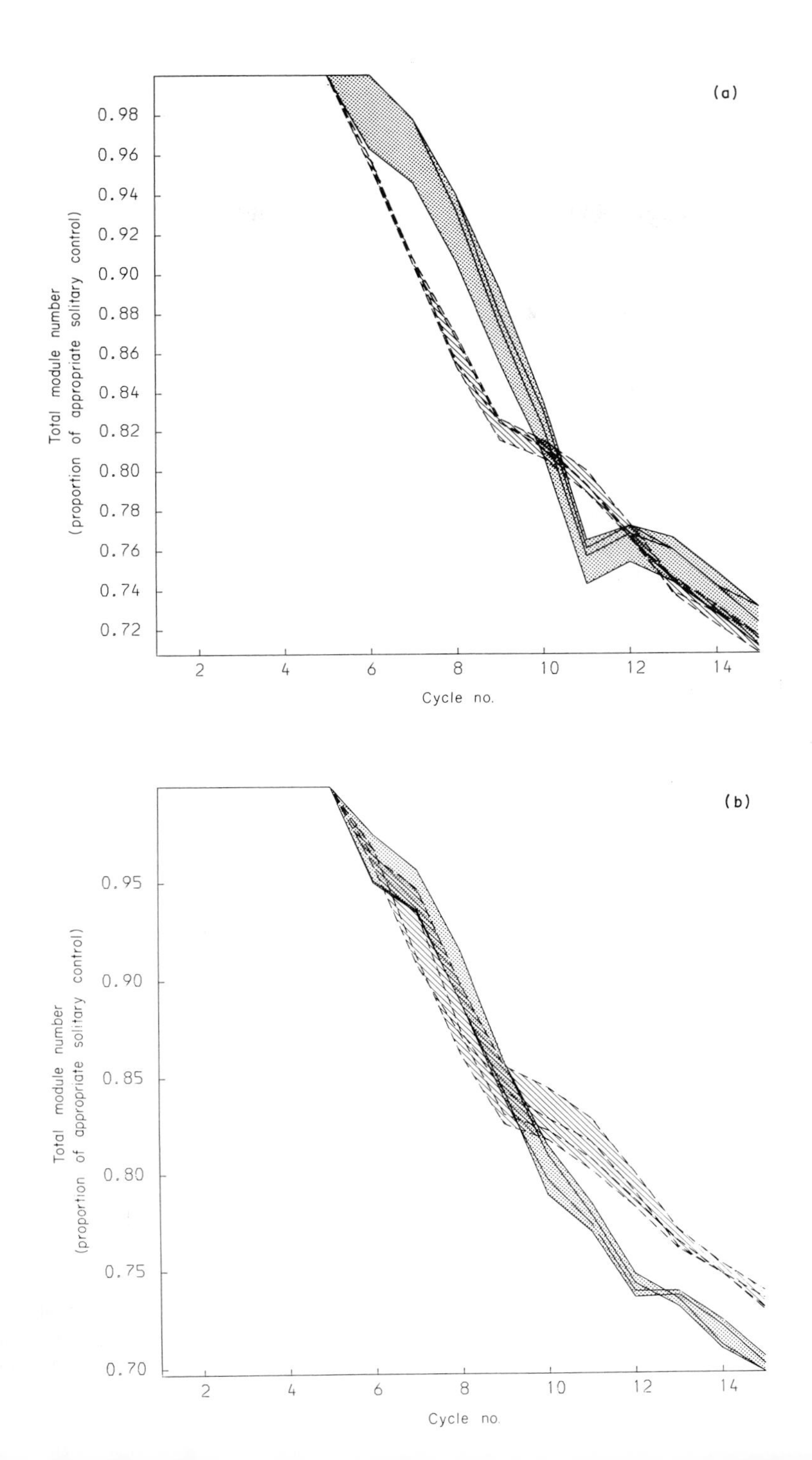
(a)
0.98
0.96
0.94
0.92
0.90
0.88
0.86
0.84
0.82
0.80
0.78
0.76
0.74
0.72
Total module number
(proportion of appropriate solitary control)
2
4
6
8
10
12
14
Cycle no.
(b)
0.95
0.90
0.85
0.80
0.75
0.70
Total module number
(proportion of appropriate solitary control)
2
4
6
8
10
12
14
Cycle no.

types than sympodial types. One possible explanation is that in linear systems based on 30°, all combinations of presentation have the same directions of radial growth established by the second cycle:

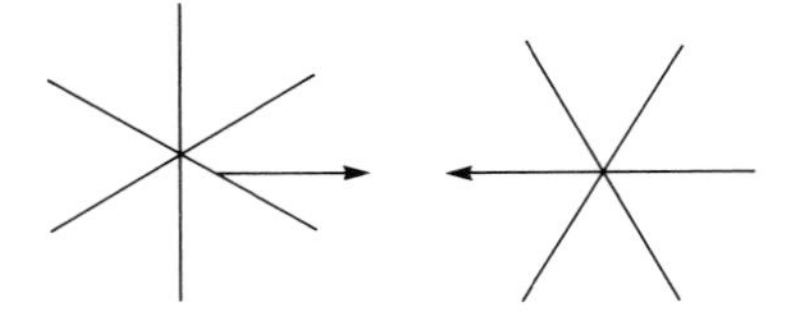

B. Different Neighbours

"I have been busily examining the little flakes. Well, they have been falling, all of them, in radial pattern, but of two kinds . . ." (Kepler 1611).

What happens when monopodial and sympodial patterns meet? One hypothesis is that the monopodial system, with its greater productivity, will suffer less when confronted by a sympodial system than by another monopodial system.

Figure 6 shows the outcome at cycle 15 of interference between a sympodial pattern and a monopodial pattern each branching at 30°. Four initial orientations are presented, as a close scrutiny of the central arms shows. Curiously, in this case it is the monopodial patterns which show a greater range of productivity depending upon their orientation (Fig. 5b). The monopodial systems are more quickly depleted than the sympodial when young. However, this situation is reversed by cycle 10.

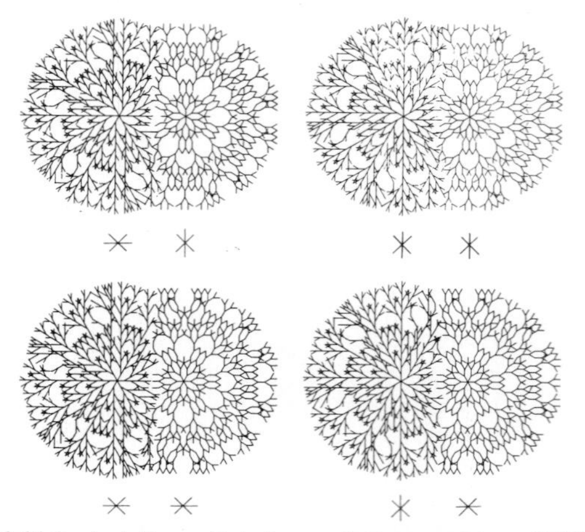

Fig. 6. Effect of initial orientation on interference between patterns of different kinds at cycle 15. Distance apart at commencement of growth, 100 units. All branching angles 30°. Monopodial patterns to the left of each colliding pair, sympodial to the right. Initial orientations as indicated. Rules of growth as in Fig. 4.

Differences in suppression of module production compared to control are not very great (0.74 for monopodial, 0.7 for sympodial). In spite of these being totally determinate patterns, with only one possible outcome to each simulation, the result of any one type of interference sequence can only be guessed at and can only be discovered by experiment. Hypothesis, experimentation, explanation can easily proceed in the wrong direction. In this instance the hypothesis is that the monopodial pattern, containing more modules at each cycle, is therefore the more robust in the face of interference. It could be argued that the sympodial pattern is of a somewhat 'phalanx' nature with its repeated dichotomies, whereas the linear component of the monopodial pattern makes it more akin to a 'guerilla' development. Then, conveniently, it can be explained that the young sympodial phalanx can resist encroachment by the monopodial guerilla but in the long-term is outflanked. If determinate patterns can allow such a carousal of argument no wonder the essence of dynamic morphology in real plants is also elusive.

V. AGE

The outcome of a meeting between two colliding clones may well depend upon their relative productivities, but it might also depend upon the timing of their merging in relation to their respective ages. It has been suggested earlier that any given module is proportionally more important to the architecture of the clone, the earlier it is formed in the sequence of astogeny. It follows that interference early in the course of developments should be more damaging than

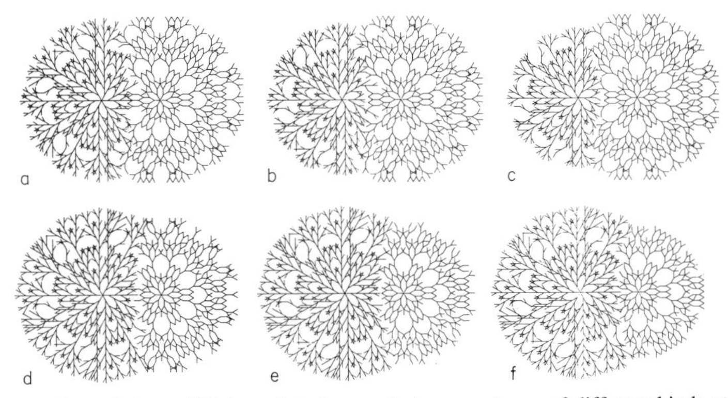

Fig. 7. Effect of time of birth on interference between patterns of different kinds at cycle 15. Distance apart at commencement of growth 100 units. All branching angles 30°. Monopodial pattern to the left, sympodial to the right. (a) Monopodial start delayed 1 cycle, (b) monopodial start delayed 2 cycles, (c) monopodial start delayed 3 cycles, (d) sympodial start delayed 1 cycle, (e) sympodial start delayed 2 cycles, and (f) sympodial start delayed 3 cycles. Rules of growth as for Fig. 4.

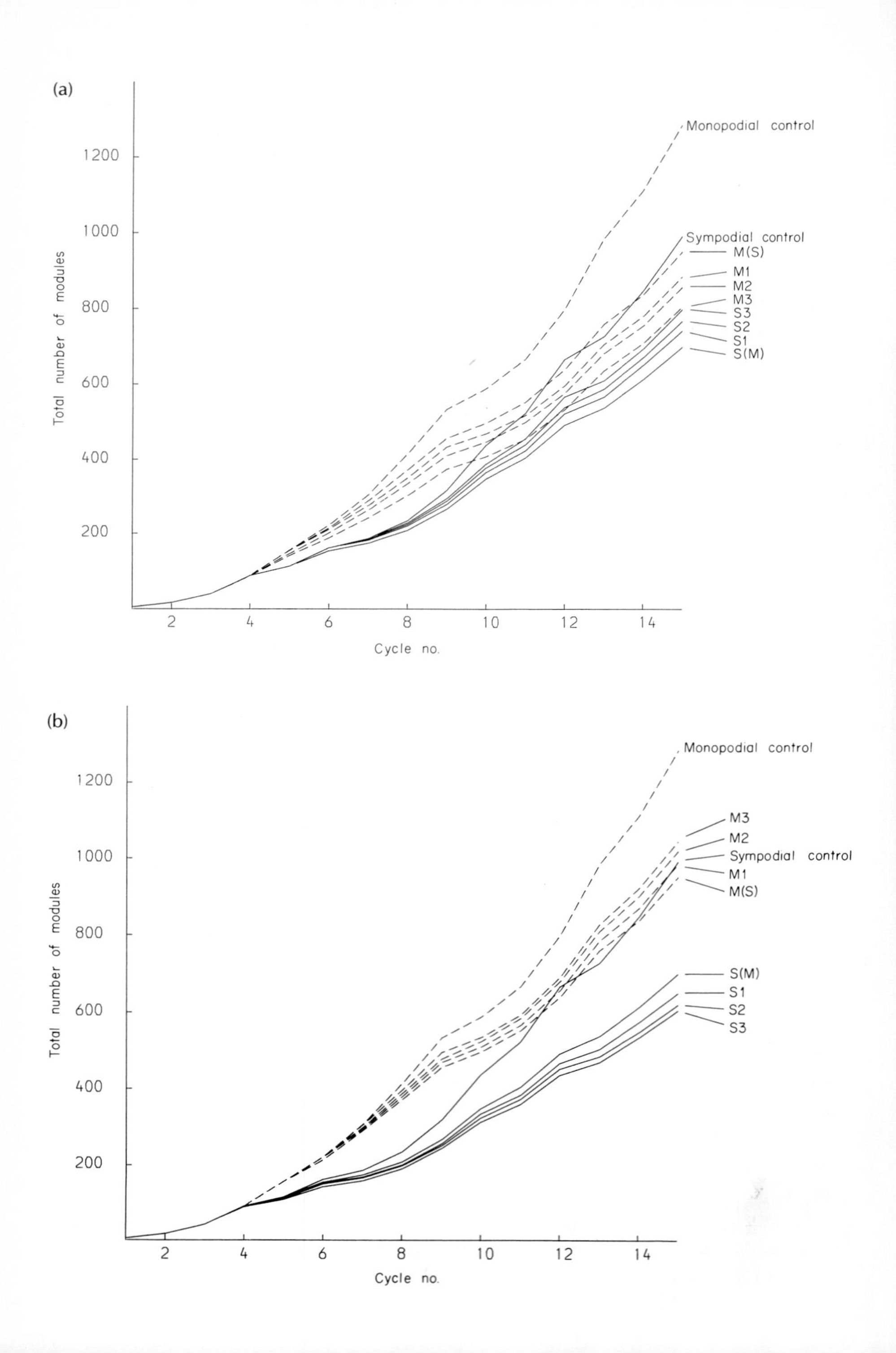
(a)
Monopodial control
Sympodial control
M(S)
M1
M2
M3
S3
S2
S1
S(M)
1200
1000
800
600
400
200
Total number of modules
2
4
6
8
10
12
14
Cycle no.
(b)
Monopodial control
M3
M2
Sympodial control
M1
M(S)
S(M)
S1
S2
S3
1200
1000
800
600
400
200
Total number of modules
2
4
6
8
10
12
14
Cycle no.

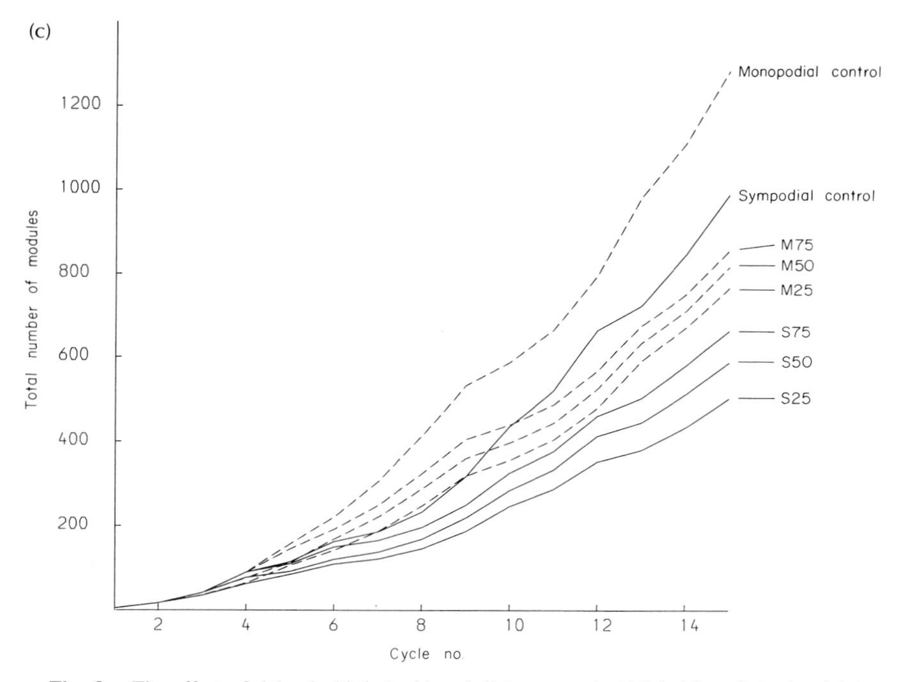

Fig. 8. The effect of delay in birth (a, b) and distance apart at birth (c) on interclonal interference in terms of total numbers of modules produced. (a) Birth of monopodial pattern delayed (see Fig. 7a, b, c). No delay: M (S), monopodial colliding with sympodial (.739); M 1, with 1-cycle delay for monopodial (.688); M 2, with 2-cycle delay for monopodial (.666); M 3, with 3-cycle delay for monopodial (.626). No delay: S (M), sympodial colliding with monopodial (.705); S 1, with 1-cycle delay for monopodial (.748); S 2, with 2-cycle delay for monopodial (.774); S 3, with 3-cycle delay for monopodial (.805). (b) Birth of sympodial pattern delayed (see Fig. 7d, e, f). As a, but sympodial delayed monopodial not delayed (.739, .766, .793, .813, .305, .656, .625, .610). (c) Proximity of birth varied. M 75, initial gap 75 units (.667); S 75, initial gap 75 units (.671); M 50, initial gap 50 units (.637); S 50, initial gap 50 units (.603); M 25, initial gap 25 units (.598); S 25, initial gap 25 units (.510). The numbers in parentheses indicate the total number of modules produced by cycle 15 as a proportion of the appropriate control.

interference occurring later. The age of a clone in terms of its susceptibility to neighbours can be measured in two ways, by its actual time of birth compared with that of its neighbour and by their distance apart. The further they are apart, the older they will be when they meet, given the constant rates of extension in the model simulation.

The results of investigating the effect of 'germination' time, and time to impact due to distance apart, are shown in Fig. 7, 8 and 9. In the simulations at cycle 15 shown in Fig. 7a–c, the start of the monopodial pattern's growth has been delayed by one, two, or three cycles. Its productivity (Fig. 8a) is pro-

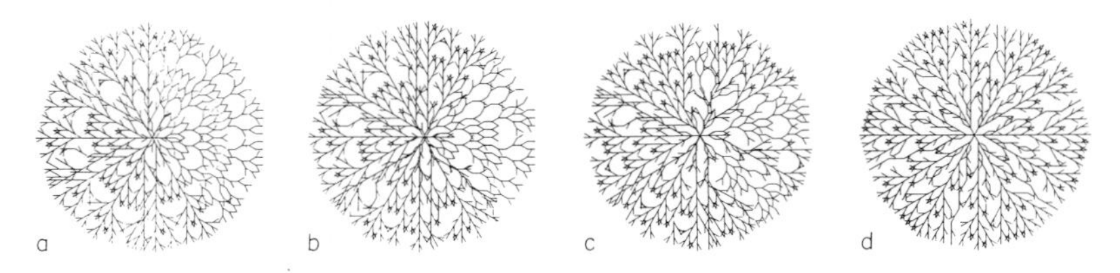

Fig. 9. Distance apart at cycle 1. a, 5 units: monopodial patterns to the left of each pair. b, 3 units: sympodial patterns to the right of each pair. c, 1 unit: all branching angles 30°. d, Same point: module length = 10 units.

gressively depressed whilst that of the sympodial neighbour is correspondingly enhanced. The numbers in brackets in Fig. 8 indicate the production at cycle 15 as a proportion of the appropriate control. With a delay of three cycles, the monopodial pattern has lost about 10%, and the sympodial has gained about 10% compared to the situation with no delay. The overall productivity of the monopodial pattern is nevertheless still greater than that of the sympodial pattern. The effect of delaying the start of the sympodial pattern in relation to the monopodial is shown in Fig. 7d,e,f and the corresponding module accumulation in Fig. 8b. The outcome is as expected: the monopodial pattern is enhanced by about 10% and the sympodial reduced by 10% compared to the situation with no delay. One pattern as defined by these rules is more affected by its relative age than the other. Is the same effect achieved by altering age at time of collision as determined by distance apart at germination? Figure 8c shows the total module number with cycle for three starting gaps, 75, 50, and 25 units (all previous simulations illustrated have commenced at 100 units apart, 1 module = 10 units). The closer the start, the sooner the interference and the greater the reduction in productivity of both components. For example, at a gap of 25 units, the linear pattern has produced 60% the number of modules of a solitary control, whereas the sympodial pattern has only managed 50% of its control. Again this monopodial pattern is superior to this sympodial pattern.

Figure 3 indicated the way a change in angle branches of 1° could make a dramatic change in architecture. This is also true of the initial distance apart of colliding clones, at least when these are very close. The starting gaps depicted in Fig. 9a, b, c are five, three and one unit. The sympodial pattern is reduced in productivity to 43%, 33% and 29% in relation to the control, respectively. It can be seen that there is a much greater degree of intermingling of the two components at one unit gap (Fig. 9b) than at three units gap (Fig. 9c). If both patterns commence to grow at the same point (Fig. 9d), something that can be tested in a simulation but not in nature, the sympodial pattern achieves a mere 10% productivity by cycle nine and then dies. The behaviour of the monopodial pattern in this particular situation is fascinating. Although it is competing for space during the early cycles (1–9) and has its productivity depressed at this time, once the sympodial neighbour has been eliminated its productivity increases and during the last six cycles (10–15) is actually *better* than the control pattern developing with no neighbour. Early pruning has apparently generated space which the later more prolific modules exploit.

There is, of course, an interaction of distance apart and time of initial development. The monopodial pattern with germination delayed by three cycles is still 'superior' to the sympodial if given time to become established (starting 100 units away, e.g. Fig. 7c). However, there is a threshold distance at a gap of 28 units distance; any closer than this, and a tardy monopodial pattern is extinguished. At 28 units gap it has just time to become precariously established before the older sympodial neighbour reaches it (see right-hand side of the title figure). At 27 units gap, merely one-tenth of a module length closer, it is engulfed and cannot grow (left-hand side of title figure).

VI. CONCLUSION

But what does it all mean? Even in these patterns, generated by very simple rules with each successful module bearing just two daughters, the permutations of astogeny are essentially limitless. Any shift of distance, timing or angle gives a new growth form. Even so, there is no chance element in the process at all: each simulation pattern is invariant given its circumstances. Here, Kepler might have said, the biologist should ponder that real organisms can also display such consistencies of form despite their potentially flexible rules of growth and the vagaries of the environment. "Let us grant that each single plant has its own principle" (Kepler 1611). So, it would appear, do dynamic patterns. "I have not yet got to the bottom of this", says Kepler, still chasing his transitory snowflakes. "In such anxious reflection as this, I crossed the bridge, embarrassed by my discourtesy in having appeared before you without a New Year's present, except in so far as I harp ceaselessly on the same chord and repeatedly bring forward Nothing". However Kepler's essay received a new lease of life when it began to snow for a second time whilst he was searching for his 'principle'. Why not 'create a snowstorm' and watch what happens? (Fig. 10).

Although a stochastic element appears in the snowstorm, there can be only one overall developmental sequence for the particular random array depicted. The monopodial individuals are expected to outsurvive the sympodial (and they do), but the details can only be speculated upon. Analysing such a snowstorm could be as difficult as understanding the interactive form of a tangle of vegetation.

But there must now be a chance; it will become possible to decipher the whole (be it a single plant or a conglomeration of plants) because it is composed of an astogenic sequence of modules, and we *are* beginning to understand modules.

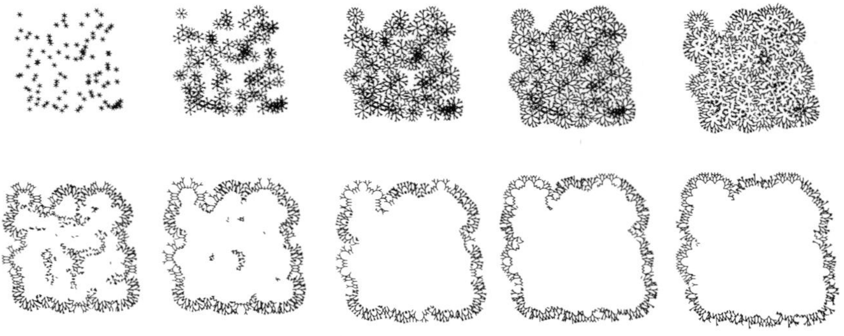

Fig. 10. Developmental sequence over 10 cycles : 100 genets initial "sowing": 50 monopodial and 50 sympodial patterns allocated at random to a 100 × 100 grid. Rules of growth and interferences as in Fig. 1–9, plus 'rotting' of old modules. Ramets dying because of interference have therefore disappeared. Survivors after 10 cycles : monopodial > sympodial.

The snowflake is a delightful object. Its attraction lies in the symmetry of its occupation of space. All organised phenomena of nature demand similar contemplation, and the problems of conception that they present are universal. Examples range from the sublime (the flake) to the unlikely: the comparison of harvesting patterns on crinoid feeding arms with roadway layouts in banana plantations (Cowen 1981); from the packing of pomegrante pips (Kepler 1611) to the symmetry of crystals. As Mackay (1975), in an essay pleading for crystallography to become the basis for a general science of regular structures, put it: "Our problem is the eternal one of the relationship between form and content in a particular field." This essay has worried itself with the insignificant minutiae of abstract pattern astogeny, but nevertheless these concepts of form are just as applicable to plants.

ACKNOWLEDGMENTS

Thanks to Dr. Brian Mayoh, Department of Computer Science, Århus, Denmark, for providing a translation (anonymous) of "De Nive Sexangula."

REFERENCES

Arber, A. (1950). *The Natural Philosophy of Plant Form.* Cambridge University Press, Cambridge.

Bell, A. D. (1984). Dynamic morphology: A contribution to plant population ecology. *Perspectives on Plant Population Ecology* (Ed. by R. Dirzo & J. Sarukhán), pp. 48–65. Sinauer Associates, Sunderland, Massachusetts.

Bell, A. D., Roberts, D. & Smith, A. (1979). Branching patterns: The simulation of plant architecture. *Journal of Theoretical Biology,* **81,** 351–375.

Corner, E. J. H. (1946). Suggestions for botanical progress. *New Phytologist*, **45**, 185–192.

Cowen, R. (1981). Crinoid arms and banana plantations: An economic harvesting analogy. *Paleobiology*, **7**, 332–343.

Cusset, G. (1982). The conceptual bases of plant morphology. *Axioms and Principles of Plant Construction* (Ed. by R. Sattler), pp. 8–86. Martinus Nijhoff/Dr. W. Junk, The Hague.

von Goethe, J. W. (1790). *Versuch Die Metamorphose Der Pflanzen Zu Erklären*. Gotha.

Greathouse, D. C. & Laetsch, W. M. (1969). Structure and development of the dimorphic branch system of *Theobroma cacoa*. *American Journal of Botany*, **56**, 1143–1151.

Hallé, F. & Oldeman, R. A. A. (1970). *Essai sur l'architecture et la dynamique de croissance des arbres tropicaux*. Collection de Monographie de Botanique et de Biolgie Végétale No. 6. Masson, Paris.

Hamshaw-Thomas, H. (1932). The old morphology and the new. *Proceedings of the Linnean Society of London*, Session 145, 1932–1933, Pt. 1, pp. 17–32.

Harper, J. L. (1978). The demography of plants with clonal growth. Structure and functioning of plant populations. *Verhandelingen der Koninklijke Nederlandse Akademie van Wetenschappen, Afdeling Natuurkunde, Reeks 2*, **70**, 27–48.

Harper, J. L. (1980). Plant demography and ecological theory. *Oikos*, **35**, 244–253.

Harper, J. L. (1981). The concept of population in modular organisms. *Theoretical Ecology: Principles and Applications* (Ed. by R. M. May), 2nd ed., pp. 57–77. Blackwell Scientific Publications, Oxford.

Harper, J. L. (1985). Modules, branches and the capture of resources. *Population Biology and Evolution of Clonal Organisms*. (Ed. by J. B. C. Jackson, L. W. Buss & R. E. Cook), Yale University Press, New Haven, Connecticut (in press).

Harper, J. L. & Bell, A. D. (1979). The population dynamics of growth form in organisms with modular construction. *Population Dynamics* (Ed. by R. M. Anderson, B. D. Turner & L. R. Taylor), *Symposium of the British Ecological Society*, **20**, pp. 29–52. Blackwell Scientific Publications, Oxford.

Holmes, A. (1979). *Henderson's Dictionary of Biological Terms*. 9th ed. Longman, London and New York.

Kepler, J. (1611). *Strena seu de nive sexangula*. Francofurti ad Molnvm, apud Godefridum Tampach.

Kepler, J. (1619). *Harmonices Mundi Libri V*. Lincii Austriae, Sumptibvs Godofredi Tampachii Bibl. Francof. excudebat Joannes Planevs.

Lang, W. H. (1916). Phyletic and causal morphology. *Report of the 85th Meeting of the British Association for the Advancement of Science*, Manchester, 1915, pp. 701–718.

Mackay, A. L. (1975). Generalized crystallography. *Izvjestaj Jugoslavenskog Centra za Kristalografiju (Zagreb)*, **10**, 15–36.

Sattler, R. (1966). Towards a more adequate approach to comparative morphology. *Phytomorphology*, **16**, 417–429.

Schneer, C. (1960). Kepler's new year's gift of a snowflake. *Isis*, **51**, 531–545.

Tomlinson, P. B. (1983). Tree architecture. (New approaches help to define the elusive biological property of tree form). *American Scientist*, **71**, 141–149.

Ulam, S. (1966). Patterns of growth of figures: Mathematical aspects. *Module, Symmetry, Rhythm* (Ed. by G. Kepes), pp. 64–74. Braziller, New York.

Webster, N. (1971). *Webster's Third New International Dictionary*. G. & C. Merriam Co.

White, J. (1979). The plant as a metapopulation. *Annual Review of Ecology and Systematics*, **10**, 109–145.

White, J. (1984). Plant metamerism. *Perspectives on Plant Population Ecology* (Ed. by R. Dirzo & J. Sarukhán), pp. 15–47. Sinauer Associates, Sunderland, Massachusetts.

Williams, D. S. (1984). *The vascular system of* Trifolium repens *L. (white clover)*. Unpublished Honours project, School of Plant Biology, University College of North Wales, Bangor, U.K.

14

The Importance of Plant Form as a Determining Factor in Competition and Habitat Exploitation

Peter H. Lovell and Patricia J. Lovell

Department of Botany
University of Auckland
Auckland, New Zealand

I. THE PLANT IN ITS ENVIRONMENT

'Plant form' is basically the three-dimensional shape of the total plant and the volume of space (air, soil or water) enclosed within its boundaries. Holland (1969) considers that plant shape is given by the geometrical distribution of its peripheral points and that plant structure is the packing of tissues into the space bounded by these points. However, it is important to consider form not only from this static viewpoint, but also from the reverse, dynamic view of function (Arber 1950). Sinnott (1960) defined form as 'the visible expression of a self-regulatory equilibrium which tends to be attained in development, maintained during life, and restored when disturbed'. Niklas and O'Rourke (1982) describe form as an interaction between the physiological requirements of growth and the inherent constraints on its mechanical design. Form is thus an organized, structural, physiological, dynamic phenomenon which is a function of space, time and environmental constraints.

The major determinants of form are the number, point of origin and length of the different orders of shoots and roots. Other features affecting it are the angles of branching; the numbers, dimensions and insertion methods of leaves; and the presence of modified shoot systems such as runners. Structural constraints may also be important; e.g. an orthotropic shoot may lack the mechanical strength to maintain its vertical growth form and bend downwards under the influence of gravity, thus having a horizontal orientation imposed upon it.

STUDIES ON PLANT DEMOGRAPHY:
A FESTSCHRIFT FOR JOHN L. HARPER

Each different type of environment exerts its own particular set of constraints on plant form. A majority of plants have one major part of their form occupying a very different type of environment from another component. The shoot system most commonly occupies an aerial environment, where sunlight is a renewed resource, and other factors such as CO_2 and O_2 are normally not limiting. Further, this environment does not exert a significant compressional effect on the plant. In contrast, some of the resources of the terrestrial part of the environment, such as minerals, are potentially much less renewable, though this is not so in all circumstances. Water may be limiting, as may oxygen. Soil forms a barrier to penetration and also exerts considerable compressional effects on plant parts. In many ways, the below-ground environment can be more limiting than the aerial one. Aquatic environments impose yet different constraints on plant form. In still water, nutrients, O_2 and CO_2 are often likely to become depleted or waste products to build up. In moving or running water the main constraint will be mechanical, with physical damage or a whole plant or part of a plant being detached (a feature exploited by certain aquatic plants). If the photosynthetic parts are submerged, light is potentially limiting.

The form of a plant at any particular point in time can be regarded as the sum of the outcome of genetic factors being expressed and their modification by environmental interactions.

The density of both shoot and root systems relates to the degree of control that the plant is exerting at present and has exerted in the past within the space it occupies, in competition with other species with aspects of growth habit in common. The 'desirability' of a space is related to its capacity to supply the requirements of the plant. It may change for any given space with time; e.g. both light quantity and quality will change if that space becomes covered by a canopy. Thus success in competition cannot be guaranteed merely by occupying a space and preventing other plants from invading it, because that space could be made less favourable as a result of other plants occupying spaces vertically above. Space above ground may be considered as a series of hierarchical strata, with occupation of a higher one bestowing a competitive advantage over plants achieving only a lower one. The ability to grow swiftly from one stratum to the next can therefore convey important strategical advantages under certain circumstances. Harper (1977) makes the very pertinent point that being high conveys no advantage in itself (more often, it is a disadvantage as it involves costs in terms of maintenance and supporting tissues); it is being higher than competing species that counts.

Alternatively, the ability to grow in a prostrate form may be very advantageous under certain circumstances. For example, many weeds of lawns and other regularly mown environments have a low, very flattened shoot form. Plants of exposed habitats subjected to frequent high winds or spray may also be favoured by having a low profile.

The situation is not merely one of plant–plant interaction or environmental action on the plant. The presence of the plant modifies the environment in both short- and long-term ways, not only by occupying it but by changing it physically and chemically. The effects often last after the death of the plant, for example, by selective removal of nutrients, by shedding parts onto and into the soil, by mechanical breaking up of the substrate, by allelopathy, by heterogeneous exploitation of resources and by many other ways. Thus the form of the current plants in an environment is in part a reflection of the legacy left by previous occupants of that space.

II. DIFFERENT FORMS

A. Cost

Different growth forms require different energy expenditures. An indeterminant orthotropic shoot system requires a substantial amount of secondary thickening if it is to remain erect. In the absence of much or any secondary tissue, it will fall over and become a prostrate 'pseudorhizomatous' shoot (Niklas & O'Rourke 1982). Even in woody plants, prostrate shoots may be more vigorous than erect shoots on the same shrub, e.g. *Arctostaphylos uva-ursi,* possibly because of the lower cost per unit length (Remphrey, Steeves & Neal 1983). Pickett and Kempf (1980) have commented that the clonal form of some shrubs may permit exploitation of a broad horizontal area relatively cheaply. Cost does not only relate to structural tissues. Orians and Solbrig (1977) point out that the costs of both production and maintenance have to be considered. Thus, although roots may cost less to produce per unit length than shoots, as they have less supportive tissue, they are still expensive because all of the carbohydrate must travel from the shoot system, and in some situations (e.g. desert plants) it may be more efficient to discard leaves or roots at various times of the year than to pay the maintenance costs (Orians & Solbrig 1977).

B. Flexibility

An individual plant is sometimes considered as a population of units, metamers or modules, their sum comprising a metapopulation (reviewed by White 1979, 1984). Existing form will be maintained by a proportional increase in the size of all existing modules, or alternatively by the formation of further metamers in the same proportions as the existing ones. If proportions are altered, or a new type of metamer is formed, then the plant undergoes a change in form, as, for example, when vegetative plants start flowering.

The genome of the plant determines the potential range of form possibilities and defines their limits. The plant has a range of detector systems for perceiving internal and environmental cues. The presence or absence of stimuli results in action or no-action responses. Flexibility is introduced by the existence of alternative responses and by interactions between form components. Responses in one part of the plant will affect growth and form of metamers elsewhere in the plant, and vice versa.

Among the alternatives available is whether to branch or not. This may be controlled by dominance from the main shoot apices and is probably affected by environmental factors (light, nutrients, etc.). Shoots can assume different forms (e.g. orthotropic and plagiotropic) and growth may be determinate or indeterminate. Shoot systems often include a mixture of these types, and in some cases, there can be qualitative change from one to another; e.g. plagiotropic rhizomes can turn upwards if the above-ground shoot system is damaged (Beasley 1970). Pfirsch and Makosso (1980) found that the node was the controlling factor for horizontal growth of stolons of *Stachys silvatica.* Isolated buds, grown *in vitro,* were always orthotropic. Diageotropic and plagiotropic shoot systems are not always readily modified, although temporary changes may be effected. Some cultivars of *Arachis hypogaea* have diageotropic side branches in addition to erect branches. The diageotropic branches become orthotropic if the plant is kept in the dark but resume a diageotropic habit when the plant is returned to the light (Zir, Halevy & Ashri 1973). If cuttings are taken from erect shoots, they produce plants with both erect and prostrate shoots, but cuttings from diageotropic shoots produce only diageotropic systems (Ashri & Goldin 1964).

Considerable flexibility of form may be achieved through differential extension of single metamers or groups of metamers. This may occur in cotyledons, hypocotyls, leaves, internodes, rhizomes, stolons, or a combination of these. Certain plants respond to a sharp vertical gradient of light. Grime and Jeffrey (1965) found that *Plantago lanceolata* showed a marked change in the shape and size of the first two leaves, in shaded conditions, producing a linear-lanceolate shape twice the length of the ovate–lanceolate leaves formed in full daylight. Hypocotyl, cotyledons and leaf three were longer than in the full daylight plants. Other species associated with dense grassland also showed increases in initial height growth when shaded, whereas species associated with bare ground or low turf did not show this ability and so failed to reach adequate light. This plasticity of form of the dense grassland species clearly confers a competitive advantage on these plants, which, on germination, may or may not be heavily shaded by plants in the next vertical stratum.

In other cases, the species possesses more than one possible response to a particular environment, but a single individual or race may be capable of only one alternative. Groups of such individuals may, for example, be prostrate in form, whereas other races may be more erect. *Poa annua* in populations on lawns consisted almost entirely of prostrate individuals, whereas those associated with flower beds were much more heterogeneous (Warwick & Briggs 1978). *Plantago major* has erect roadside races and prostrate races in lawns (Warwick & Briggs 1980). Under standard growing conditions which included a cutting regime at 20 mm above ground level, the populations from the lawns produced several thousand seeds per plant, whereas those from the roadside populations were severely damaged and lost all reproductive structures. The prostrate race was at a great advantage in lawn habitats in comparison with the erect type. However, the reverse applies in the roadsides, where over-shading occurs. Thus the species has been able to exploit two different environments as a result of having two different genotypes (Warwick & Briggs 1980).

In some cases, elongation of parts and change in form may occur only after a certain amount of time has elapsed, or a particular state, such as a minimum size, has been attained. In *Dipsacus,* the rosette forms a flowering stalk only after reaching a critical size, developmental stage being more important than plant age (Werner & Caswell 1977). Similarly, in *Viola blanda* (Thompson & Beattie 1981) a characteristic minimum weight is required before either stolons or sexual reproductive structures are produced. For a perennial plant (or a biennial which might become perennial if the vegetative stage is prolonged) an unsuccessful attempt to produce seed is far worse than not making the attempt, as resources have been wasted on the effort. Therefore, it is probably very important to have some mechanism for assessing the chances of success and initiating flowering only when the chances are good. For an annual like *Soliva,* success is determined entirely by whether seeds are produced or not, so the attempt must always be worthwhile. Thus whereas in a perennial, one might expect factors such as size to affect the decision whether to become reproductive or not and for this information to override environmental cues, these would not be expected to be overridden in an annual.

III. THE PROSTRATE HABIT

A. Advantages and Disadvantages

A prostrate habit permits maximum ground cover at minimum cost as the requirement for supporting tissue is reduced. The energy input into secondary thickening of a prostrate shrub will be far less than for a tree supporting an equivalent amount of foliage (Pickett & Kempf 1980). Wind speeds are less, close to the ground, and so transpiration losses and possible foliage damage will be reduced in low-growing individuals. The prostrate habit (plagiotropic shoots, pseudorhizomes, stolons, runners, rhizomes) encourages vegetative spread because these organs are prolific producers of adventitious roots (Lovell & White 1986), essential processes for the establishment of independent ramets. There is an inverse relationship between ramet number and the likelihood of death of the genet as long as the death of an individual ramet does not affect the entire network, if the ramets are still interconnected (Cook 1979). However, a prostrate habit means that the plant normally carries the risk of being overtopped and shaded.

B. The Effect of Internode Length and Vigour of Development at Nodes

Harper (1977 p. 774) expressed this theme very succinctly when he observed that "The form of an individual genet determines the way in which it meets neighbours". He contrasts the 'guerilla' strategy (plants with long internodes and wandering form) with those of a 'phalanx' strategy (with compact structures). Thus, clonal plants exhibit a range of strategies from very conservative with short internodes and overlapping rosettes as in *Soliva valdiviana* (see Section III,C) to highly exploratory e.g. *Ranunculus repens* (Ginzo & Lovell 1973a, b; Lovett Doust 1981b). It seems uncertain how, or whether, clonal plants assess or control density (Pitelka 1984). Hutchings and Barkham (1976) suggest that ramets need to be sufficiently close together to resist invasion but not so close as to inhibit the growth of individuals. Pitelka (1984) observes that this may apply to individuals operating a conservative strategy but not to those with a "guerilla" approach, in which individuals may be quite widely spaced.

C. Conservative Strategies

A high density of shoots, stems or leaves entails a substantial input of material per unit volume of space occupied. The consequences may be lack of resources for extension of the plant system, and an increased degree of self-shading but greater resistance to competition for occupied space from other species.

In a clonal plant there may be multiple root systems exploring a small volume of soil. *Soliva valdiviana* and *S. pterosperma* are annual rosette plants which produce short stolons after the formation of a central, terminal capitulum. Further rosettes, capitula and stolons may be produced by the first-order stolons. Adventitious roots may develop but they are not very vigorous. The rosettes overlap and the ramets and parent form a tight clone, giving rise to a dense, circular patch in a uniform environment (Fig. 1). They commonly occur in lawns, and under a mowing regime in competition with other species their growth form is much less uniform. Individual branch systems may have 10 times as many rosettes as others on the same plant. In summer the patches of *Soliva*

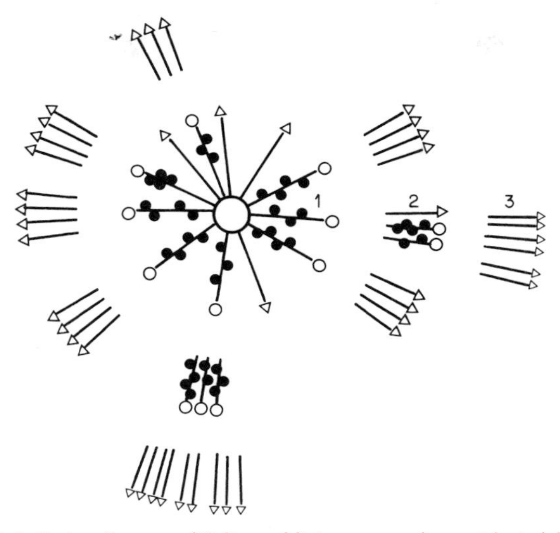

Fig. 1. An 'exploded' plan diagram of *Soliva valdiviana* grown in a nutrient-rich environment in the absence of competition. Large circle, central capitulum; small circles, lateral capitula; closed circles, point of origin of the next order laterals; small triangles, vegetative apices. Three orders of laterals are present, numbered 1, 2 and 3. Not to scale.

die, and these areas tend to remain uncolonized during this relatively dry period. The cycle begins again with the first rains in autumn (Johnson & Lovell 1980). This is essentially a replacement strategy, that is, the maintenance of a safe site for the next season, but dispersal may be achieved by carriage of the sharp-pointed cypselas away from the site by animals.

Veronica filiformis, another lawn weed, is similar to *Soliva* in that it may have a circular outline in the absence of competition. It is a perennial, however, and has longer internodes and a rather greater rate of spread during the growing season than *Soliva*. Its growth form is largely prostrate but the young part of the shoot is orthotropic and has a high chance of severance during mowing. The shoot also tends to fragment, and because it roots vigorously at the nodes the separated pieces can establish rapidly. These may merge with the parent if they are deposited close to it (Harris & Lovell 1980a, b). Once again, when the environment is more heterogeneous the growth form become more variable. If the substrate is too dry, rooting will not occur and branching is inhibited. The ease of fragmentation and the vigour of adventitious rooting enable *V. filiformis* to flourish in lawns by vegetative spread (Thaler 1951; Harris & Lovell 1980b); sexual reproduction does not occur in *V. filiformis* in Britain (Bangerter & Kent 1957, 1962).

Species such as *Solanum tuberosum* and *Oxalis latifolia* are 'delayed conservatives'. The new plants of *Solanum* are displaced from the parent on underground stolons, but their growth is delayed for several months whilst they overwinter as tubers. The initial growth of the new plant will be dependent on the reserves laid down in the tuber, which will have been determined almost entirely by the conditions experienced by the parent plant rather than those at the site of the daughter tubers, except perhaps for soil compaction, which could impede tuber bulking. The new plants will be affected by the microenvironment around the tuber once growth resumes.

Oxalis latifolia produces two orders of bulbs during a growing season in Auckland. Many of the primary bulbs develop close to the parent bulb and produce leaves and secondary bulbs whilst still attached to the parent. Secondary bulbs remain dormant until the next year (Beath 1981). This conservative strategy results in a modest extension of the site occupied by the parent during the first season with a dense production of petioles and leaves and a substantial residue of bulbs occupying a somewhat larger area for subsequent years.

D. Exploratory, 'Guerilla' Strategies

Prostrate plants with relatively long internodes and rapid production of nodes have the potential to explore territory farther from the parent either by lateral runners, underground rhizomes or by both above- and below-ground rhizomes (as in *Pennisetum clandestinum*). Initially the growth of a lateral is dependent entirely on the already established plant for its resources. If it remains below ground (as a subterranean rhizome) it will always be dependent on other parts for a carbon supply, but root production at nodes may enable a rhizome to respond to local conditions, as shown for *Agropyron repens* (McIntyre 1976). The pattern of growth may be greatly affected by environmental conditions. First-order stolons of *Ranunculus repens* achieved the same total length in low- and high-nutrient conditions, but branching was severely limited at low-nutrient levels (Ginzo & Lovell 1973a) Thus, the strategy for maximum distance from the parent is maintained with minimum energy expenditure on branching. This may be a favourable response when the parent is in a nutrient-poor environment.

In a homogeneous environment the best strategy may be to place ramets at a sufficient distance from the parent and from each other that intra-genet competition is minimized. This situation could be achieved by organized, predictable patterns of branching and growth giving rise to evenly dispersed ramets (Bell & Tomlinson 1980). To some extent this does occur in aquatic environments. Ramets of the water fern *Salvinia* tend to be of uniform size and arranged in regular patterns (Room 1983). However, in terrestrial situations it is unlikely that the environment will be uniform; it will vary with respect to light, nutrients, water relations, degree of competition and obstacles such as stones. Schellner, Newell and Solbrig (1982), working with 16 populations of *Viola blanda, V. pallens* and *V. incognito,* found that ramets that were 1 year old or more were

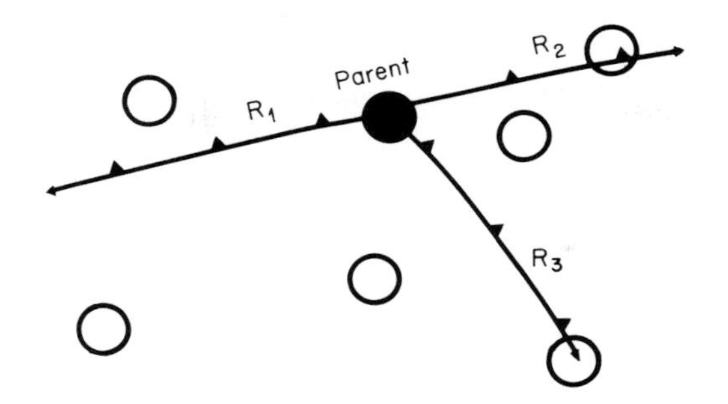

Fig. 2. Spread through a heterogeneous environment by a clonal species which has a 'guerilla' strategy. Closed circle, parent; triangles, nodes: open circles, favourable micro-habitats in an area which is generally less favourable. Runners, R_1, R_2 and R_3.

either aggregated or randomly distributed in space. Favourable sites in the forest floor environment where these species occur are unevenly distributed, on a small scale. They suggest that random scattering of stolons in space and time could be a good strategy for placing new ramets in favourable microsites. This raises the interesting question of the way that species with an exploratory strategy place ramets in favourable microsites in a patchy environment. Ramets develop as a result of adventitious root and shoot formation at some or all nodes. In a patchy environment (Fig. 2) some rhizomes (R_1) may, by chance, not encounter any favourable microsites. Others (R_2) may pass through a favourable site, having located a node in the site, giving the potential for ramet development there. If the mean length of the favourable microsites is of the same order or smaller than the mean length of internode the rhizome could pass across a microsite and fail to utilize it. The boundaries between favourable and unfavourable sites may not be abrupt. A feature of critical importance is whether the apex of a stolon or rhizome can 'recognise' a favourable site, and, if it can, whether the newest internode length can be modified so that a node is located on the site. The problem of recognition will differ, depending on whether the stolon is above or below ground and also whether the favourable characters relate to the substrate (water, nutrient) or to light (high-irradiance patches). If the node is placed adjacent to the microsite a ramet may be able to grow into it.

If a node is located on a favourable microsite then it is important that the ramet be able to respond and not be conditioned in its development by the parent. There are interactive effects within clones. Lovett Doust (1981a) found that when stolon connections were severed in clones of *Ranunculus repens,* biomass allocations were altered, and clones sometimes grew less well than clones with intact stolons. Similarly, Hartnett and Bazzaz (1983) found that ramets of *Solidago canadensis* showed reduction in growth, survivorship and flowering when stolon connections were severed experimentally. This was particularly significant for

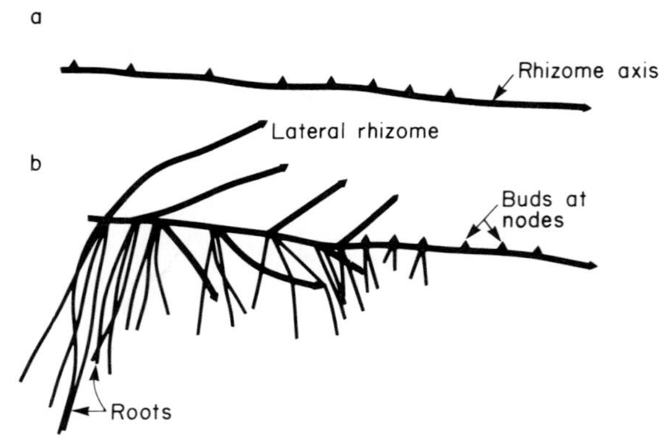

Fig. 3. Rhizomes of *Agropyron repens*. a, Low availability of water. b, High availability of water. [Adapted from McIntyre (1976).]

young ramets. Shaded ramets were supported initially by non-shaded ramets to which they were connected, but this support was probably temporary. A similar, temporary supply of assimilates to shaded tillers of *Lolium multiflorum* has also been observed (Ong & Marshall 1979). Newell (1982) found that photoassimilates were translocated within clones of *Viola blanda*. However, despite physiological interactions there is also evidence that individual ramets can respond to localized environmental conditions. Schellner, Newell and Solbrig (1982) found large differences in size between ramets of *Viola* and attributed these to microenvironmental differences. McIntyre (1976) emphasizes the importance of water availability and shows that in *Agropyron repens* the degree of inhibition of buds on the rhizome is inversely related to internal water content of the segment of rhizome bearing the bud. When the supply of water to individual rhizomes was increased vigorous root production occurred at the nodes and was associated with loss of inhibition of the bud (Fig. 3). Under low-nutrient conditions a bud on a rhizome of *Agropyron* is more likely to develop as a rhizome than as an erect shoot (McIntyre 1972). After root formation has taken place a locally higher level of nitrate or phosphate will induce more vigorous branching in the part of the root system that comes into contact with it (Drew & Saker 1978). This will not only increase the surface area available for absorption but will also give rise to higher numbers of root tips, which are regions of cytokinin production. There is increasing evidence for functional relationships between root systems and shoot systems (Chung, Rowe & Field 1982), and this balance appears to be controlled at least in part by cytokinin movement from root tips to the shoot system. Richards and Rowe (1977a, b) have shown a relationship between the number of root tips and the number of leaves and have found that cytokinins can substitute for parts of the root system in root-pruned plants. This effect of cytokinins on branching and leaf number may not be direct but may work via other growth regulators (Tucker 1981).

Even when physiological connections are still intact, evidence from translocation studies (Ginzo & Lovell 1973b; Schellner, Newell & Solbrig 1982) suggests that the effect of a vigorous root system of a ramet is most likely to be felt either by the shoot system of that ramet or possibly by young material on the stolon, rather than older parts, i.e. to increase the vigour of the ramet at that site. This could lead to increased branching, thus generating more ramets close to the vigorous individual (the aggregations noted by Schellner, Newell & Solbrig 1982). The important role of new root systems in increased vigour has also been noted for *Ammophila breviligulata,* which flourishes on shifting dune systems (Disraeli 1984).

When a clone produces ramets they can respond to irradiance or to the environment around their roots. The more heterogenous the environment the greater the variation in vigour and size that might be expected, even allowing for physiological interactions within the clone. Substantial variation in ramet size occurs in terrestrial plants, e.g. *Viola* (Schellner, Newell & Solbrig 1982). There is also variation in vigour of branch systems in non-clonal plants. In *Muehlenbeckia* the variation does not fit easily into any clearly defined dominance pattern, and it is harder to attribute localized environmental effects as causal factors. The onset of a new branch system occurs when the above-ground parts have been dormant during winter, but it is not clear why a new shoot should develop, instead of buds from the already existing systems. Possibly the new shoot is favoured by its proximity to the root system, or perhaps it could be affected by a specific part of the root system. Evidence for links between specific parts of shoot and root systems has been shown for some species (White 1979).

The differential vigour of ramets can be retained after severance from the parent. Even uniform ramets do not perform identically. Hunt (1984) grew, under standard conditions, a series of cohorts of three-leaved ramets of *Carex flacca* cloned from a single genet taken from a grassland in north Derbyshire. He found that the range of variation in mean relative growth rate was of the same order as that for 'between-genet' variation for *Trifolium repens* (Burdon & Harper 1980). Hunt suggests that variation in vigour between populations, between genets and between ramets may be of the same order despite the differences in the degree of genetic uniformity. Further comparative studies on differential vigour of ramets and genets under a range of conditions will help us to understand the extent of the retention of 'individuality'.

REFERENCES

Arber, A. (1950). *The Natural Philosophy of Plant Form.* Cambridge University Press, Cambridge.

Ashri, A. & Goldin, E. (1964). Vegetative propagation in peanut breeding. *Crop Science,* **4,** 110–111.

Bangerter, E. B. & Kent, D. H. (1957). *Veronica filiformis* Sm. in the British Isles. *Proceedings of the Botanical Society of the British Isles,* **2,** 197–217.

Bangerter, E. B. & Kent, D. H. (1962). Further notes on *Veronica filiformis. Proceedings of the Botanical Society of the British Isles,* **4,** 384–398.

Beasley, C. A. (1970). Development of axillary buds from Johnsongrass rhizomes. *Weed Science,* **18,** 218–222.

Beath, G. C. (1981). *The growth and reproductive strategy of* Oxalis latifolia *H. B. K.* M.Sc. thesis, University of Auckland.

Bell, A. D. & Tomlinson, P. B. (1980). Adaptive architecture in rhizomatous plants. *Botanical Journal of the Linnean Society,* **80,** 125–160.

Burdon, J. J. & Harper, J. L. (1980). Relative growth rates of individual members of a plant population. *Journal of Ecology,* **68,** 953–957.

Chung, G. C., Rowe, R. N. & Field, R. J. (1982). Relationship between shoot and roots of cucumber plants under nutritional stress. *Annals of Botany (London),* **50,** 859–861.

Cook, R. E. (1979). Asexual reproduction: A further consideration. *American Naturalist,* **113,** 769–772.

Disraeli, D. J. (1984). The effect of sand deposits on the growth and morphology of *Ammophila breviligulata. Journal of Ecology,* **72,** 145–154.

Drew, M. C. & Saker, L. R. (1978). Nutrient supply and the growth of the seminal root system in barley. III. Compensatory increases in growth of lateral roots, and in rates of phosphate uptake, in response to a localized supply of phosphate. Journal of Experimental Botany, **29,** 435–451.

Ginzo, H. D. & Lovell, P. H. (1973a). Aspects of the comparative physiology of *Ranunculus bulbosus* L. and *Ranunculus repens* L. I. Response to nitrogen. *Annals of Botany (London),* **37,** 753–764.

Ginzo, H. D. & Lovell, P. H. (1973b). Aspects of the comparative physiology of *Ranunculus bulbosus* L. and *Ranunculus repens* L. *II. Carbon dioxide assimilation and distribution of photosynthates. Annals of Botany (London),* **37,** 765–776.

Grime, J. P. & Jeffrey, D. W. (1965). Seedling establishment in vertical gradients of sunlight. *Journal of Ecology,* **53,** 621–642.

Harper, J. L. (1977). *Population Biology of Plants.* Academic Press, London.

Harris, G. R. & Lovell, P. H. (1980a). Adventitious root formation in *Veronica* spp. *Annals of Botany (London),* **45,** 459–468.

Harris, G. R. & Lovell, P. H. (1980b). Localized spread of *Veronica filiformis, V. agrestis* and *V. persica. Journal of Applied Ecology,* **17,** 815–826.

Hartnett, D. C. & Bazzaz, F. A. (1983). Physiological integration among intraclonal ramets in *Solidago canadensis. Ecology,* **64,** 779–788.

Holland, P. G. (1969). The maintenance of structure and shape in three mallee eucalypts. *New Phytologist,* **68,** 411–421.

Hunt, R. (1984). Relative growth rates of cohorts of ramets cloned from a single genet. *Journal of Ecology,* **72,** 299–305.

Hutchings, M. J. & Barkham, J. P. (1976). An investigation of shoot interactions in *Mercurialis perennis* L., a rhizomatous herb. *Journal of Ecology,* **64,** 723–743.

Johnson, C. D. & Lovell, P. H. (1980). Germination, establishment and spread of *Soliva valdiviana* (Compositae). *New Zealand Journal of Botany,* **18,** 487–493.

Lovell, P. H. & White, J. (1986). Anatomy of adventitous rooting. *New Root Formation in Plants and Cuttings* (Ed. by M. B. Jackson). Martinus Nijhoff, The Hague (in press).

Lovett Doust, L. (1981a). Intraclonal variation and competition in *Ranunculus repens. New Phytologist,* **89,** 495–502.

Lovett Doust, L. (1981b). Population dynamics and local specialisation in a clonal perennial (*Ranunculus repens*). I. The dynamics of ramets in contrasting habitats. *Journal of Ecology,* **69,** 743–755.

McIntyre, G. I. (1972). Studies on bud development in the rhizome of *Agropyron repens.* II. The effect of the nitrogen supply. *Canadian Journal of Botany,* **50,** 393–401.

McIntyre, G. I. (1976). Apical dominance in the rhizome of *Agropyron repens:* The influence of water stress on bud activity. *Canadian Journal of Botany,* **54,** 2747–2754.

Newell, S. J. (1982). Translocation of ^{14}C-assimilate in two stoloniferous *Viola* species. *Bulletin of the Torrey Botanical Club,* **109**, 306–317.

Niklas, K. J. & O'Rourke, T. D. (1982). Growth patterns of plants that maximize vertical growth and minimize internal stresses. *American Journal of Botany,* **69,** 1367–1374.

Ong, C. K. & Marshall, C. (1979). The growth and survival of severely-shaded tillers in *Lolium perenne* L. *Annals of Botany (London),* **43,** 147–155.

Orians, G. H. & Solbrig, O. T. (1977). A cost-income model of leaves and roots with special reference to arid and semi-arid areas. *American Naturalist,* **111,** 677–690.

Pfirsch, E. & Makosso, T. (1980). Effet des noeuds sur la direction de croissance des bourgeons axillaires des stolons de *Stachys silvatica. Canadian Journal of Botany,* **58,** 466–470.

Pickett, S. T. A. & Kempf, J. S. (1980). Branching patterns in forest shrub and understory trees in relation to habitat. *New Phytologist,* **86,** 219–228.

Pitelka, L. F. (1984). Application of the $-3/2$ power law to clonal herbs. *American Naturalist,* **123,** 442–449.

Remphrey, W. R., Steeves, T. A. & Neal, B. R. (1983). The morphology and growth of *Arctostaphylos uva-ursi* (bearberry): An architectural analysis. *Canadian Journal of Botany,* **61,** 2430–2450.

Richards, D. & Rowe, R. N. (1977a). Effects of root restriction, root pruning and 6-benzylaminopurine on the growth of peach seedlings. *Annals of Botany (London)* **41,** 729–740.

Richards, D. & Rowe, R. N. (1977b). Root-shoot interactions in peach: the function of the root. *Annals of Botany (London),* **41,** 1211–1216.

Room, P. M. (1983). 'Falling apart' as a life style: The rhizome architecture and population growth of *Salvinia molesta. Journal of Ecology,* **71,** 349–365.

Schellner, R. A., Newell, S. J. & Solbrig, O. T. (1982). Studies on the population biology of the genus *Viola.* IV. Spatial pattern of ramets and seedlings in three stoloniferous species. *Journal of Ecology,* **70,** 273–290.

Sinnott, E. W. (1960). *Plant Morphogenesis.* McGraw-Hill, New York.

Thaler, I. (1951). Morphologisches uber *Veronica filiformis* Smith und ihre Verwandten. *Phyton,* **3,** 216–226.

Thompson, D. A. & Beattie, A. J. (1981). Density-mediated seed and stolon production in *Viola* (Violaceae). *American Journal of Botany,* **68,** 383–388.

Tucker, D. J. (1981). Axillary bud formation in two isogenic lines of tomato showing different degrees of apical dominance. *Annals of Botany (London),* **48,** 837–843.

Warwick, S. I. & Briggs, D. (1978). The genecology of lawn weeds. I. Population differentiation in *Poa annua* L. in a mosaic environment of bowling green lawns and flower beds. *New Phytologist,* **81,** 711–723.

Warwick, S. I. & Briggs, D. (1980). The genecology of lawn weeds. V. The adaptive significance of different growth habit in lawn and roadside populations of *Plantago major* L. New Phytologist, **85,** 289–300.

Werner, P. A. & Caswell, H. (1977). Population growth rates and age versus stage-distribution models for Teasel (*Dipsacus sylvestris* Huds.). *Ecology,* **58,** 1103–1111.

White, J. (1979). The plant as a metapopulation. *Annual Review of Ecology and Systematics,* **10,** 109–145.

White, J. (1984). Plant metamerism. *Perspectives on Plant Population Ecology* (Ed. by R. Dirgo & J. Sarukhán), pp. 15–47. Sinauer Associates, Sunderland, Massachusetts.

Zir, M., Halevy, A. H. & Ashri, A. (1973). Phytohormones and light regulation of the growth habit in peanuts (*Arachis hypogaea* L.). *Plant and Cell Physiology,* **14,** 727–735.

15

Modular Demography and Form in Silver Birch

Michelle Jones

School of Plant Biology
University College of North Wales
Bangor, Gwynedd, Wales

I. INTRODUCTION

By comparison with plants, most animals have very fixed patterns of growth and development and their bodies consist of a limited number of parts and organs (often highly specialized), which are laid down in the embryo and have no subsequent ability to increase. A much more plastic mode of development is exhibited by plants. Growth proceeds by the continuous production of subunits which within a species are relatively constant in size and appearance (Harper & Bell 1979). These subunits (which are usually considered to comprise the leaf–node–internode–axillary meristem groupings on a shoot) have been termed metamers by White (1979, 1984). The work of Hallé and Oldeman (1979) and Hallé, Oldeman and Tomlinson (1978) focussed attention on the module. A *module* is defined as the product of a single apical meristem and thus differs from a metamer, but as White (1984) points out, *module* has been used more generally by plant ecologists to refer to any subunit of the plant body. The modular construction of plants means that, broadly speaking, whereas the size of an animal is a function of the *size* of its parts, the size of a plant is a function of the *number* of its parts: all rabbits have four legs, two eyes and one tail, but all oak trees do not have the same number of leaves or branches.

This crucial difference between modular and non-modular organisms has far-reaching influences on the ways in which we can interpret them demographically. Much demographic theory, originally developed for man and higher animals, needs modification before it can be applied to plants (Harper & White 1974; White 1979, 1980, 1984; Hickman 1979). For instance, the range in size of individual plants of the same chronological age within a species can cover two to three orders of magnitude. This phenomenon is frequently seen in dense stands (Koyama & Kira 1956; Obeid, Machin & Harper 1967; White & Harper 1970), in which pronounced size hierarchies develop with the passage of time. The variation in size of similarly-aged animals in a population is minute by comparison. Although a simple count of the number of animals in a population may be expected to give a reasonable amount of information on the size and activities of that population, a census of a plant population is much less informative because of the great range of variation in size (and hence biomass and reproductive output) of individuals.

A full appreciation of the demographic processes taking place in a population of plants therefore requires a 'hierarchical' approach (Harper & Bell 1979; Harper 1978, 1980; Hickman 1979). The population must be considered on at least two levels: that of the individual genotype, or *genet,* and that of the module. Other elements in the hierarchy of construction may also be important in particular studies (White 1979, 1984); Tomlinson (1982) has commented that the varieties of constructional elements in plants are still inadequately known.

The importance of this is illustrated experimentally in a study by Kays and Harper (1974). *Lolium perenne* was grown from seed, and the birth and death rates of whole plants (genets) and tillers (modules) were monitored. They found that the density of modules over time reached a point at which it became independent of the density of genets. This came about through total death of some genets (that is, all their modules died), while surviving genets had varying numbers of modules. Had this sward produced seed, the seed progeny would have carried a biased representation of genes from the parental population caused (a) by the elimination of some genets and (b) by the differential production of modules (and hence flowers and seed) by the surviving genets.

The fitness of a genet is therefore highly dependent on the demography of its modules; an understanding of the modular dynamics of an individual and the way they are affected by changes in the environment, such as the presence of neighbours, is the starting point for the understanding of the demography of genetically separate individuals (White 1980).

Having drawn attention to the relevance of processes taking place at the modular level, we can go on to examine two interesting implications resulting from the application of modular concepts to plant ecology. The first is that growth, which proceeds by the accumulation of modules, can be considered as a demographic process (Harper 1980). This approach has been used by Kobayashi (1975), who modelled the growth of sunflower stands as populations of internodal segments; Maillette (1982a, b), who studied the dynamics of the bud populations of trees; Porter (1983a, b), who analysed the modular dynamics of fuchsias; and Bazzaz and Harper (1977), who used leaf demography to describe the growth of plants of *Linum usitatissimum*. The second, and perhaps more neglected implication of using modular demography relates to form. As growth proceeds and is characterized by the birth and death of modules, the locations of these births and deaths will determine the form and shape of the whole plant (Harper & Bell 1979). The form of a tree is the outcome of the demography of its modules. These may in turn be influenced by the environmental conditions to which the individual branches are subjected. This highlights another of the differences between animals and plants. The mobility of animals enables them to react quickly to deterioration in their environment: they can simply move away. In plants this option is much more restricted or does not exist at all. The only forms of action open to plants are either growth or the discarding of parts, both of which involve changes in the size and shape of the organism (Arber 1950; Harper 1980). Parts of the same genet may therefore develop in different ways if they grow into different micro-environments. The work reported here is an attempt to investigate this possibility.

II. THE DEMOGRAPHY OF FORM IN SILVER BIRCH

An experiment was set up in spring 1980, when 4-year-old silver birch trees (*Betula pendula* Roth.) were planted out in an area of parkland near to the School of Plant Biology, University College of North Wales. At this site there were nine trees, arranged in three groups of three (Fig. 1a). The crowns of the trees were symmetrical at this stage and within a group the trees were planted in a triangular configuration, so that each tree had some of its branches meeting those of its neighbours, while others, on the opposite side of the trunk, could grow out into open space. Distances between the bases of the trees in a group were about 20 cm, which ensured that branches from neighbouring trees were intermingling from the start of the experiment. Distances between the groups were 3–4 m. Trees were chosen as units of study because it was thought that changes in their modular dynamics would have easily discernible and persistent effects on form.

An initial census of the number of buds (condensed modules) and their positions on the branches was made at the start of the experiment. Subsequent recordings were made at the end of each growing season, in autumn 1980, 1981 and 1982. A plan of each tree at the previous recording occasion provided the basis for each new record made. The states of all the buds previously present was noted, together with the positions of any new buds produced. These raw data were used to derive the following basic variables for each primary branch (those growing directly from the trunk): (a) total number of buds (living and dead), (b) number of living buds, (c) number of secondary branches, (d) number of tertiary and higher-order branches.

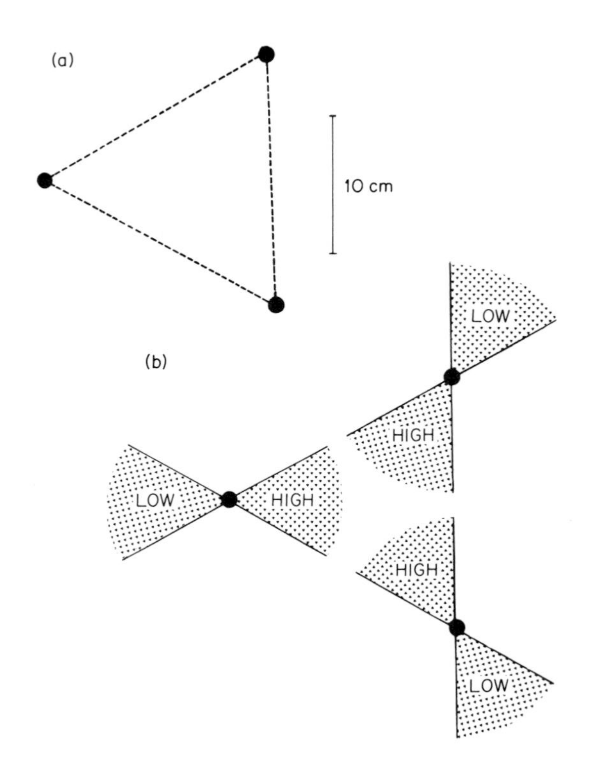

Fig. 1. a, Layout of one group of trees. Points 1, 2, and 3 represent the positions of the bases of the trees. b, Shaded portions illustrate for each tree the locations of the high- and low-interference zones considered in this paper; zones are shown diagrammatically and do not necessarily represent the actual extent of the area covered during growth.

In this chapter I shall confine myself to a comparison of branches from two areas of the tree crown, as illustrated in Fig. 1b. Branches growing directly into the triangle formed by the bases of the three trees in a group make up the 'high-interference' zone, and branches existing in the opposite 60° segment, at the other side of the tree, comprise the 'low-interference' zone. Branches of the same age, that is those emerging in the same season, generally form a vertical series on the trunk, with younger branches above and older ones below. Branch age and position have a strong influence on growth; the effect of interference is superimposed on these. It is therefore necessary to categorize branches into age classes. This is easily done as shoot growth in silver birch is sympodial, and the division between annual increments is clearly marked by the position of the previous year's aborted terminal bud.

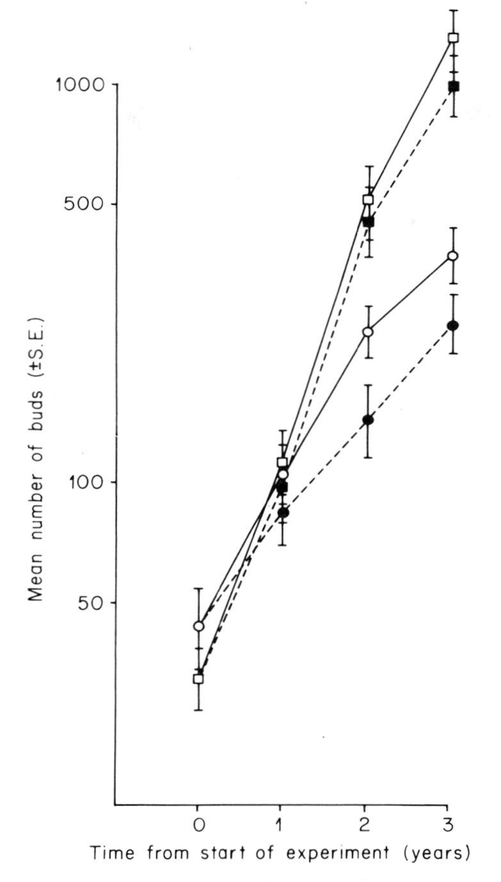

Fig. 2. Total number of buds in branches of each interference zone with time. Points plotted are the means of nine trees. □——□ total buds (living and dead) in the low-interference zone; ■- - - -■ living buds in the low-interference zone; ○——○ total buds in the high-interference zone; ●- - - -● living buds in the high-interference zone.

A. Growth in the Bud Population

1. *For All Branches*

Figure 2 illustrates the increase in numbers of buds in the two interference zones over time. There is very little difference between them during the first year of the experiment, but beyond this the declining rate of growth in the high-interference zone is very clear. Furthermore, the discrepancy between total number of buds and number of living buds is much larger for branches in the

TABLE I.

Total Bud Numbers (Mean Values) in High- and Low-interference Zones of a Silver Birch Tree Crown[a]

	Years from start of experiment			
	0	1	2	3
Number of buds in high-interference zone (*a*)	43.67	103.33 (84.67)	237.22 (143.11)	372.89 (248.78)
Number of buds in low-interference zone (*b*)	32.33	112.11 (97.33)	513.56 (455.44)	1281.89 (992.44)
Ratio (*b*/*a*)	0.74	1.08 (1.15)	2.16 (3.18)	3.44 (3.99)

[a] The total number of buds (living and dead) is given; figures in brackets are for living buds only.

high-interference area, suggesting that these suffer higher rates of mortality as well. The differences between the numbers of buds in each zone are compared directly in Table I. The ratio of the numbers of living buds in the low-interference zone to those in the high-interference zone changes from approximately 1 after 1 year of the experiment to 4 by the end of year 3. Clearly the growth processes occurring on the two sets of branches during this time must be different.

2. *Branches of Different Ages*

Branches of different ages occupy different positions on the trunk. First-year branches form a series at the top of the tree, with older branches situated lower and lower in the crown. All the branches produced in one growing season can be considered to form a cohort, with the growth of these cohorts followed over time.

Figure 3 illustrates changes in the number of buds in each cohort of branches over the course of the experiment. Cohorts of high- and low-interference branches are plotted seperately, and their patterns of growth are obviously different.

There is a pronounced decline in the growth rate of high-interference branches; total bud numbers in the two oldest cohorts do not appear to increase appreciably after the second year of the experiment. This condition is in contrast to the situation for the low-interference branches, where growth is maintained throughout the experiment, with only a slight decrease in rate in the oldest cohort in the final year. (Figs. 3a and b). Figure 3c and d show that the cohort differences in number of living buds are even more marked. Although the trends for number of living buds are much the same as those for total numbers in the low-interference

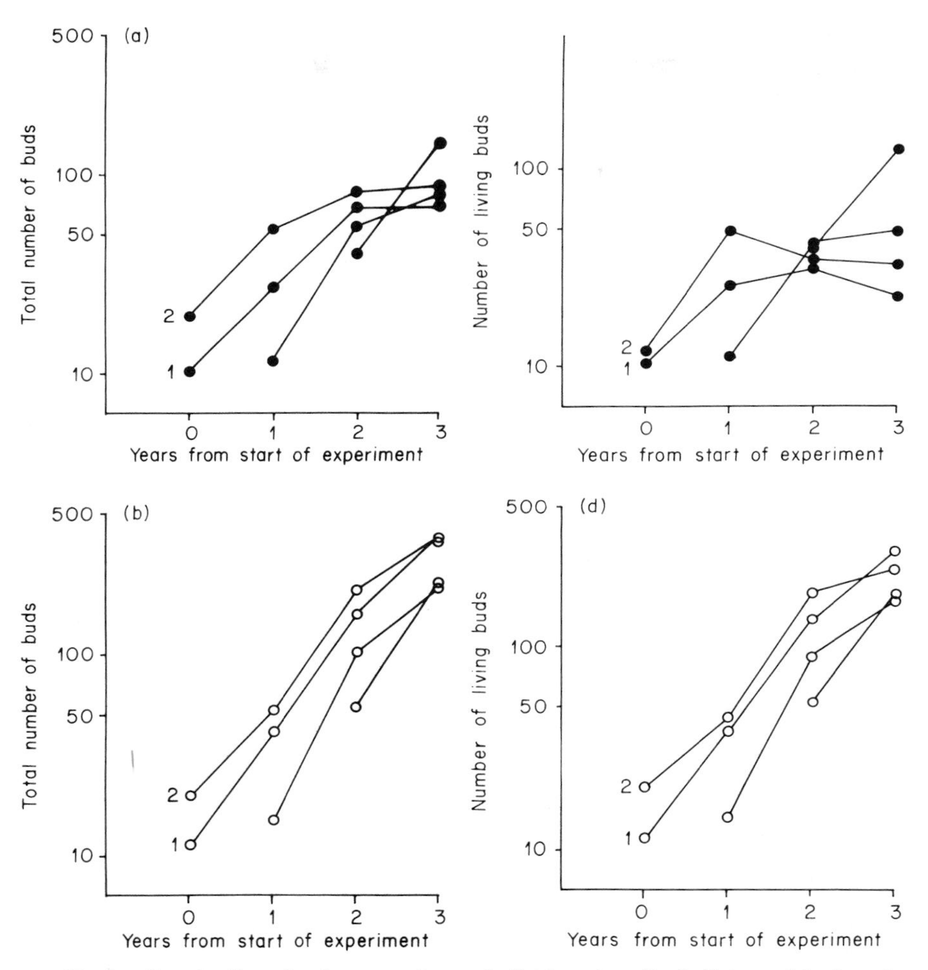

Fig. 3. Growth of branch cohorts over time. a, b, Total number of buds (living and dead). c, d, Number of living buds. Oldest cohorts start at left: numbers represent their ages at the start of the experiment. ●——● branches in the high-interference zone; ○——○ branches in the low-interference zone.

branches (i.e. sustained increase in numbers over time), for the older high-interference cohorts living bud numbers tend to fall after the first 1–2 years of the experiment.

These plots also indicate that differences in growth between high- and low-interference branches are relatively small at first (initial numbers in cohorts of both groups are similar), but that they increase with time. Thus the tree tends to develop asymmetrically in terms of bud distribution, not only from one side to the other (as the figures in Table I indicate), but from top to bottom as well. Although there may be four times as many buds in total on the branches of the low-interference zone, this does not mean that low-interference branches of all ages are four times larger than their high-interference counterparts. Figure 4 illustrates this. The effects of interference on branch growth and form appear to be strongly developed only below a certain level in the canopy (in branches older than 2 years).

(a) At start of experiment (Year 0)

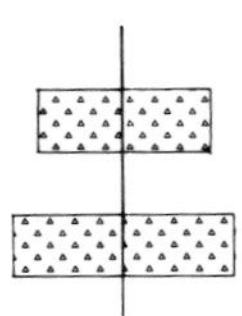

(b) Year 1

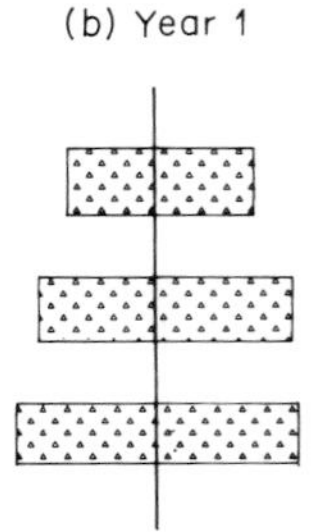

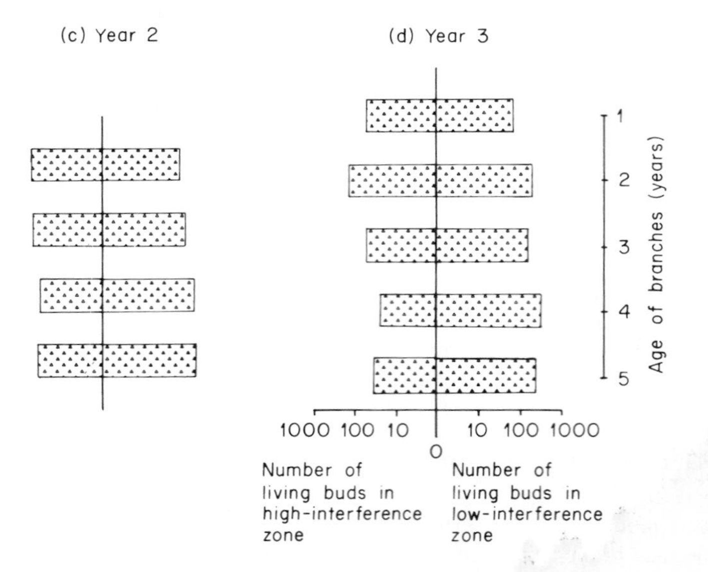

Fig. 4. Distribution of living buds within a tree. The numbers shown at each level are numbers per branch age cohort, not numbers per unit depth or area.

Up to this point the results have been largely concerned with changes in bud numbers over the tree as a whole, and the way these may affect form in a very broad sense (Table I and Fig. 4). Now I shall discuss in greater detail the dynamics of growth for branches of one cohort.

B. Cohort Analysis

The cohort studied comprises branches which were 1 year old at the start of the experiment. These branches were most suitable for analysis because they were both old enough to have been growing throughout the 3 years of the experiment and sufficiently small at the start that there was a minimum of past growth history to influence their performance.

Once again branches were split into high- and low-interference groups, and the results from these two groups compared. Table II gives a condensed summary of births and deaths of buds. The rate of increase in the living bud population is very similar for both interference classes in the first year, but by the third year this value has fallen below 1.0 for the high-interference cohort, indicating that the number of buds produced is less than the number dying. Not surprisingly, overall

TABLE II.

Flux in the Bud Numbers in High- and Low-interference Branches of the Same Cohort

	High-interference branches	Low-interference branches
Number of buds at start of year 1 (*a*)	59	74
Number of buds produced during year 1 (*b*)	194	226
Number of buds dying during year 1 (*c*)	17	8
Number of living buds at end of year 1 (*d*)	177	218
Rate of increase in the living bud population (d/a)	3.00	2.95
Mortality (c/b)	0.09	0.04
Number of buds at start of year 2 (*a*)	177	218
Number of buds produced during year 2 (*b*)	443	1060
Number of buds dying during year 2 (*c*)	238	109
Number of living buds at end of year 2 (*d*)	205	951
Rate of increase in the living bud population (d/a)	1.16	4.36
Mortality (c/b)	0.54	0.10
Number of buds at start of year 3 (*a*)	205	951
Number of buds produced during year 3 (*b*)	234	2248
Number of buds dying during year 3 (*c*)	78	311
Number of living buds at end of year 3 (*d*)	156	1937
Rate of increase in the living bud population (d/a)	0.76	2.04
Mortality (c/b)	0.33	0.14

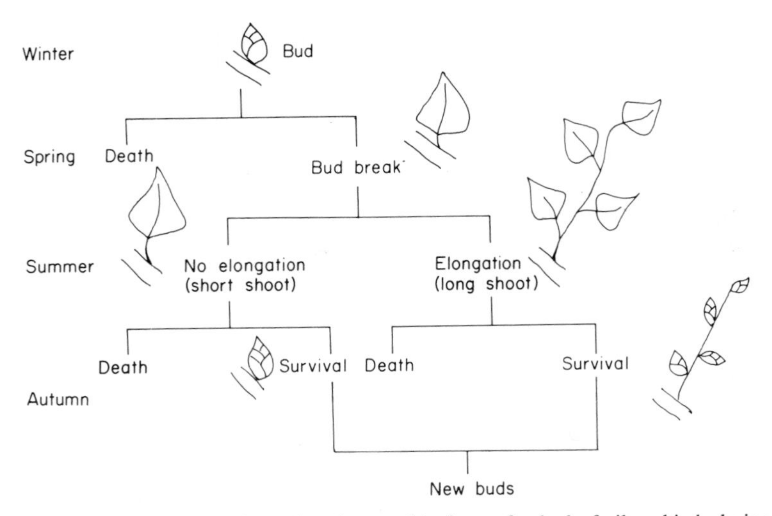

Fig. 5. Simplified scheme illustrating the possible fates of a bud of silver birch during one growing season. [Adapted from Maillette (1982), with permission from Blackwell Scientific Publications, Ltd.]

levels of mortality are very much higher in this group, especially in the second year.

A bud present at the start of the growing season faces a number of possible fates (Maillette 1982a) (Fig. 5). It may grow out and produce daughter buds (on a long shoot), replace itself (on a short shoot), or die. The fates of all the buds in the population during the growing season thus determine the number of buds that go on to form the population in the following year. Death of buds often occurs after outgrowth has taken place, in which case any daughters produced die also. Unfortunately as census of the bud population was made only once a year, at the end of each growing season, it was not always possible to say precisely when the deaths of particular buds occurred. A proportion of the dead buds recorded in any year was composed of buds on long shoots which had not been present in the previous year. Obviously these were the offspring of buds which had grown out and subsequently died. Of the dead short-shoot buds, however, where evidence of current-year growth was much more difficult to establish, it was often not clear whether a bud had died before the start of the growing season, without making any growth, or whether death had taken place during the time of active growth. Figure 6 shows the fates of buds in the high- and low-interference cohorts over the 3 years of the experiment.

Table III gives the proportions of buds undergoing each of the fates. The proportion of buds producing long shoots was similar in both interference zones for the first 2 years of the experiment, but in year two there was a very high level of mortality (48%) suffered by the newly produced buds on these shoots in the high-interference zone. By contrast only 7% of their counterparts in the low-

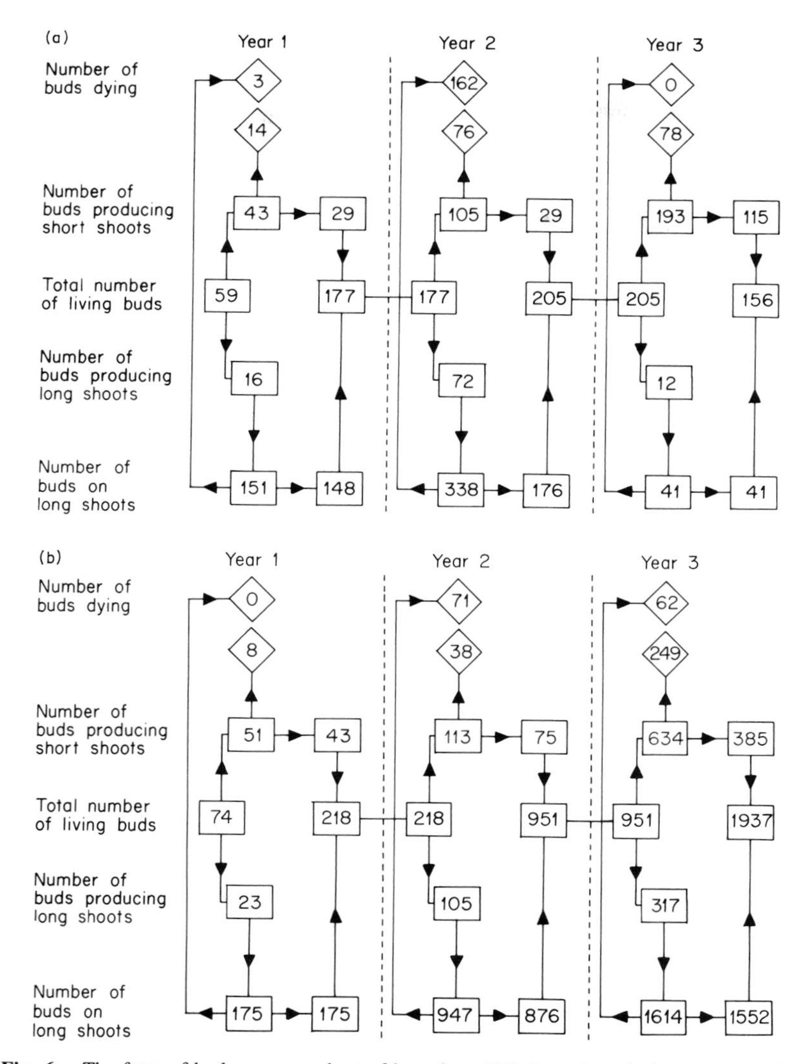

Fig. 6. The fates of buds on one cohort of branches. (All short-shoot buds are assumed to die during the growing season.) a, Branches in the high-interference zone. b, Branches in the low-interference zone.

interference zone died. In the following year, a very small proportion (6%) of high-interference buds produced offspring, and although there were no deaths among the latter, continuing mortality in the rest of the population meant that there was a decrease in the total living bud numbers in this cohort (from 205 to 156). In the low-interference zone, 33% of the buds underwent long-shoot production and survival of daughters was very high. The sustained production of

TABLE III.

Bud Fates

Fate	Buds present at start of growing season		Buds produced during growing season	
	High-interference zone	Low-interference zone	High-interference zone	Low-interference zone
In year 1 of experiment				
Long-shoot production	0.27	0.31	0.00	0.00
Short-shoot production / survival	0.49	0.58	0.98	1.00
Death	0.24	0.11	0.02	0.00
In year 2 of experiment				
Long-shoot production	0.41	0.48	0.00	0.00
Short-shoot production / survival	0.16	0.35	0.52	0.93
Death	0.43	0.17	0.48	0.07
In year 3 of experiment				
Long-shoot production	0.06	0.33	0.00	0.00
Short-shoot production / survival	0.56	0.41	1.00	0.96
Death	0.38	0.26	0.00	0.04

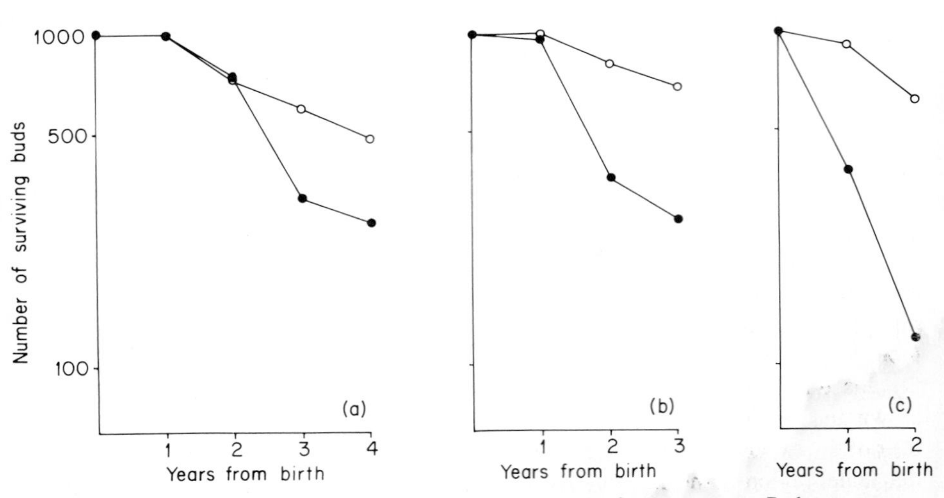

Fig. 7. Survivorship curves of buds in the high- and low-interference zones. a, Buds present at the start of the experiment (year 1). b, Buds entering the population during year 1 which were not direct descendants of (a). c, Buds entering the population during year 2 which were not direct descendants of (a) or (b). ●——● buds in the high-interference zone; ○——○ buds in the low-interference zone.

long shoots in relatively high proportions by buds in the low-interference zone and their generally good survivorship led to a continual increase in the bud population over the course of the experiment. Survivorship curves for the buds are plotted in Fig. 7. A bud is considered to have survived to a given year if it has a living direct descendant in that year. Because buds producing short shoots have only one descendant in a year, but buds which produce long shoots have many, complications in the construction of survivorship curves arise. In this paper therefore the direct descendants of buds producing long shoots are defined as the buds at the apices of daughter long shoots. All other long-shoot buds produced in a year form a new group, which has its own survivorship curve. The generally lower levels of survivorship in the high-interference cohort are evident.

III. CONCLUSIONS

Clearly neighbours have a strong influence on branch growth processes. Buds on branches growing in the low-interference zone have higher probabilities of surviving and of producing daughters than their counterparts on branches of the same age in the centre of a tree group. It is these fundamental differences in modular dynamics which affect branch size and shape and, in so doing, underlie the differences in bud distribution (form) from one side of the tree to the other. It is appropriate then to consider what lies behind the changes in modular dynamics themselves.

The branch cohort whose dynamics were studied in detail was 1 year old at the start of the experiment. It was situated at the top of the canopy, the branches were young and small in size, and contact with neighbours was minimal. Growth of branches in the low- and high-interference zones followed much the same pattern (Tables II and III). By the end of the first year of the experiment the branches of this cohort had attained a high level of growth and there was a new set of branches situated above them. In the second year of the experiment, initial growth of branches was again very similar in both zones: comparable proportions of buds grew out and produced daughters. However, despite this, by the end of the growing season the total number of living buds carried forward in the following year was considerably different. This difference was largely due to high mortality amongst the newly produced buds on the high-interference branches. By this time there would have been two groups of branches higher up in the crown and, as Fig. 4 shows, there is relatively little difference, even in the third year of the experiment, between high- and low-interference zone bud numbers in this upper region of the canopy. We might reasonably assume, therefore, that the bud (and hence, in summer, leaf) numbers here would be two to three times greater in the centre of a group, that is, in the high-interference zone, than they would be in the low-interference zone of any one tree, and that the light climate would deteriorate accordingly. As a result of this, many leaves existing lower in the canopy may be unable to support themselves and die before the end of the summer, together with the developing buds in their axils. The number of buds in the lower canopy in the centre of a group is therefore very small. The canopy of a tree group, which at the start of the experiment resembled three separate trees in its bud distribution, thus begins to look increasingly like that of a single tree, with buds concentrated on the periphery.

Maillette's innovative study (1982b) on growth in the bud population of mature silver birch trees revealed that there were substantial differences in bud behaviour with canopy position. Buds in the top canopy (on the leader and all branches produced by the leader during the previous year) had a greater chance of producing long shoots than those lower in the crown. The work reported here indicates that comparable changes in growth can occur across a canopy when the environment differs from one side to the other. Buds respond to their local environment, causing changes in dynamics and form on a local scale.

ACKNOWLEDGMENTS

I would like to express my thanks to Prof. J. L. Harper for his support and encouragement. The research was carried out while the author was receiving a grant from the Natural Environment Research Council.

REFERENCES

Arber, A. (1950). *The Natural Philosophy of Plant Form.* Cambridge University Press, Cambridge.

Bazzaz, F. A. & Harper, J. L. (1977). Demographic analysis of the growth of *Linum usitatissimum. New Phytologist,* **78,** 193–207.

Hallé, F. & Oldeman, R. A. A. (1970). *Essai sur l'architecture et la dynamique de croissance des arbres tropicaux.* Masson, Paris.

Hallé, F., Oldeman, R. A. A., & Tomlinson, P. B. (1978). *Tropical Trees and Forests: An Architectural Analysis.* Springer-Verlag, Berlin.

Harper, J. L. (1978). The demography of plants with clonal growth. *Structure and Functioning of Plant Populations* (Ed. by A. H. J. Freysen & J. W. Woldendorp), pp. 27–48. North-Holland, Amsterdam.

Harper, J. L. (1980). Plant demography and evolutionary theory. *Oikos,* **35,** 244–253.

Harper, J. L. & Bell, A. D. (1979). The population dynamics of growth form in organisms with modular growth. *Population Dynamics* (Ed. by R. M. Anderson, B. D. Turner & L. R. Taylor), Symposium of the British Ecological Society, 20, pp. 29–52. Blackwell Scientific Publications, Oxford.

Harper, J. L. & White, J. (1974). The demography of plants. *Annual Review of Ecology and Systematics,* **5,** 419–463.

Hickman, J. C. (1979). The basic biology of plant numbers. *Topics in Plant Population Biology* (Ed. by O. T. Solbrig, S. Jain, G. B. Johnson & P. H. Raven), pp. 232–263. Columbia University Press, New York.

Kays, S. & Harper, J. L. (1974). The regulation of plant and tiller density in a grass sward. *Journal of Ecology,* **62,** 97–105.

Kobayashi, S. (1975). Growth analysis of plants as an assemblage of internodal segments—a case of sunflower plants in pure stands. *Japanese Journal of Ecology,* **25,** 61–70.

Koyama, H. & Kira, T. (1956). Intraspecific competition among higher plants. VIII. Frequency distribution of individual plant weight as affected by the interaction between plants. *Journal of the Institute of Polytechnics, Osaka City University, Series D,* **7,** 73–94.

Maillette, L. (1982a). Structural dynamics of silver birch. I. Fates of buds. *Journal of Applied Ecology,* **19,** 203–218.

Maillette, L. (1982b). Structural dynamics of silver birch. II. A matrix model of the bud population. *Journal of Applied Ecology,* **19,** 219–238.

Obeid, M., Machin, D. & Harper, J. L. (1967). Influence of density on plant to plant variation in Fiber flax, *Linum usitatissimum* L. *Crop Science,* **7,** 471–473.

Porter, J. R. (1983a). A modular approach to analysis of plant growth. I. Theory and principles. *New Phytologist,* **94,** 183–190.

Porter, J. R. (1983b). A modular approach to analysis of plant growth. II. Methods and results. *New Phytologist,* **94,** 191–200.

Tomlinson, P. B. (1982). Chance and design in the construction of plants. *Axioms and Principles of Plant Construction* (Ed. by R. Sattler), pp. 162–183. Martinus Nijhoff/Dr. W. Junk, The Hague.

White, J. (1979). The plant as a metapopulation. *Annual Review of Ecology and Systematics,* **10,** 109–145.

White, J. (1980). Demographic factors in populations of plants. *Demography and Evolution of Plant Populations* (Ed. by O. T. Solbrig), pp. 21–48. Blackwell Scientific Publications, Oxford.

White, J. (1984). Plant metamerism. *Perspectives on Plant Population Ecology* (Ed. by R. Dirzo & J. Sarukhán), pp. 15–47. Sinauer Associates, Sunderland, Massachusetts.

White, J. & Harper, J. L. (1970). Correlated changes in plant size and number in plant populations. *Journal of Ecology,* **58,** 467–485.

16

*Modular Demography and Growth Patterns of Two Annual Weeds (*Chenopodium album *L. and* Spergula arvensis *L.) in Relation to Flowering*

Lucie Maillette[1]

Centre de recherches écologiques de Montréal
Montreal, Quebec, Canada

I. INTRODUCTION

Plants grow by the progressive accumulation of repeated elements: leaves, buds, internodes, branches, and flowers which together make up modules. These modules are assembled according to specific architectural patterns of meristem behaviour (Hallé, Oldeman & Tomlinson 1978), phyllotaxy (Jean 1983), branching angles, leaf display and bifurcation ratios (see review by Fisher 1984), which together contribute to the particular shape of a plant. In most plants the number of elements is not fixed; it changes with time, because of growth and senescence processes, and from plant to plant because of plasticity and genetic variability. Changes in the number of parts are caused by demographic events, births, and deaths; a plant can indeed be viewed as a population of metameric or modular units, a metapopulation (White 1979; see White 1984 for terminology).

Bazzaz and Harper (1977) pioneered modular demography with a study of *Linum usitatissimum* in which they showed that leaf birth and death rates were affected by plant density and light conditions. Botanists have been particularly interested in demography of leaves because of their photosynthetic function. Demographic studies have included leaves of herbs (e.g. Abul-Fatih & Bazzaz

[1]Present address: Centre d'études nordiques, Université Laval, Sainte-Foy, Quebec, Canada G1K 7P4.

STUDIES ON PLANT DEMOGRAPHY:

1980; Garbutt & Bazzaz 1983; Huiskes & Harper 1979; Lovett Doust 1981) and trees (e.g. Boojh & Ramakrishnan 1982; Kikuzawa 1983), as well as conifer needles (Hadley & Smith 1983; Maillette 1982a). Other components have also been considered: buds (Maillette 1982b), rhizomes (Bell 1976), shoots (White 1980) and branches (McGraw & Antonovics 1983). Modular demography is particularly well suited for small populations of plants in which repeated destructive harvests are not possible; it allows monitoring of overall fecundity of individuals, and results can be fitted into predictive models (McGraw & Wulff 1983; see also Harper 1980; Hunt & Bazzaz 1980). Modular demography also makes it possible to link ecophysiology with population biology by assessing the impact of physiological processes on plant fecundity and survival via monitoring single plant growth (McGraw & Wulff 1983).

This study was undertaken primarily to explore the growth of weeds infesting wheat (*Triticum aestivum*) crops in Quebec, Canada. These applied results will be published elsewhere. The project also had a more academic interest: the study of modular dynamics of annuals examined in relation to flowering. Two weeds with contrasting growth form, erect *Chenopodium album* and procumbent *Spergula arvensis,* were grown in pure stands. Their performances were assessed by making censuses of plants and component metamers and by monitoring leaf size and canopy development. Their overall patterns of growth could thus be compared, in particular, their responses to flowering.

II. EXPERIMENTAL PROCEDURE

A. Species Studied

Chenopodium album (lamb's quarter, Chenopodiaceae) is a branched erect annual. Flowers are perfect, wind-pollinated and grouped in paniculate spikes or cymes found on shoot tips and leaf axils (Fernald 1950). It grows well on a wide range of soil types and pH (Williams 1963). It can quickly invade newly ploughed fields through banks of seeds that can remain dormant in the soil for 30–40 years (Holm *et al.* 1977). It shows a large degree of plasticity, ranging in height from 2–3 cm to 3 m in the most favourable circumstances. Its 'growth strategy' is that of a competitive–ruderal (Grime 1979). It has an almost universal distribution and is particularly successful in major crops of the temperate region, such as potatoes. It is one of the 10 worst weeds in the world, not only reducing yields through competition with crops but also acting as an intermediary host for a number of plant diseases (Holm *et al.* 1977). The biology of the species has been reviewed by Williams (1963) and by Bassett and Crompton (1978).

Spergula arvensis (corn spurrey, Caryophyllaceae) is a small annual. It has weak ascending shoots branching at the lowest internodes (up to 14 such branches) and small filiform leaves growing in whorls (New 1961). Its perfect flowers occur in terminal cymes (Marie-Victorin 1964). Individuals studied all exhibited a low spreading habit. Seeds can germinate even after a very long dormancy, up to 2000 years (Odum 1965). Corn spurrey is cosmopolitan, although restricted to the temperate zone, and it prefers acid soils of pH 4.5–5.0 (New 1961). It has been characterized by Trivedi & Tripathi (1982) as a ruderal restricted to recently disturbed habitats. It is not as serious a weed as *C. album*, but it can significantly reduce yields in some annual crops. A review of the biology of corn spurrey has been presented by New (1961).

B. Experiments

All seed material for the experiments came from La Pocatière Experimental Farm of Agriculture Canada, 100 km east of Quebec City (70°02′ W, 47°27′ N). Seeds were collected in 1975 and subsequently kept in a seed granary at 5°C. In both species germination rates of 71% were obtained when seeds were tested, a few days prior to the start of the experiments.

All experiments were carried out simultaneously in a cool greenhouse of the Montréal Botanic Garden during the summer of 1980. Seeds were sprinkled on damp Perlite and left to germinate in an incubator from 28 May until needed. Seedlings were transplanted from 5 to 13 June into aluminium trays (35 cm × 50 cm × 8 cm), filled with a mixture of peat moss, sand, garden compost and loam (pH 6.2). Some seedlings were kept in the incubator to replace those that did not survive the first 2 weeks after transplantation. In the greenhouse plants were watered as needed and treated against pests as necessary. Liquid fertilizer (20 : 20 : 20; 12.5 ml/litre) was added on 28 July. Experiments were terminated in the first week of September.

Templates were used during transplantation to ensure even spacing between plants. There were four density treatments: D1: 4 plants per tray (23 per square metre), D2: 20 plants per tray (114 per square metre), D3: 100 plants per tray (571 per square metre), D4: 500 plants per tray (2857 per square metre).

1. Demography

The experiment on demography included four density treatments, D1, D2, D3 and D4, with three replicates per treatment, except *Chenopodium album* D4, which had no replicates at the highest density because of a shortage of seedlings. Trays were placed in a randomized design. Each week the number of surviving plants was recorded as well as the number of leaves per cohort on selected individuals in each tray. All plants at densities D1 and D2 and 20 plants randomly chosen in both D3 and D4 were monitored. The youngest leaf of each cohort was marked with a colour-coded ring cut from a plastic straw. The experiment was started on 12–13 June and lasted 12 weeks. Towards the end of summer some D1 plants had reached such a large size that records of leaf births and deaths could only be taken every other week.

2. *Architecture*

The experiment on plant architecture used two density treatments, D1 and D3, with two replicates per treatment in a randomized design. Each week component elements of all D1 plants and of 20 randomly selected D3 plants per tray were examined. These included internodes (number and length), branches, leaves and flowers (numbers). The experiment was started on 13 June 1980 and lasted 12 weeks. On 3 September, plants were harvested, dried at 60°C for 48 hr and weighed. Monitored plants were weighed individually. Pearson linear correlation coefficients were calculated between dry weight and weekly architectural data with batch system SPSS^x (computer program for statistical analyses).

3. *Canopy Development*

For the experiment on canopy development plants were grown at two density levels, D1 and D3, with three repetitions per treatment, except for two repetitions of D3 with *Chenopodium album*. Each tray was filled with soil, planted with seedlings and divided into 70 squares, 5×5 cm^2 each, with fishing line and sticky tape. Seven squares per tray were chosen at random and marked. Each week the number of leaves found above each chosen square was recorded in 5-cm layers: this gave the number of leaves in a cubic volume, 5 cm on edge. The experiment was started on 5 June and lasted 13 weeks. At the end of the experiment, plants were harvested, dried 48 hr at 60°C and weighed.

III. RESULTS

A. Demography

1. *Plant Demography*

Most plant mortality occurred within the first month after transplantation. There were also some later deaths at high density, especially for *S. arvensis*. Mortality increased with density: final survivorship percentages for D1, D2, D3 and D4 were respectively 83, 90, 86, and 75 for *C. album* and 100, 84, 78 and 63 for *S. arvensis*. In D1 *C. album*, 2 plants died not long after transplantation, out of 12 that had been planted; that explains the low survivorship value.

TABLE I.

Leaf Demography: Linear Regressions Calculated between Final Number of Leaves and Density of Surviving Plants of *Chenopodium album* and *Spergula arvensis* Planted in Pure Stands at Four Density Levels[a]

	C. album		*S. arvensis*	
Variable[b]	Regression line	r	Regression line	r
Log(births)	$-0.78 \log x + 3.2$	-0.992	$-0.68 \log x + 3.5$	-0.995
Log(deaths)	$-0.28 \log x + 1.8$	-0.979	$-0.47 \log x + 2.8$	-0.994
Log(current number)	$-1.00 \log x + 3.4$	-0.999	$-0.83 \log x + 3.5$	-0.990

[a] Planted at 23 per square metre, 114 per square metre, 571 per square metre and 2857 per square metre.

[b] Births = cumulative number of leaf births per plant; deaths = cumulative number of leaf deaths per plant; current number = total number of leaves present per plant at the end of the experiment; x = density of survivors. r = coefficient of determination.

2. *Leaf Demography*

Absolute numbers of leaf births and deaths decreased exponentially with increasing density (Table I). Linear regressions of log of leaf number on log of final density yielded coefficients of determination ranging from −0.979 to −0.999. For both species regression lines for leaf births and deaths were not parallel, with proportionately more deaths at high density. The same trend is also apparent in the relative changes in the number of leaf births and deaths (Fig. 1a,b). At high density, fewer leaves were produced and proportionately more died sooner, thus reducing the actual number of leaves present on the plants. Leaf mortality increased with time.

B. Plant Architecture

In both species shoot elongation followed a sigmoid pattern: rapid at first, it later slowed down and finally reached a plateau (Fig. 2). The number of leaves increased in a similar way. At high density, growth stopped sooner, plants were smaller, and they had fewer leaves, branches and flowers. The difference in size between D1 and D3 plants widened markedly after the onset of flowering (Fig. 2). *Spergula arvensis* had much more leaves but fewer branches and flowers than *C. album*.

The 'flowering effort' was estimated by dividing the number of flowers alive on each plant at time t by the total length of shoot alive on each plant at time t (Fig. 3). Flowering effort was at first greater in D3 plants, but later the trend was reversed. In *S. arvensis* the flowering effort was bimodal: the first peak coin-

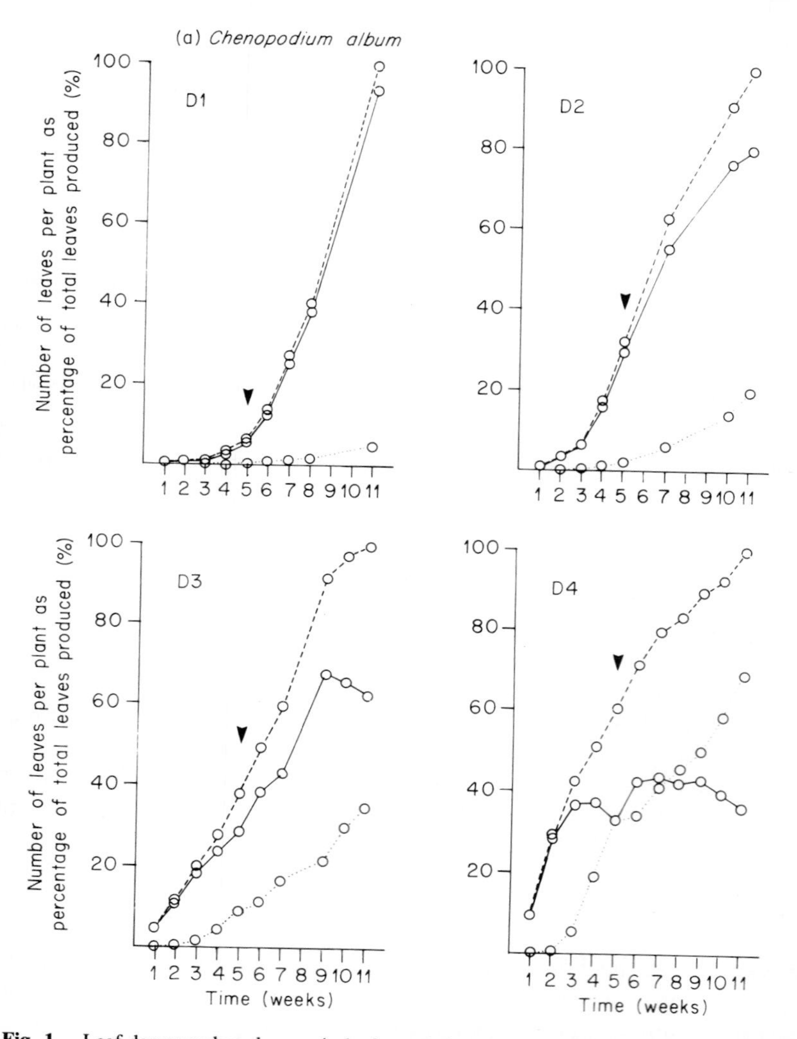

Fig. 1. Leaf demography: changes in leaf population size of plants growing in pure stands in a greenhouse. a, *Chenopodium album*. b, *Spergula arvensis*. Four planting densities: D1 = 4 per tray, D2 = 20 per tray, D3 = 100 per tray, D4 = 500 per tray. Week 1 = 12 June 1980. ——Total number of leaves present on plants. - - - - Cumulative number of births. ···· Cumulative number of deaths. 100% = Total number of leaf births. ▼ Beginning of flowering.

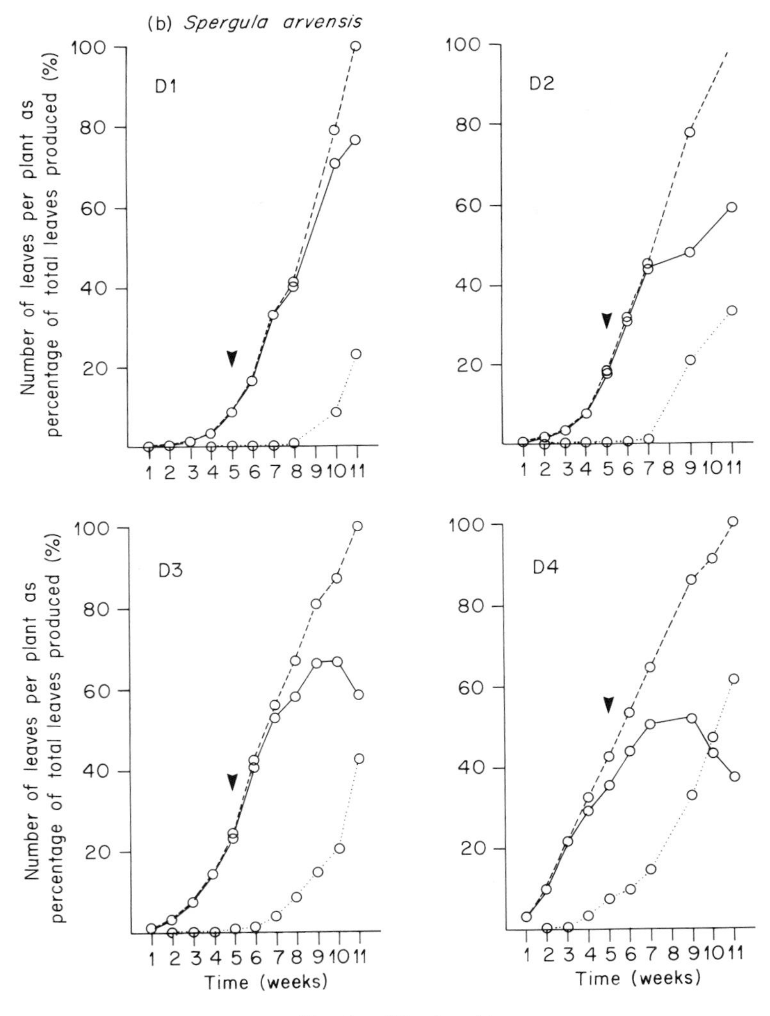

Fig. 1. *(Continued.)*

cided with the end of elongation of the main stem and the second one with a similar event on lateral branches. Only one peak was recorded in *C. album* before harvest. In both species flowering started on the main axis and later spread out to lateral branches.

Pearson linear correlation coefficients were calculated between final dry weights and architectural data for each sampling date (Table II). At first there was little or no correlation between final biomass and growth parameters. The

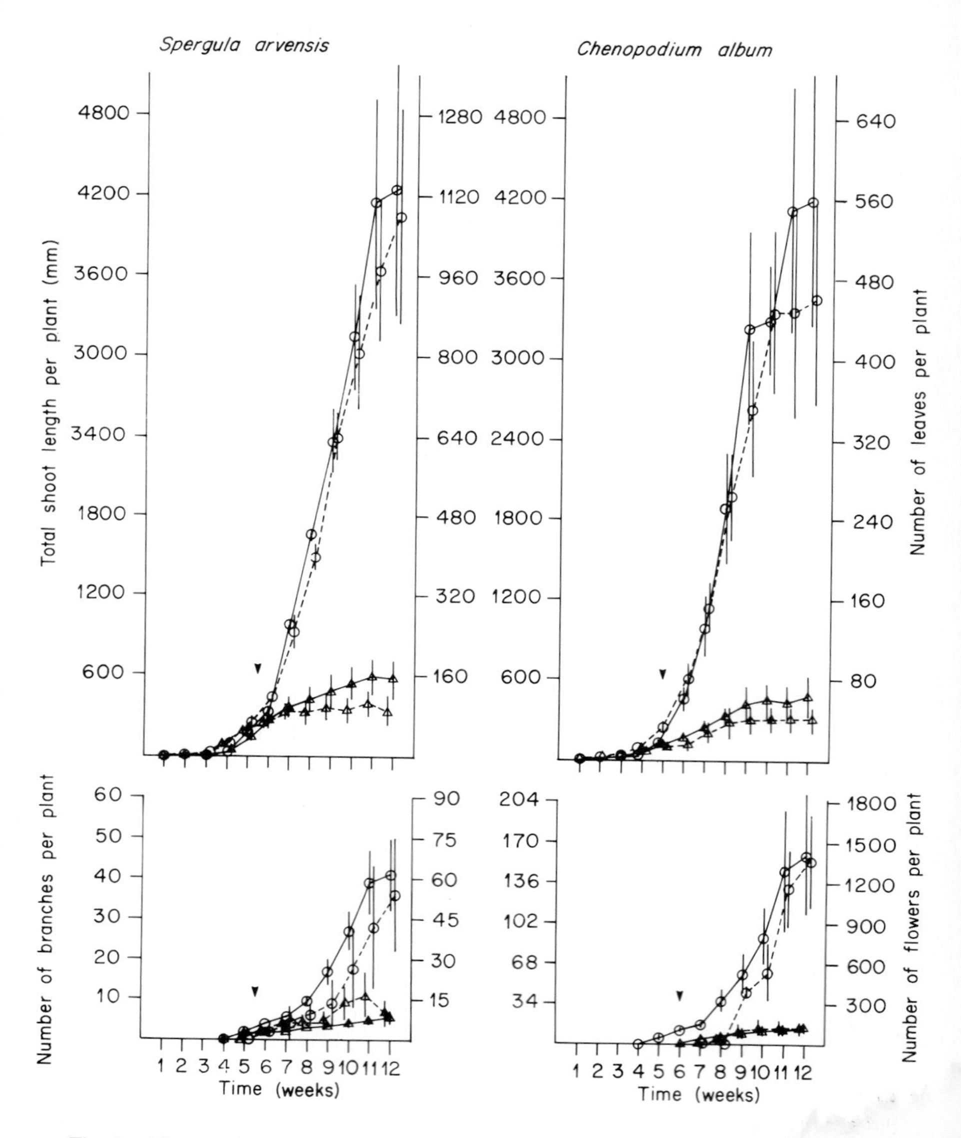

Fig. 2. Plant architecture: changes in population size of plant parts of *Chenopodium album* and *Spergula arvensis* growing in pure stands. ○, D1 = 4 per tray (23 per square metre). △, D3 = 100 per tray (571 per square metre). ——, Total shoot length per plant or number of branches per plant. - - - -, Number of leaves per plant or number of flowers per plant. ▼, Beginning of flowering. Vertical lines, standard deviation. Week 1 = 13 June 1980. Sample sizes: for *C. album* D1: $n = 7$, D3: $n = 39$; for *S. arvensis* D1: $n = 8$, D3: $n = 42$.

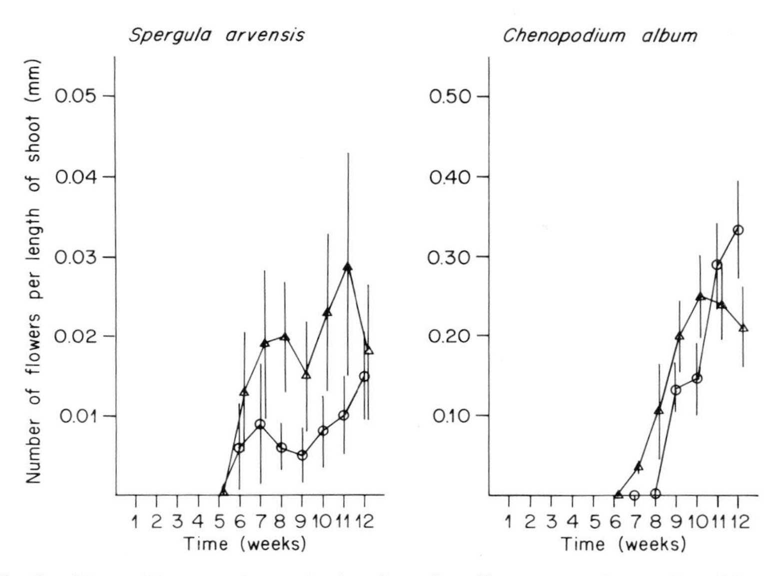

Fig. 3. Plant architecture: changes in plant flowering effort, expressed as number of flowers per millimetre of shoot length, for pure stands of *Chenopodium album* and *Spergula arvensis* growing in a greenhouse. Week 1 = 13 June 1980. ○, D1 = 4 plants per tray (23 per square metre). △, D2 = 100 plants per tray (571 per square metre). Vertical lines = standard deviation.

correlation for all vegetative indices increased sharply during weeks five and six, when flowering started. Significant correlations between numbers of flowers and dry weights appeared only during the last few weeks of the experiment.

C. Canopy Development

The canopy development of *C. album* and *S. arvensis* is shown in Fig. 4. *Spergula arvensis* produced small linear leaves in a dense mat near the ground. At high density its canopy was at first much more crowded and also taller than at low density; later it progressively grew flatter and thinner. Total biomass at high density was significantly higher than at low density (F-test, D1 = 14.6 g, D3 = 19.4 g, $p < .05$). *Chenopodium album* grew 15 cm taller than *S. arvensis* at D3. At high density, *C. album* too produced more leaves, especially at low levels, from weeks three to seven. Later on, as plants grew taller, leaves became less concentrated. There were no significant differences between D1 and D3 total biomass (D1 = 26.8 g, D3 = 25.0 g). At the end of the experiment there was a reduction in the number of leaves for both species.

TABLE II.

Plant Architecture: Correlation Coefficients between Final Biomass (13th Week) and Architectural Components of Plants of *Chenopodium album* and *Spergula arvensis* Growing in Pure Stands in a Greenhouse[a]

	Time (weeks)											
Architectural component	1	2	3	4	5	6	7	8	9	10	11	12
Chenopodium album												
Total shoot length	0.116	0.065	0.105	−0.181	0.342 *	0.931 ***	0.953 ***	0.969 ***	0.981 ***	0.977 ***	0.985 ***	0.984 ***
Number of internodes	nc	nc	−0.093	−0.016	0.740 ***	0.932 ***	0.953 ***	0.975 ***	0.960 ***	0.969 ***	0.947 ***	0.957 ***
Number of leaves	0.075	−0.211	−0.217	0.271 *	0.910 ***	0.918 ***	0.940 ***	0.923 ***	0.942 ***	0.981 ***	0.930 ***	0.953 ***
Number of branches	nc	nc	nc	−0.057	0.815 ***	0.775 ***	0.686 ***	0.750 ***	0.828 ***	0.960 ***	0.853 ***	0.869 ***
Number of flowers	nc	nc	nc	nc	nc	−0.109	−0.179	−0.063	0.633 ***	0.813 ***	0.879 ***	0.889 ***
Sample size	41	41	41	41	41	41	41	41	41	38	41	41
Spergula arvensis												
Total shoot length	0.176	−0.031	−0.122	−0.276 *	0.349 *	0.694 ***	0.895 ***	0.953 ***	0.953 ***	0.955 ***	0.920 ***	0.907 ***
Number of internodes	nc	nc	−0.052	0.054	0.522 ***	0.741 ***	0.882 ***	0.928 ***	0.928 ***	0.922 ***	0.888 ***	0.875 ***
Number of leaves	0.176	0.308 *	0.181	0.201	0.301 *	0.708 ***	0.869 ***	0.955 ***	0.970 ***	0.957 ***	0.958 ***	0.934 ***
Number of branches	nc	nc	nc	−0.028	0.315 *	0.807 ***	0.880 ***	0.856 ***	0.863 ***	0.886 ***	0.882 ***	0.898 ***
Number of flowers	nc	nc	nc	nc	−0.090	−0.056	−0.007	0.112	0.223	0.243	0.388 **	0.668 ***
Sample size	43	43	43	43	43	43	43	43	43	43	43	43

[a] D1 and D3 data pooled together for each calculation. Plants dead before the end of the experiment were not included. Week 1, 13 June 1980. Significance levels: *, $p < 0.05$; **, $p < 0.01$; ***, $p < 0.001$; nc, could not be computed.

IV. DISCUSSION

A. Demography

Plasticity seemed more important than mortality in the response of plants to increased density. In spite of a 125-fold range in planting density there was only a 37% range in observed mortality. Such low mortality rates have also been observed in field populations of *C. album* (Lapointe *et al.* 1984). Plasticity manifested itself in several ways. At high density, there were fewer leaves produced (Table I and Fig. 1), as in *Ambrosia trifida* (Abul-Fatih & Bazzaz 1980), but unlike the latter, *C. album* and *S. arvensis* lost proportionately more leaves at high density than at low density. Fruit set rather than flowering may have been related to an increase in leaf deaths, but the trend is not absolutely clear-cut because prereproductive deaths also occurred (Fig. 1).

Increased density was accompanied by increased mortality of plants and leaves. Population regulation thus took place at two levels: that of the population of individuals and that of the metapopulation of leaves, as observed also by Abul-Fatih and Bazzaz (1980). The two levels of population regulation were related in a log–log fashion (Table 1). Since leaf number and plant biomass are closely correlated (Table II) such a situation may be a reflection of the self-thinning rule which negatively relates the log of plant density to the log of mean biomass over a period of time, in populations where density-dependent mortality is occurring (Kira, Ogawa & Sakazaki 1953; Harper 1977; White 1981).

B. Architecture

The growth of *C. album* and *S. arvensis* is of a mixed nature. Individual shoots are hapaxanthic and not indeterminate, but since not all shoots flower in synchrony and numbers of metamers per shoot are variable, plant size is also not determinate. Both species displayed sigmoid patterns of shoot elongation and leaf and branch production (Fig. 2). At low density, plants were bigger, and disparity between D1 and D3 plants increased markedly when flowering started. Each D1 plant had produced several branches and branch primordia prior to flowering. These branches grew until fully developed, when their apices turned into inflorescences. On the other hand, D3 plants were little or not branched at all, and once the apex of the main stem had been transformed elongation quickly stopped. Thus D1 plants grew for a longer period of time and reached larger sizes than D3 plants.

Sigmoid functions have been interpreted as signs of density-dependent regulation of numbers (Hutchinson 1978) and of competition among members of a population or a metapopulation (White 1980). In the case of *C. album* and *S. arvensis* monocarpy and hapaxanthy as well as competition probably influenced

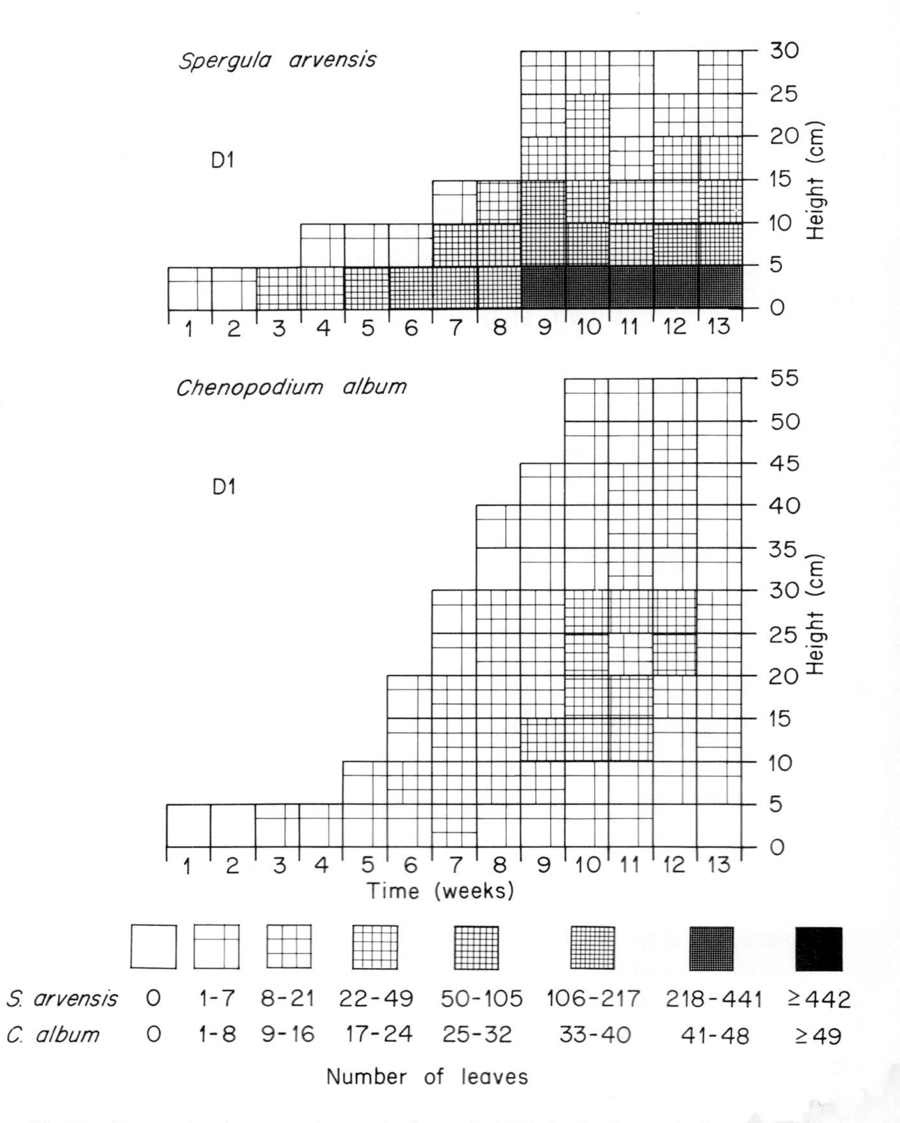

Fig. 4. Canopy development: changes in the vertical distribution leaves in the canopy of *Chenopodium album* and *Spergula arvensis* growing in pure stands in a greenhouse. Each square represents the number of leaves found in 21 cubes 5 × 5 × 5 cm^3. Week 1 = 5–9 June 1980.

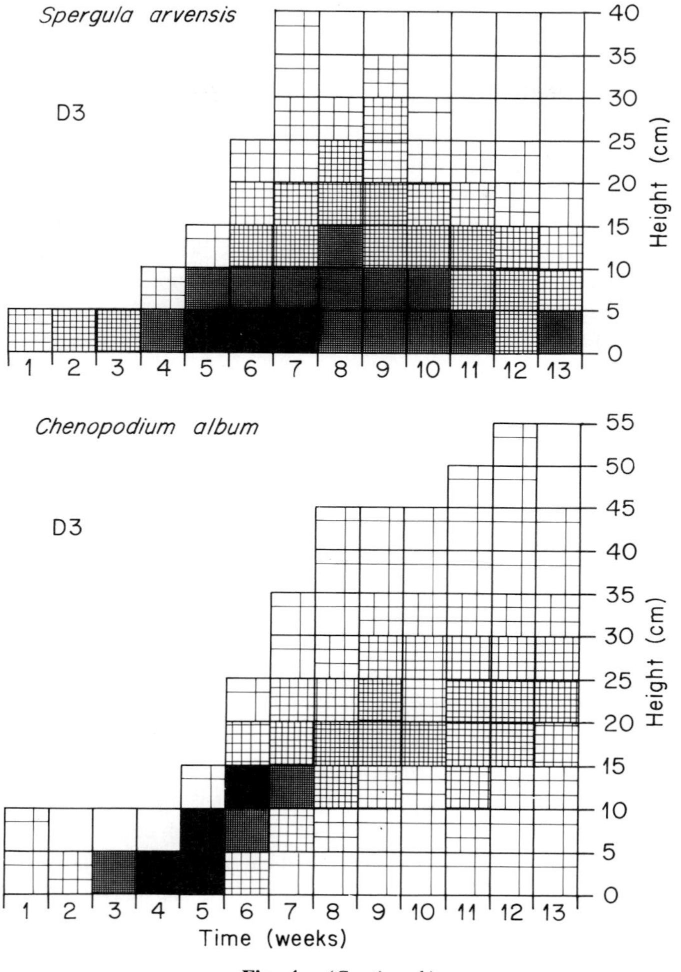

Fig. 4. (*Continued.*)

the shape of the curve but competition alone has to account for a large extent of the final levels of the asymptotes. Competition could well be the common mechanism of regulation of populations and metapopulations that gave rise to the log–log relationships seen in Table I.

Flowering was continuous in *C. album* and bimodal in *S. arvensis* (Fig. 3). This may be due to the differences in the morphology, and presumably the hormonal traffic, of the two species. The canopy of *C. album* is conical with a regular distribution of branches along the main axis, whereas the canopy of *S.*

arvensis is more shrub-like with most of its branches borne at the level of the first few nodes on the main stem.

Low-density plants produced more flowers per plant than high-density plants in both species (Fig. 2). Similar observations were made by Lemieux, Deschênes and Morisset (1984) on field populations of *S. arvensis* where increasing density meant fewer flowers and seeds per plant. For both species, however, the 'flowering effort' (Fig. 3) seemed as important in D1 plants as in D3 plants although there was a delay in D1 plants, probably due to branch development, as mentioned earlier. For the period of study *C. album* produced 20–25 times as many flowers as *S. arvensis* (Fig. 2). However actual seed production may not have been much different since *S. arvensis* produces 14–33 seeds per capsule (Lemieux, Deschênes & Morisset 1984) and *C. album*, only 1 seed per flower.

Reproduction had an impact on the overall growth of metapopulations. Before the start of flowering on week five, there was little or no correlation at all between vegetative growth parameters and final plant size, expressed in biomass (Table II). When reproduction started correlations increased suddenly to very highly significant levels. Floral initiation ended the juvenile, plastic and indeterminate phase of growth and started the reproductive, determinate phase of growth.

C. Canopy Development and Leaf Size

Chenopodium album and *S. arvensis* generated very different leaf canopies (Fig. 4). In spite of those differences there were similarities in the canopy development of both species: initially crowded canopies which later thinned out at high density and sparse leaf cover gradually increased at low density. At the end of the experiment leaf density was only slightly higher in D3 than in D1 treatments: plants with a 25-fold difference in planting density produced very similar leaf distributions. Actual leaf area may have been even more similar because D3 leaves were significantly smaller than D1 leaves (L. Maillette, unpublished data). Final canopy occupation may have been related to the 'carrying capacity' of the substrate.

V. CONCLUSIONS

Annuals devote a large proportion of their resources to reproduction (Ogden 1968 in Harper 1977). *Chenopodium album* allocates up to 68% of its dry weight to reproduction (Warwick & Marriage 1982; Fukuda & Hayashi 1982) and *S. arvensis,* up to 63% (Trivedi & Tripathi 1982). It is not surprising that reproduction had such a significant impact on various growth patterns (Table II). Furthermore, *C. album* and *S. arvensis* survive from year to year largely because of their ability to establish seed banks that can survive long periods of dormancy in the soil. They can colonize newly disturbed habitats, grow, reproduce and replenish their seed banks and wait until the next set of favorable conditions arises. Both are successful weeds, especially *C. album*. Both belong to a group of species sometimes called 'fugitives' or 'opportunists' (Harper 1977).

Yet *C. album* and *S. arvensis* are very different plants. The former looks like a miniature tree, and the latter is a dense tangle of filiform leaves. Nevertheless, both species generate their canopies through similar growth patterns: similar sigmoid growth functions (Fig. 2), similar plant and leaf population regulation (Table I), similar patterns of canopy development (Fig. 4). Both species also react to density with some plant mortality and a large degree of plasticity in the number and size of component elements (Figs. 1 and 2). Morphological development processes thus appear to be tied with 'life-style' (for lack of a better word), regardless of final shape in these two weeds.

ACKNOWLEDGMENTS

The author wishes to thank the Montreal Botanic Garden for their technical support. She is very grateful for the patient work of dedicated assistants: C. Abord-Hugon, L. Dudley, V. Jarry and I. Saucier. Drs. Pierre Legendre of University of Montreal and Jean-Marc Deschênes of Canada Agriculture made useful suggestions on the statistical analysis of the data. Drs. Serge Payette and Pierre Morisset of Laval University made helpful comments on the manuscript. L. Bélisle assisted with the computing. The study was supported by a grant from the 'Fonds annuel de soutien' of the University of Montreal.

REFERENCES

Abul-Fatih, H. A. & Bazzaz, F. A. (1980). The biology of *Ambrosia trifida* L. IV. Demography of plants and leaves. *New Phytologist,* **84**(1), 107–112.

Bassett, I. J. & Crompton, C. W. (1978). The biology of Canadian weeds. 32. *Chenopodium album* L. *Canadian Journal of Plant Science,* **58**(4), 1061–1072.

Bazzaz, F. A. & Harper, J. L. (1977). Demographic analysis of the growth of *Linum usitatissimum. New Phytologist,* **78,** 193–207.

Bell, A. D. (1976). Computerized vegetative mobility in rhizomatous plants. *Automata, Languages, Development* (Ed. by A. Lindermayer & G. Rozenberg), pp. 3–14. North-Holland, Amsterdam.

Boojh, R. & Ramakrishnan, P. S. (1982). Growth strategy of trees related to successional status. *Forest Ecology and Mangement,* **4,** 359–386.

Fernald, M. L. (1950). *Gray's Manual of Botany.* 8th ed. American Book Company, New York.

Fisher, J. B. (1984). Tree architecture: Relationship between structure and function. *Contemporary Problems in Plant Anatomy* (Ed. By R. A. White & W. C. Dickison) pp. 541–589. Academic Press, Orlando, Florida.

Fukuda, H. & Hayashi, I. (1982). Ecology of dominant plant species of early stages in secondary succession: On *Chenopodium album* L. *Japanese Journal of Ecology,* **32,** 517–526.

Garbutt, K. & Bazzaz, F. A. (1983). Leaf demography, flower production and biomass of diploid and tetraploid populations of *Phlox drummondii* Hook. on a soil moisture gradient. *New Phytologist*, **93**, 129–141.

Grime, J. P. (1979). *Plant Strategies and Vegetation Processes*. Wiley, New York.

Hadley, J. L. & Smith, W. K. (1983). Influence of wind exposure on needle dessication and mortality for timberline conifers, Wyoming. *Arctic and Alpine Research*, **15**(1), 127–136.

Hallé, F., Oldeman, R. A. A. & Tomlinson, P. B. (1978). *Tropical Trees and Forests*. Springer-Verlag, Berlin.

Harper, J. L. (1977). *Population Biology of Plants*. Academic Press, London.

Harper, J. L. (1980). Plant demography and ecological theory. *Oikos*, **35**, 244–253.

Holm, L. G., Plucknett, D. L., Pancho, J. V. & Herberger, J. P. (1977). *The World's Worst Weeds*. University Press of Hawaii, Honolulu.

Huiskes, A. H. L. & Harper, J. L. (1979). The demography of leaves and tillers of *Ammophila arenaria* in a dune sere. *Oecologia Plantarum*, **14**(4), 435–446.

Hunt, R. & Bazzaz, F. A. (1980). The biology of *Ambrosia trifida* L. V. Response to fertilizer, with growth analysis at the organismal and sub-organismal level. *New Phytologist*, **84**(1), 113–121.

Hutchinson, G. E. (1978). *An Introduction to Population Ecology*. Yale University Press, New Haven, Connecticut.

Jean, R. V. (1983). *Croissance végétale et morphogénèse*. Masson, Paris.

Kikuzawa, K. (1983). Leaf survival of woody plants in deciduous broad-leaved forests. I. Tall trees. *Canadian Journal of Botany*, **61**, 2133–2139.

Kira, T., Ogawa, H. & Sakazaki, N. (1953). Intraspecific competition among higher plants. I. Competition-density-yield inter-relationships in regularly dispersed populations. *Journal of the Institute of Polytechnics, Osaka City University, Series D*, **4**, 1–16.

Lapointe, A.-M. Deschênes, J.-M., Gervais, P. & Lemieux, C. (1984). Biologie du chénopode blanc (*Chenopodium album* L.) influence du travail du sol sur la levée et celle de la densité du peuplement sur la croissance. *Canadian Journal of Botany*, **62**, 2587–2593.

Lemieux, C., Deschênes, J.-M. & Morisset, P. (1984). Compétition entre la spargoute des champs (*Spergula arvensis* L.) et la sétaire glauque (*Setaria glauca* (L.) Beauv.). II. Production de graines. *Canadian Journal of Botany*, **62**, 1852–1857.

Lovett Doust, L. (1981). Population dynamics and local specialization in a clonal perennial (*Ranunculus repens*). II. The dynamics of leaves and a recriprocal transplant-replant experiment. *Journal of Ecology*, **69**, 757–768.

Maillette, L. (1982a). Needle demography and growth dynamics of Corsican pine. *Canadian Journal of Botany*, **60**(2), 105–116.

Maillette, L. (1982b). Structural dynamics of silver birch. 1. The fates of buds. *Journal of Applied Ecology*, **19**, 203–218.

Marie-Victorin, F. (1964). *Flore laurentienne*. Presses de l'Université de Montréal, Montréal.

McGraw, J. B. & Antonovics, J. (1983). Experimental ecology of *Dryas octopetala* ecotypes. II. A demographic model of growth, branching and fecundity. *Journal of Ecology*, **71**(3), 899–912.

McGraw, J. B. & Wulff, R. D. (1983). The study of plant growth: A link between the physiological ecology and population biology of plants. *Journal of Theoretical Biology*, **103**, 21–28.

New, J. K. (1961). Biological flora of the British Isles, *Spergula arvensis* L. (*S. sativa* Boenn., *S. vulgaris* Boenn.). *Journal of Ecology*, **49**(1), 205–215.

Odum, S. (1965). Germination of ancient seeds. Floristical observations and experiments with archaeologically dated soil samples. *Dansk Botanisk Arkiv*, **24**(2), 1–70.

Trivedi, S. & Tripathi, R. S. (1982). The effects of soil texture and moisture on reproductive strategies of *Spergula arvensis* L. and *Plantago major* L. *Weed Research*, **22**, 41–49.

Warwick, S. I. & Marriage, P. B. (1982). Geographical variation in populations of *Chenopodium album* resistant and susceptible to atrazine. II. Photoperiod and reciprocal transplant studies. *Canadian Journal of Botany*, **60**, 494–504.

White, J. (1979). The plant as a metapopulation. *Annual Review of Ecology and Systematics,* **10,** 109–145.

White, J. (1980). Demographic factors in populations of plants. *Demography and Evolution in Plant Populations* (Ed. by O. T. Solbrig), pp. 21–48. University of California Press, Berkeley.

White, J. (1981). The allometric interpretation of the self-thinning rule. *Journal of Theoretical Biology,* **89,** 475–500.

White, J. (1984). Plant metamerism. *Perspectives on Plant Population Ecology* (Ed. by R. Dirzo & J. Sarukhán), pp. 15–47. Sinauer Associates, Sunderland, Massachusetts.

Williams, J. T. (1963). Biological flora of the British Isles, *Chenopodium album* L. *Journal of Ecology,* **51**(3), 711–725.

17

A Modular Approach to Tree Production

Miguel Franco[1]

School of Plant Biology
University College of North Wales
Bangor, Gwynedd, Wales

I. MODULAR CONSTRUCTION AND THE GROWTH OF TREES

The idea that a plant can be considered as an assemblage of constructionally simple, basic units is not new and it has its roots in several, sometimes independent schools of thought (Cusset 1982). This conception, however, has been borrowed and revitalized in recent years by plant population biologists seeking a more tangible 'individuality' in plants which does not show the tremendous variation (in size, reproductive output, etc.) shown by individual plants. It was in one of his less quoted papers that Harper (1968) first proposed the study of individual plants as populations of repeating units, in particular the unit formed by the leaf and associated internode and axillary meristem. Later (Harper 1977) he hinted again at this proposition but did not explore it in detail. In more recent years the idea has been developed rapidly (Harper 1981). It is not the purpose of this paper to discuss the scope of this approach, as White has already done (1979, 1980, 1984), but rather to formulate some of the questions (and I am afraid only a few answers) which can be raised using this methodology when one studies trees.

It has recently been suggested (Waring, Thies & Muscato 1980; Waring, Newman & Bell 1981; Satoo & Madgwick 1982; Waring 1983) that canopy leaf area and growth efficiency are the parameters most likely to permit a biologically meaningful description of forest production. This approach parallels the classical field of plant growth analysis (Evans 1972; Hunt 1978a). In spite of the argu-

[1]Present address: Departamento de Ecología, Instituto de Biologia, Universidad Nacional Autonóma de México, 04510 Mexico City, Mexico.

STUDIES ON PLANT DEMOGRAPHY:
A FESTSCHRIFT FOR JOHN L. HARPER

ments already raised against the demography of modules, which may themselves be variable in size (Hunt 1978b), I would like to present a heuristic model of tree growth which could be used as an idealistic, hypothesis-generator model.

Two central concepts are relevant to the approach presented here: the pipe model (Shinozaki *et al.* 1964a, b) and the integrated physiological unit (Watson & Casper 1984). The *pipe model* theory is a restatement of an old observation on the hydraulic architecture of trees (Zimmermann 1983). It is based on the empirical evidence that the total amount of foliage at and above a certain height of the trunk is proportional to its cross-sectional area at that same height. This leads to the observation that an individual tree (or even the whole forest) can be seen as an assemblage of individual units, each of them consisting of a unit amount of leaves and a long and thin conductive–supportive pipe of constant thickness. A dynamic version of the pipe model would suggest that these pipes are accumulated through a demographic-type process. However, in order for the proportionality between foliage and cross-sectional area to be held constant, a tall tree would have to produce longer (heavier) pipes than a short one, unless the proportionality were achieved through the use of some of the old pipes. Without any *a priori* assumptions about the possible variation in this proportionality, this dynamic version will convey the necessary heuristic idealization of an individual module.

In a recent paper Watson and Casper (1984) have proposed, on the basis of morphological and physiological evidence, the use of the term *integrated physiological unit* (IPU) for individual 'modules' with more or less independent carbon economy. They claim that most plants do not behave as completely integrated entities but more as populations of relatively independent physiological units.

In this paper I shall argue in favor of a modular approach for the description and quantification of tree growth by making use of a loose combination of these two concepts into what will be called a *structural unit* (SU).

II. DEFINITION OF CONCEPTS

In order to develop an appropriate argument for the growth of tree populations in terms of modular structure, I shall first define some concepts.

Total mass, or simply mass, is the total amount of dry matter, both living and dead, present at any given moment.

Biomass is only that part of the total tissue mass which still shows some kind of metabolic activity expressed as respiration. As will be discussed, this is assumed to be proportional to leaf area.

Necromass is defined as that part of the total mass that has been accumulated throughout the years and which does not show any sign of metabolic activity. This is somehow related, but not necessarily equal, to heartwood.

The three of them may be measured either on an individual or a population basis.

I have avoided the use of the term *biomass* when refering to mass, both living and dead combined. This distinction is usually not made by foresters or production ecologists. Existing models to quantify tree growth make use of variables which combine both dead and living tissues. Since dead tissues do not contribute to growth, I believe they should be excluded from realistic biological models of tree growth. All trees continuously shed leaves, twigs and whole branches, and these are grouped together under the term *litterfall,* not as part of the total mass (biomass in the traditional sense). Since nobody would consider it convenient to include these figures in the definition of a predictor variable of forest growth, I cannot see any reason to include accumulated dead wood material.

III. A SIMPLE MODEL OF MASS ACCUMULATION

If light limits the size of the leaf population so that it cannot exceed the carrying capacity of the environment, one could more conveniently define the total amount of foliage present at a certain time (say LAI at the end of the growing season) as a measure of population size. The basic dynamics of the total amount of leaves per unit area as a forest develops can then be described as a birth and death process resembling logistic growth:

$$\frac{df}{dt} = (b - d)f \left[\frac{F - f}{F}\right] \tag{1}$$

where:

f = amount of foliage per unit area; it can be expressed as leaf numbers, leaf area or leaf weight.

b = leaf birth rate; mean amount of leaves produced by a unit amount of leaves per unit of time.

d = leaf death rate; mean amount of leaves lost per unit amount of leaves per unit of time.

F = maximum amount of foliage that can be supported.

t = time.

The solution of the logistic equation is

$$f = \frac{F}{1 + e^{a-(b-d)t}} \tag{2}$$

where:

e = the base of natural logarithms and a is a constant.

Now, let us assume that structural units of construction of biomass (SUs) exist and each one of them consists of a distal (photosynthetic) part (PP) and a proximal (conductive and supportive) part (CSP). As more SUs are accumulated, some of them, presumably the ones at the bottom of the canopy, start to die at the distal end. The PP will be shed first. Part of the CSP (e.g. twigs) may also be shed (and this typically later than leaves) but in trees most of it, being part of the trunk and major branches, is reatined and remains active for a few years, forming part of the sapwood. Eventually, the CSP will also die and become part of the central core or heartwood.

The process of accumulation of CSPs is proportional to the total amount of leaves present (PPs) and to their average birth rate, but is independent of leaf mortality rate; once formed, CSPs are not shed by the tree and persist irrespective of leaf longevity. Leaf mortality rate only affects the growth of the wood mass inasmuch as fewer leaves are available to produce copies of themselves. Each leaf comes attached to a 'body' of a certain size. Whether it is the 'parent' leaf which produces the 'daughter' leaf's body or it is the 'daughter' leaf itself is just a semantic problem which needs not concern us here. From this point of view a leaf is the means to produce more SUs, not a means to produce wood. The distinction is important because I believe this semantic difference may allow a better understanding of the basic process of tree growth.

Following the preceding argument, the change in the amount of CSPs (wood) can be expressed in differential form as

$$\frac{dw}{dt} = Cbf \tag{3}$$

where

w = amount of supportive–conductive parts or wood mass.

C = constant of proportionality between photosynthetic and supportive–conductive parts of the SU.

Substituting (2) in (3) and integrating from zero to t, one obtains:

$$w_t = w_0 + \frac{CbF}{r}\,(rt + \ln[(1 + e^{a-rt})/(1 + e^{a})] \tag{4}$$

where:

ln represents Naperian or natural logarithms and $r = b - d$.

This simple idealization of mass accumulation in forest trees produces a sigmoid increase in foliage while wood mass initially increases following a concave curve and, once maximum LAI has been reached, wood mass increases linearly; that is, annual wood production becomes constant. The objective of any person engaged in the management of forests would then be to bring their production as close as possible to this ideal pattern.

The general trend of a gradual increase followed by a plateau in current annual wood increment deserves a closer examination from a mechanistic point of view. If parameter b were a function of soil fertility so that at constant maximum LAI richer sites had a higher foliage turnover rate, according to this model wood production would then be proportionally higher. In fact there is evidence that some interaction between light and soil fertility occurs so that at same light intensity level richer soils seem to support a higher maximum LAI (Satoo & Madgwick 1982).

The plateau in current annual wood increment may be a short or a long one. Nonetheless, it will eventually give place to a drop in mass accumulation in the forest. Two main factors account for this decrease:

1. Regardless of whether branch weight has been taken into account or not, there is always some loss of wood material through litter fall and death of 'whole' trees. Yield tables do not normally consider dead trees unless some sort of thinning practice is performed and included in the predicted yield. One should also be aware that suppressed trees may be ecologically dead (or literally almost 100% necromass) and the decision on whether to include them in the predicted yield or not will certainly produce different results.

2. The second possibility is a true drop in net primary production as a consequence of either deprivation of some essential mineral resource (assuming light and water supply remain constant) or some impediment in the physiological efficiency of the plant itself. These two cases may be viewed, respectively, as a lessening of the leaf birth rate and a decrease in the value of C, the constant of proportionality in equations (3) and (4). I shall come back to these issues in the following section but in order to do so I shall use a different version of the model.

Temperate trees do not grow continuously but produce annual flushes which can better be described by a discrete model. For illustration consider the following case. Cohorts of leaves are produced in annual flushes, and their numbers (area or weight) decline thereafter, following a particular survivorship curve (f_x). Old meristems produce new ones at an average, age-specific rate (m_x).

If one let m_0 be the ratio $f_{0,t+1}/f_{0,t}$, one would avoid the need of quantifying any physiological activity to measure the contribution that each meristem makes to future growth. Meristems produced in year $t + 1$ come from meristems produced the previous year; thus m_x has only one positive value m_0; that is, only meristems produced the previous season 'reproduce'.

This process can be summarized in the following flux:

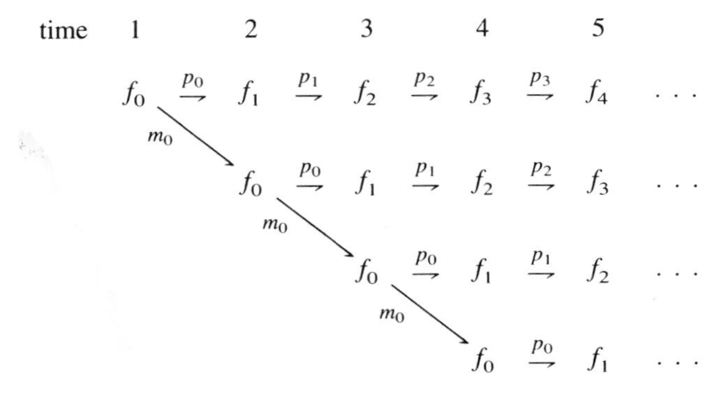

The behaviour of the leaf population can be simulated by making use of a Leslie-type matrix system (Leslie 1945, 1948; see also Maillete 1982; McGraw & Antonovics 1983) of the following form:

$$\begin{bmatrix} f_0 \\ f_1 \\ f_2 \\ \vdots \\ f_n \end{bmatrix}_{t+1} = \begin{bmatrix} m_0 & m_1 & m_2 & \ldots & m_n \\ p_0 & 0 & 0 & \ldots & 0 \\ 0 & p_1 & 0 & \ldots & 0 \\ \vdots & \vdots & \vdots & & \vdots \\ 0 & 0 & 0 & \ldots & 0 \end{bmatrix} \begin{bmatrix} f_0 \\ f_1 \\ f_2 \\ \vdots \\ f_n \end{bmatrix}_t \quad (5)$$

where:

p_x = the probability that a leaf aged x at time t will be alive in the age group $x + 1$ at time $t + 1$.

m_x = mean amount of foliage produced in the interval t to $t + 1$ by a unit amount of leaf aged x.

f_x = amount of foliage aged x present per unit area.

The total amount of foliage present at time $t + 1$ is obviously

$$F_{t+1} = \sum_x f_{x,t+1} \quad (6)$$

One might expect the parameters m_x and p_x to be a decreasing function of density (LAI) and an increasing function of soil fertility or introduce some sort of random environmental variation from year to year.

The total amount of wood mass (W) at time $t + 1$ is the cumulative sum of the wood net primary production (w_i) times the proportion still attached to the tree (s_i). The latter is equal to 1 minus the percentage lost through litter fall and decomposition ($s_i = 1 - d_i$):

$$W_{t+1} = s_0 w_0 + s_1 w_1 + s_2 w_2 + \ldots + s_t w_t + s_{t+1} w_{t+1} \quad (7)$$

As before, the growth of the wood mass would be proportional to the amount of new leaves produced times a constant of proportionality between proximal and distant part of the SU, or

$$w_{t+1} = C m_0 f_{0,t} = C f_{0,t+1} \quad (8)$$

Assuming there is no loss of leaf and wood material during the year these are produced, it is clear from this equation that C is the ratio wood net primary production (w_{t+1}) to foliage net primary production ($f_{0,t+1}$).

IV. TEST OF THE MODEL USING FORESTRY DATA

The crucial assumption in the application of the model is that the parameter C either is constant or varies in a predictable manner. Thus, the first problem to be solved is whether there is any evidence from forestry literature data which shows some general pattern in the range of values the parameter C can take. This should be in accordance with patterns of resource allocation, although it must be appreciated that the classical way in which resource allocation is measured does not allow direct comparison, because it is normally recorded without consideration of the dynamic aspect of the leaf population mentioned before.

Cannell's compilation on World Forest Productivity Studies (Cannell 1982) is a valuable source of relevant data. From all the information presented, I selected 477 stands on the basis that they should contain information on both standing crop (his 'biomass') and net primary production (NPP) of at least leaves and main stem and of branches if possible. I discarded those plots in which foliage NPP was assumed to be some proportion of the total foliage mass present.

With this information and assuming

1. The rate of foliage production remains approximately constant for a few years,
2. The proportion wood NPP to foliage NPP also remains constant during this time interval,
3. Leaf survivorship is of a perfect Type I; that is, no leaf mortality occurs until the end of its predicted lifetime for all cohorts,

the following were calculated:

$$\text{Leaf lifetime} = \frac{\text{foliage mass}}{\text{foliage NPP}}$$

$$\text{Biomass} = \text{leaf lifetime (foliage NPP + stem NPP + branch NPP)}$$

$$\text{Total mass} = \text{foliage mass + stem mass + branch mass}$$

$$\text{Necromass} = \text{total mass} - \text{biomass}$$

In a few cases biomass exceeded total mass. In such cases biomass was made equal to total mass, and consequently necromass was zero.

Although both total mass and necromass increased with age or size, however measured, of the average tree, biomass rarely exceeded 100 tons/ha (Fig. 1). Biomass behaved in essentially the same way as leaf area index, increasing rapidly with age; LAI rarely exceeded a value of 10. So the percentage of necromass could vary from 0 to nearly 100% of the total mass (Fig. 2).

It can be argued that these assumptions underestimate the amount of living

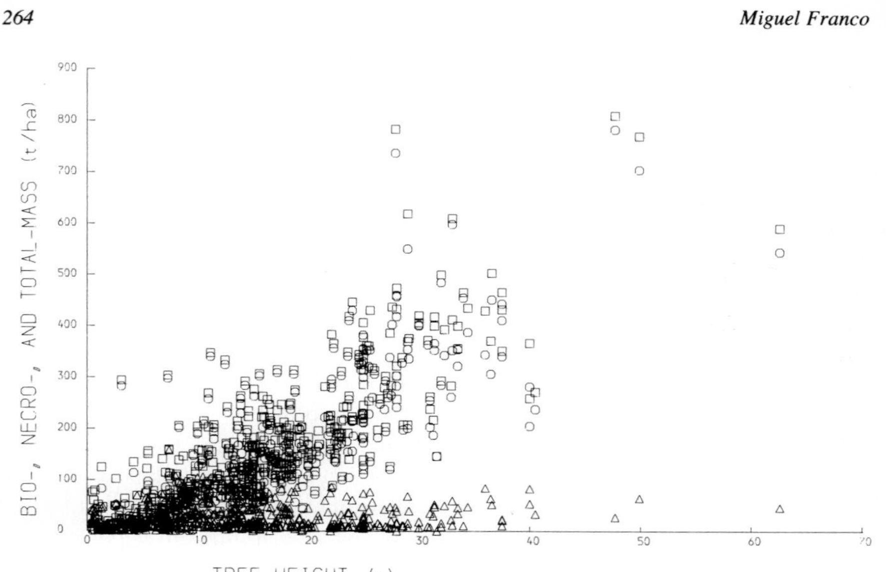

Fig. 1. Amount of biomass (△), necromass (○), and total mass (□), plotted against tree height, of 432 sample forests worldwide. Height, instead of age, was used as the independent variable because trees accumulate mass in proportion to the volume they occupy, and regardless of the time that it takes to fill that volume. Plots without information on height were excluded. (Calculated from data reported in Cannell [1982].)

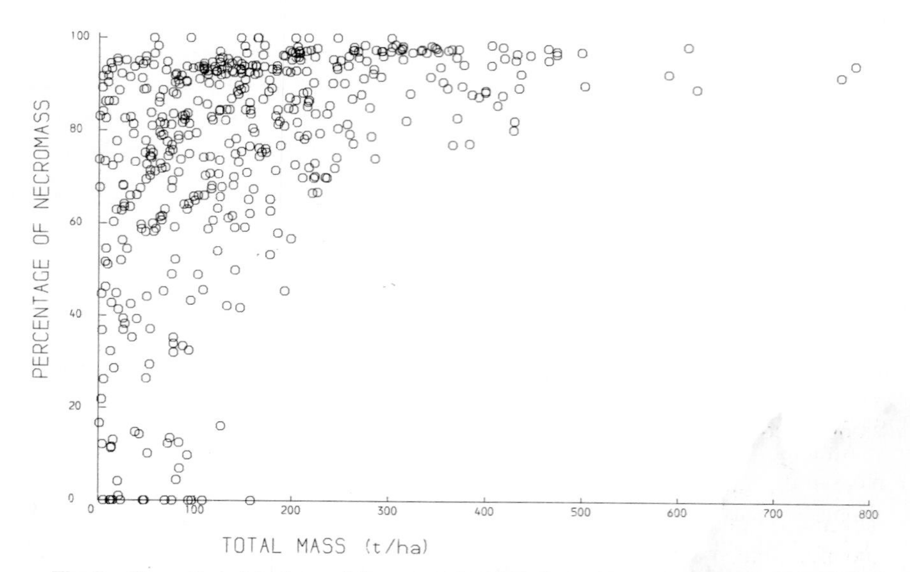

Fig. 2. Proportion of dead material present, plotted against total mass accumulated in 477 forest stands. (Calculated from data reported in Cannell [1982].)

tissues or biomass for stands composed of big trees, because sapwood volume is not proportional to leaf mass. Actually sapwood cross-sectional area is proportional to leaf mass (Grier & Waring 1974; Waring *et al.* 1977; Whitehead 1978; Snell & Brown 1978; Long, Smith & Scott 1981; Waring, Newman & Bell 1981; Waring, Schroeder & Oren 1982). Were one to weigh the 'livingness' of a tissue by its metabolic activity (respiration), sapwood activity drops dramatically from the last two or three rings towards the pith (Möller 1945; Möller, Müller & Nielsen 1954; Yoda *et al.* 1965; Negisi 1970, 1977). Kozlowski (1971 p. 12) points out that of all the cells comprising sapwood only about 10% are alive. Similarly, at least for ring-porous species, some essential physiological activities such as water conduction are performed preferentially or exclusively through the last few rings (Huber 1935; Kozlowski 1961; Zimmermann 1971, 1983).

Since not all the studies summarized by Cannell (1982) reported data on bark, this factor was excluded from the analysis. Its inclusion would simply increase the value of the parameter C proportionally. The parameter C was calculated as follows:

$$C = \frac{\text{stem NPP} + \text{branch NPP}}{\text{foliage NPP}}$$

As expected, this wood–foliage productivity ratio did not show any correlation whatsoever with variables such as age, density or height of trees. When it was plotted against biomass, however, the trend shown in Fig. 3 emerged. The range of values for the regression went from less than 1 for plots with a total biomass between 0 and 10 tons per hectare, to near 6 for stands with a total biomass around 150 tons per hectare. Most values however were between 0.75 and 3.0 ($\bar{X} = 1.90$; $s = 1.13$). There seems to be a minimum value in this ratio which increases as biomass accumulates. This has been represented as a dotted line parallel to the calculated regression line in Fig. 3.

As to the variation around the line, it can be explained by the following factors:

1. It must be stressed that both foliage NPP and wood NPP are not direct measurements but have been estimated through several, sometimes completely different methods, which are not quite appropriate for testing the hypothesis proposed here accurately. In particular, the use of allometric regressions to estimate some of these figures represents only a crude guess about them.

2. In some cases wood NPP represented stem NPP only.

3. There may be some differences between species, although with the present information this may be difficult to test. Dividing the data set into three groups (conifer, temperate broad-leaved and tropical species) did not show any difference between them. (I was rather surprised, indeed, to find such a consistent trend, despite its variability, from such a range of species, ages, sizes, etc.!).

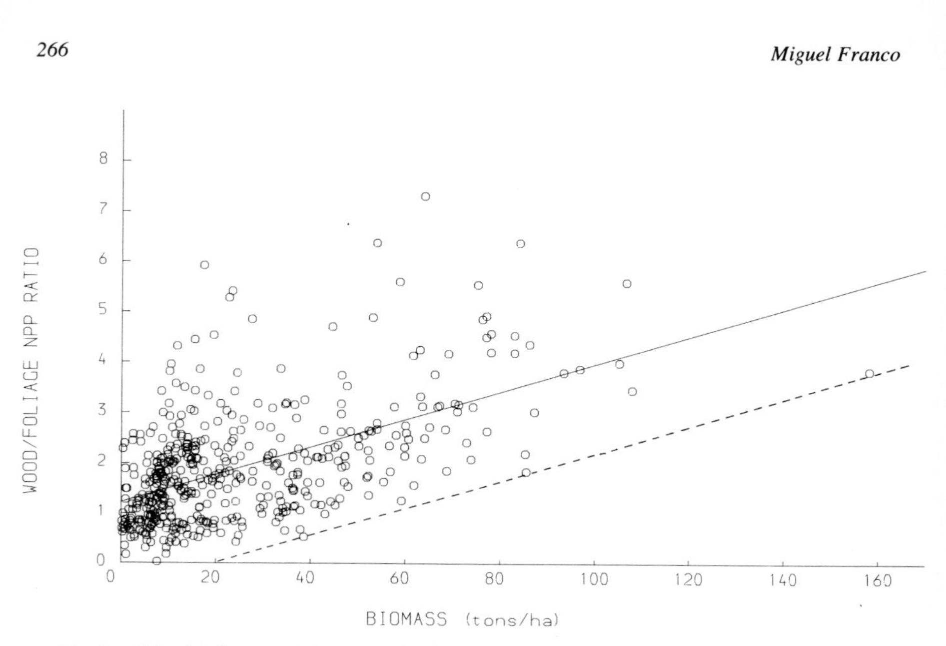

Fig. 3. Wood/foliage net primary production ratio, as related to the total amount of living mass, of 477 forest stands ($Y = 1.236 + 0.027X$; $r = 0.55$; $n = 477$; $p \ll 0.001$). The dotted line represents a guess of the minimum values this relationship can take. (Calculated from data in Cannell [1982].)

4. It is likely that differences in growing conditions affect this idealized allocation pattern.

To illustrate this last point I shall use four examples on the effect of fertilizers on the productivity of tree populations (Table I).

The first example shown represents stands of *Eucalyptus globulus* with four levels of fertilizer at two different ages. In this case, foliage NPP increased with both amount of fertilizer and age of the trees. *C*, however, did not seem to show any response to the application of nitrogen and phosphorus, although it seemed to decrease from age 4 to age 9.5 years old. Thus, wood NPP behaved essentially

---

$^{a',b',c',d'}$ Nine-and-one-half-year-old trees with treatment described in preceding note.

^{e} Forty-four-year-old trees growing on good-quality sites.

^{f} Forty-four-year-old trees growing on medium-quality sites.

^{g,h,i,j,k} Thirty-nine-year-old trees which received the same amount of fertilizer at age 36 plus increasing amounts in the order $g < h < i < j < k$ in the 3 subsequent years.

^{l} Sixty-year-old stand in an infertile soil.

^{m} Fifty-two-year-old stand in an intermediate site.

^{n} Fifty-two-year-old stand in a good site.

^{o} All data as reported in Cannell (1982, pp. 9–10, 48–49, 246, 268). The original sources are described therein.

TABLE I.

Parameters Determining Forest Growth in Four Species of Trees under Different Soil Fertility Levels[o]

Foliage NPP $f_{0,t}$ t/ha/y	Wood NPP w_t t/ha/y	Relative allocation C g/g	Efficiency of wood production w_t/F g/g
	Eucalyptus globulus		
0.79	2.02	2.56	1.836[a]
2.10	4.20	2.00	1.680[b]
2.85	5.72	2.01	1.733[c]
4.02	8.37	2.08	1.610[d]
2.60	3.40	1.31	0.850[a']
3.80	6.20	1.63	1.216[b']
5.00	7.80	1.56	1.164[c']
5.90	8.20	1.39	1.242[d']
	Pinus banksiana		
1.36	3.06	2.25	0.638[e]
1.36	3.16	2.32	0.702[e]
1.43	3.00	2.10	0.588[e]
1.45	3.04	2.10	0.633[e]
0.92	1.90	2.07	0.559[e]
1.08	2.19	2.03	0.592[e]
1.71	3.51	2.05	0.450[f]
1.74	3.53	2.03	0.484[f]
1.69	2.88	1.70	0.406[f]
1.87	3.11	1.66	0.451[f]
1.64	2.25	1.37	0.388[f]
1.58	2.13	1.35	0.374[f]
	Pinus nigra var. *maritima*		
3.18	5.37	1.69	0.726[g]
4.01	7.10	1.77	0.724[h]
4.97	8.93	1.80	0.757[i]
5.37	9.35	1.74	0.698[j]
5.46	7.60	1.39	0.514[k]
	Populus grandidentata		
0.96	1.69	1.76	1.760[l]
1.76	4.77	2.71	2.710[m]
2.41	7.65	3.17	3.174[n]

[a,b,c,d] Four-year-old trees with increasing levels of fertilizer (N and P) in the order $a < b < c < d$ applied during the first 2 years of development.

as foliage NPP. If efficiency is expressed as the ratio of wood NPP to total amount of foliage present, no pattern is apparent. This could be interpreted as a consequence of the stands' still building up their canopies and so increasing in wood mass in proportion to the total amount of live leaves. Were the canopies at similar, maximum levels of foliage mass, stands with higher values of NPP would be more 'efficient'.

The second example concerns a series of 44-year-old plots of *Pinus banksiana* which seemed to respond to variations in fertility through changes in the parameter *C:* this was higher in good- than in medium-quality sites. Contrary to expectation, foliage NPP was lower in good- than in medium-quality sites. This might be an indication that nutrient capture in these stands may be a slow process. Nutrients already captured are utilized in the production of supporting tissues, which might give an advantage in the competition for light, but nutrient accumulation has presumably slowed down considerably. The ratio of wood NPP to total amount of foliage was, however, higher as a consequence of good sites' supporting fewer leaves than medium-quality ones. One still wonders whether these plots bear an amount of leaves close to maximum LAI.

The third example, *Pinus nigra* var. *maritima,* suggested that for 39-year-old stands, leaf NPP increased with fertility level, C remained approximately constant (Table I), with one exception, the plot with the higher concentration of fertilizer. No apparent trend was seen in the ratio of wood NPP to total amount of foliage. Although no data are given which might suggest a trend in time of $f_{0,t}$, if this were proportional to m_0, the example would seem to coincide with the predictions concerning the relationship between soil fertility and both C and m_0 made previously.

The last example, *Populus gradidentata,* showed that both foliage NPP and wood NPP increased with soil fertility (Table I). As wood NPP increased in a bigger proportion than foliage NPP, C also increased with soil fertility. In the case of deciduous species the parameter C is equal to the ratio of wood NPP to foliage mass and the ratio of foliage NPP to foliage mass is always equal to 1. This suggests that the latter might be better defined by the ratio $f_{0,t+1}/f_{0,t}$. The information given did not permit the estimation of this value, however.

A more complete study, well suited for illustrating the approach proposed, is Sprugel's investigation on the wave-regenerated balsam fir forests of North America (Table II). Foliage mass increased very rapidly in the first 15 years, reaching its maximum value before the population was 20 years old. This, and evidence on self-thinning, indicated that the forest is at carrying capacity most of the time. Although foliage NPP increased with time, wood NPP first increased and then decreased. This was interpreted as a gradual reduction in both C and m_0 as the forest aged. Because foliage mass was maximal and tended to remain constant, the ratio of wood NPP to foliage mass followed the same pattern as wood NPP. Efficiency from the point of view of this ratio increased until the age of 30–40 years and then declined as the stand aged. It is interesting to note that

TABLE II.

Variation in the Growth Parameters of *Abies balsamea* throughout Its Life-span[a]

Age x years	Foliage mass F t/ha	Wood mass t/ha	Foliage NPP $f_{0,t}$ t/ha/y	Wood NPP w_t t/ha/y	Relative allocation C g/g	Growth ratio[b] m_0 g/g	Efficiency of wood production w_t/F g/g/y	Production rate $f_{0,t}/F$ g/g/y	Biomass t/ha
10	13.0	16.5	2.05	3.40	1.66	1.44	0.262	0.158	35.56
20	15.0	26.5	2.95	4.25	1.44	1.14	0.283	0.197	36.61
30	15.0	56.0	3.35	4.45	1.33	1.07	0.297	0.223	34.93
40	14.0	74.0	3.60	4.30	1.19	1.04	0.307	0.257	30.72
50	14.5	87.5	3.75	4.10	1.09	1.04	0.283	0.259	30.36
60	15.5	99.0	3.90	3.95	1.01	—	0.255	0.252	31.20

[a] From data in Sprugel (1984).

[b] No attempt was made to adjust for mean annual values of the growth ratio, but calculations were done as if changes occur abruptly every 10 years.

efficiency as measured by both C and m_0 decreased monotonically with age. Wood NPP at early stages was small because less foliage was produced. At later stages wood NPP was small as a consequence of reduced C and m_0. It is also interesting to observe that C in this study did not increase with age, since biomass had reached its maximum value early in the life of the forest. On the contrary, once maximum biomass had been attained, C decreased monotonically. The gradual increase in foliage NPP can be interpreted as a consequence of nutrient accumulation during the whole lifetime of the forest, even though this proceeded at slower rates at older ages.

More substantial information is necessary to test whether the partition of the components of primary production into the ones put forward in this paper merits a closer examination. Also, it would be of interest to investigate the consequences of the model on the interpretation of competition effects at the individual tree level.

ACKNOWLEDGMENTS

Some of the ideas presented in this paper were conceived while working under the direction of Jim Sarukhán in Mexico and were further developed during my studies in Bangor with Professor J. L. Harper. Additionally I thank Dr. J. White for his thorough criticism of the original manuscript.

This work was supported by the British Council and the Universidad Nacional Autónoma de México.

REFERENCES

Cannell, M. G. R. (1982). *World Forest Biomass and Primary Production Data.* Academic Press, London.

Cusset, G. (1982). The conceptual bases of plant morphology. *Axioms and Principles of Plant Construction* (Ed. by R. Sattler), pp. 8–86. Martinus Nijhoff/Dr. W. Junk, The Hague.

Evans, G. C. (1972). *The Quantitative Analysis of Plant Growth.* Blackwell Scientific Publications, Oxford.

Grier, C. C. & Waring, R. H. (1974). Conifer foliage mass related to sapwood area. *Forest Science,* **20,** 205–206.

Harper, J. L. (1968). The regulation of numbers and mass in plant populations. *Population Biology and Evolution* (Ed. by R. C. Lewontin), pp. 139–158. Syracuse University Press, Syracuse, New York.

Harper, J. L. (1977). *Population Biology of Plants.* Academic Press, London.

Harper, J. L. (1981). The concept of population in modular organisms. *Theoretical Ecology: Principles and Applications* (Ed. by R. M. May), pp. 53–77. Blackwell Scientific Publications, Oxford.

Huber, B. (1935). Die physiologische Bedeutung der Ring- und Zerstreutporigkeit. *Berichte der Deutschen Botanischen Gesellschaft,* **53,** 711–719.

Hunt, R. (1978a). *Plant Growth Analysis.* Edward Arnold, London.

Hunt, R. (1978b). Demography versus plant growth analysis. *New Phytologist,* **80,** 269–272.

Kozlowski, T. T. (1961). The movement of water in trees. *Forest Science*, **7**, 177–192.
Kozlowski, T. T. (1971). *Growth and Development of Trees*. Vol. 1. Academic Press, New York.
Leslie, P. H. (1945). On the use of matrices in certain population mathematics. *Biometrika*, **33**, 183–212.
Leslie, P. H. (1948). Some further notes on the use of matrices in population biology. *Biometrika*, **35**, 213–245.
Long, J. N., Smith, F. W. & Scott, D. R. M. (1981). The role of Douglas-fir stem sapwood and heartwood in the mechanical and physiological support of crowns and development of stem form. *Canadian Journal of Forest Research*, **11**, 459–464.
Maillette, L. (1982). Structural dynamics of silver birch. II. A matrix model of the bud population. *Journal of Applied Ecology*, **19**, 219–238.
McGraw, J. B. & Antonovics, J. (1983). Experimental ecology of *Dryas octopetala* ecotypes. II. A demographic model of the growth, branching and fecundity. *Journal of Ecology*, **71**, 899–912.
Möller, C. M. (1945). Untersuchungen über Laubmenge, Stoffverlust und Stoffproduktion des Waldes. *Forstlige Forsgsøvaesen i Danmark*, **17**, 1–287.
Möller, C. M., Müller, D. & Nielsen, J. (1954). Respiration in stems and branches of beech. *Forstlige Forsgsøvaesen i Danmark*, **21**, 273–301.
Negisi, K. (1970). Respiration in non-photosynthetic organs of trees in relation to dry matter production of forests. *Journal of the Japanese Forest Society*, **52**, 331–345.
Negisi, K. (1977). Respiration in forest trees. *Primary Productivity of Japanese Forests - Productivity of Terrestrial Communities* (Ed. by T. Shidei & T. Kira), JIBP Synthesis, 16, pp. 86–99. University of Tokyo Press, Tokyo.
Satoo, T. & Madgwick, H. A. I. (1982). *Forest Biomass*. Martinus Nijhoff/Dr. W. Junk Publishers, The Hague.
Shinozaki, K., Yoda, K., Hozumi, K. & Kira, T. (1964a). A quantitative analysis of plant form - the pipe model theory. I. Basic analysis. *Japanese Journal of Ecology*, **14**, 97–105.
Shinozaki, K., Yoda, K., Hozumi, K. & Kira, T. (1964b). A quantitative analysis of plant form - the pipe model theory. II. Further evidence of the theory and its application in forest ecology. *Japanese Journal of Ecology*, **14**, 133–139.
Snell, J. A. K. & Brown, J. K. (1978). Comparison of tree biomass estimators - dbh and sapwood area. *Forest Science*, **24**, 455–457.
Sprugel, D. G. (1984). Density, biomass, productivity, and nutrient-cycling changes during stand development in wave-regenerated balsam fir forests. *Ecological Monographs*, **54**, 165–186.
Waring, R. H. (1983). Estimating forest growth and efficiency in relation to canopy leaf area. *Advances in Ecological Research*, **13**, 327–354.
Waring, R. H., Gholz, H. L., Grier, C. C. & Plummer, M. L. (1977). Evaluating stem conducting tissue as an estimator of leaf area in four woody angiosperms. *Canadian Journal of Botany*, **55**, 1474–1477.
Waring, R. H., Thies, W. G. & Muscato, D. (1980). Stem growth per unit of leaf area: A measure of tree vigor. *Forest Science*, **26**, 112–117.
Waring, R. H., Newman, K. & Bell, J. (1981). Efficiency of tree crowns and stem wood production at different canopy leaf intensities. *Forestry*, **54**, 129–137.
Waring, R. H., Schroeder, P. E. & Oren, R. (1982). Application of the pipe model theory to predict canopy leaf area. *Canadian Journal of Forest Research*, **12**, 556–560.
Watson, M. A. & Casper, B. B. (1984). Morphogenetic constraints on patterns of carbon distribution in plants. *Annual Review of Ecology and Systematics*, **15**, 233–258.
White, J. (1979). The plant as a metapopulation. *Annual Review of Ecology and Systematics*, **10**, 109–145.
White, J. (1980). Demographic factors in populations of plants. *Demography and Evolution in Plant Populations* (Ed. by O. T. Solbrig), pp. 21–48. Blackwell Scientific Publications, Oxford.

White, J. (1984). Plant metamerism. *Perspectives on Plant Population Ecology* (Ed. by R. Dirzo & J. Sarukhán), pp. 15–47. Sinauer Associates, Sunderland, Massachusetts.

Whitehead, D. (1978). The estimation of foliage area from sapwood basal area in Scots pine. *Forestry,* **51,** 137–149.

Yoda, K., Shinozaki, K., Ogawa, H., Hozumi, K. & Kira, T. (1965). Estimation of the total amount of respiration in woody organs of trees and forest communities. *Journal of Biology, Osaka City University,* **16,** 15–26.

Zimmermann, M. H. (1971). Transport in the xylem. *Trees: Structure and Function* (M. H. Zimmermann & C. L. Brown), pp. 169–220. Springer-Verlag, New York.

Zimmermann, M. H. (1983). *Xylem Structure and the Ascent of Sap*. Springer-Verlag, Berlin.

IV

Plant Interference: The Effects of Neighbours

18

Plant Responses to Crowding

A. R. Watkinson

School of Biological Sciences
University of East Anglia
Norwich, England

I. INTRODUCTION

Some of the best documented generalizations of plant demography concern the relationships between density and the size of plants. Much of the early interest in yield–density relationships centred around how crop yield responded to density, but an understanding of how plants respond to the proximity of neighbours is also essential if one is to understand how population size is regulated and the abundance of plants determined. In this respect it was Harper (1961) who realized the importance of the vast quantity of data that related yield to density in the literature of agriculture and forestry. Most of the data that had been collected by agronomists and foresters, however, related to stands of single species, a situation that is far removed from the complexities of natural communities. It remained an act of faith of population biologists that the observations on yield–density relationships in monoculture were relevant to understanding the dynamics of populations in multi-species mixtures. One of the major aims of this paper is to examine our current understanding of yield–density relationships in monocultures and then to relate this information to the population regulation of annual plants in the field. The other is to examine which factors are important in determining the yield of individual plants in populations: it is, of course, the study of variance rather than mean yield that is most relevant to understanding evolution (Harper 1981).

II. YIELD–DENSITY RELATIONSHIPS IN MONOCULTURE

Two important generalizations have emerged from the study of yield–density relationships in monocultures. The first relates to the negatively density–dependent relationship between plant size and density whilst the second relates to the density–dependent mortality of plants that results from self-thinning. Both are largely a consequence of the modular nature of plant growth (Harper 1981).

In monocultures the total yield of a population at any given time is restricted by the availability of resources such that there is a reciprocal relationship between mean plant dry weight (w) and density (N)of the form

$$w = w_m(1 + aN)^{-b} \tag{1}$$

where w_m, a and b are parameters (Watkinson 1980). The number of plants that can survive to a given time in a monoculture is, however, restricted at high densities by self-thinning (Yoda *et al.* 1963) such that the relationship between the initial density of plants (N_i) and the density of surviving plants (N) can be described by the equation

$$N = N_i(1 + mN_i)^{-1} \tag{2}$$

where m^{-1} is a parameter that represents the maximum density of plants that the habitat can support. The value of m changes with time as the plants grow larger, such that

$$w = cm^k \tag{3}$$

where w is the mean shoot dry weight of the surviving plants and c and k are constants (Firbank 1984). Since m^{-1} represents the maximum density of plants that the habitat can support, the relationship between the number of surviving plants and the mean shoot dry weight of the survivors through time can be expressed by the equation

$$w = cN^{-k} \tag{4}$$

This relationship represents a boundary condition and truncates the reciprocal Eq. (1) such that densities higher than the point of intersection cannot be realized because of the effects of density-dependent mortality. As the value of k in Eq. (4) typically has a value of approximately 1.5, it is generally referred to as the $-3/2$ power law, or self-thinning rule.

The biological interpretation of the parameters in Eqs. (1) and (4) and some indication of the way the parameters vary with conditions are both of considerable interest if one is to gain a fuller understanding of plants' responses to crowding. In Eq. (1) w_m can be interpreted biologically as the dry matter production of an isolated plant, a as the area required to achieve a yield of w_m and b as the effectiveness with which resources are taken up from that area (Watkinson 1980, 1984). The parameter a can also be considered as the ecological neighbourhood area (*sensu* Antonovics & Levin 1980) since only when the neighbourhood areas of individual plants overlap will yield be density-dependent.

All three of the parameters in Eq. (1) increase with time. One would expect w_m and a to increase because the area of resources required to achieve a yield of

w_m will inevitably increase as the plants grow larger. However, although it has previously been assumed that b is a constant, a detailed series of experiments on *Vulpia fasciculata* (Watkinson 1984) has shown that b increases with time towards a value of unity. An increase in the value of b indicates that the resources within the space available to each plant are more fully utilized with time and presumably reflects the proliferation of the root and shoot systems within the volume of space occupied by a plant. It is only when the value of $b = 1$ that the resources within the space available to each plant are fully utilized and the commonly observed 'law of constant final yield' is found over a wide range of densities (e.g. Shinozaki & Kira 1956).

Equation (1) quantifies the extent to which plants sown at high densities interfere with each other's growth earlier in their development than plants sown at lower densities, which do so only when they have grown larger. The addition of nutrients to plants sown over a range of densities might well be expected to increase their growth rate and thus the rate at which the effects of crowding occur but it also affects the intensity of interference. Two series of experiments on *Vulpia fasciculata* have shown that over a series of harvests

$$w_m = Ca^{Db} \tag{5}$$

for a given set of nutrient conditions where w_m, a and b are as previously defined and C and D are parameters (Watkinson 1984). The addition of nutrients results in an increase in the value of C. Such an increase means that a larger weight can be supported by a given area because the resources within that area are greater. The corollary is that a smaller area is required to support a given weight when nutrients are added. In lowering the area of resources (a^b) needed for a given value of w_m the addition of nutrients delayed the onset of interference at high densities. The data of Harper (1961) on two species of *Bromus* are also entirely consistent with the hypothesis that the addition of nutrients decreases the area of resources required to support a given value of w_m and delays the onset of interference.

III. SIZE HIERARCHY DEVELOPMENT

In considering the effects of crowding within monocultures I have so far dealt with mean plant size and ignored the variation that occurs among individuals within populations. Frequency distributions of seed weight within monospecific populations are generally normally distributed (Obeid, Machin & Harper 1967; Rabinowitz 1979) but within a population individual plants have frequency distributions of weight that are asymmetric and positively skewed. Such hierarchies with few large individuals and numerous smaller ones are established soon after seedling emergence and appear to become more positively skewed as seedlings grow. Simple exponential growth is sufficient to cause such shift from a symmetrical to a skewed distribution if individuals differ slightly in their relative growth rates (Koyama & Kira 1956), although other factors may also play a role. In high-density stands where interference occurs between individuals asymmetric or one-sided competition may further exaggerate the difference between individuals, leading to dominance and suppression, eventually to such an extent that some individuals are lost from the population (e.g. Aikman & Watkinson 1980; Ford & Diggle 1981; Cannell, Rothery & Ford 1984). Depending on the exact nature of the survivorship-probability function, density-dependent mortality could then further exaggerate the skewness of the population with time or counteract it (Begon 1984). Simulation models that incorporate variation in intrinsic growth rate together with one-sided competition (Aikman & Watkinson 1980) produce a hierarchy development through time that closely mimics hierarchy development in real monocultures (Ford 1975).

It is often argued that the skewness of a population will increase not only with time but with density (e.g. Obeid, Machin & Harper 1967; Ford 1975). There are, however, other examples, in which skewness does not increase significantly with density (e.g. Koyama & Kira 1956; Turner & Rabinowitz 1983). Clearly it is not possible to infer the extent of intraspecific competition within populations from the shape of the frequency distribution alone. It should also be noted at this point that there are problems associated with using skewness to evaluate size hierarchies. Weiner and Solbrig (1984) have recently criticised the concept of skewness as a measure of size hierarchies within populations and suggested that it is the concept of size inequality which corresponds more closely to the notion of size hierarchy. The Gini coefficient (G), a hierarchy parameter developed from economic distribution theory, provides a useful quantification of inequality

$$G = \frac{\sum_{i=1}^{n} \sum_{j=1}^{n} |x_i - x_j|}{2n^2\bar{x}} \tag{6}$$

in a population of n individuals where the mean size of plants whose sizes are referred to as $x_1, x_2, x_3 \ldots, x_n$, is $\bar{x}$ (Weiner & Solbrig 1984). The Gini coefficient varies between 0 (absolute equality) and a theoretical maximum of 1 in an infinite population in which all individuals but one are infinitesimally small. So far the coefficient has been calculated for too few populations but it will be interesting to see whether it gives a clearer picture of size hierarchy development within populations than the measure of skewness.

Experimental studies have shown that a large number of factors may influence the survival and performance of individual plants within a population. These include the size and spatial arrangement of neighbours, seed size and quality, herbivore and pathogen activity, germination time and intrinsic growth rate. Since a number of these factors are under genetic control, variation in the survival and reproductive performance of plants may well reflect differences in individual fitness (Burdon, Marshall & Brown 1983), although this need not always be the case (Gottlieb 1977). Of the environmental factors influencing individual plant yield particular attention has been focused on the spatial arrangement of neighbours, perhaps because of the correlation between mean plant yield and density. A number of measures such as polygon area and the number of neighbours within a given area have been used as estimates for the level of crowding but generally only about 20% of the variation in individual plant yield can be accounted for on the basis of the size and distance of neighbouring plants (e.g. Mead 1966; Liddle, Budd & Hutchings 1982). This is not surprising, since models based on the degree of local crowding assume that competition is two-sided, with resources in areas of overlap being shared equally between neighbouring plants. This assumption is, however, only compatible with the idea that competition can be described in two dimensions, whereas in reality it is a spatial process in which the performance of any individual depends on the structure of its canopy and root system in relation to neighbours as well as its distance from them. Clearly those plants that emerge first and have the fastest intrinsic growth rate are likely to gain a disproportionate share of the resources within a population (Ross and Harper 1972; Watkinson, Lonsdale & Firbank 1983). A simulation model of growth and survival developed by Firbank (1984) neatly illustrates the relative importance of local crowding and emergence time in determining individual plant yield. A low variation in emergence time resulted in a close correlation between individual plant yield and the number of close neighbours (Fig. 1), whereas a high variance resulted in a much greater variation in the yield of plants at a given level of crowding.

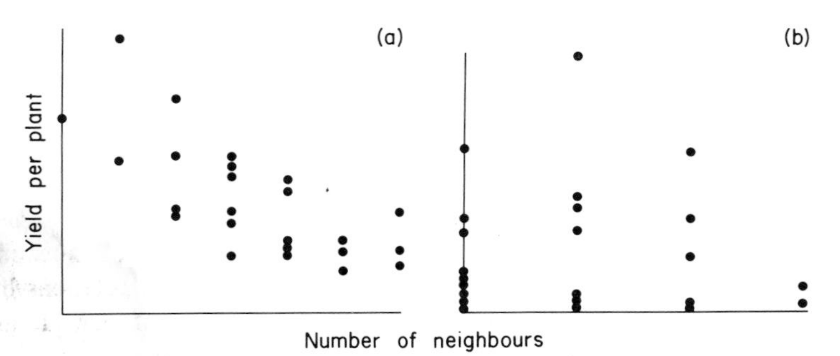

Fig. 1. The relationship between the yield of individual plants and the number of neighbours within a given neighbourhood in simulated monocultures where interference is assumed to be partially one-sided and the variance in emergence time is (a) low and (b) high. [From Firbank (1984).]

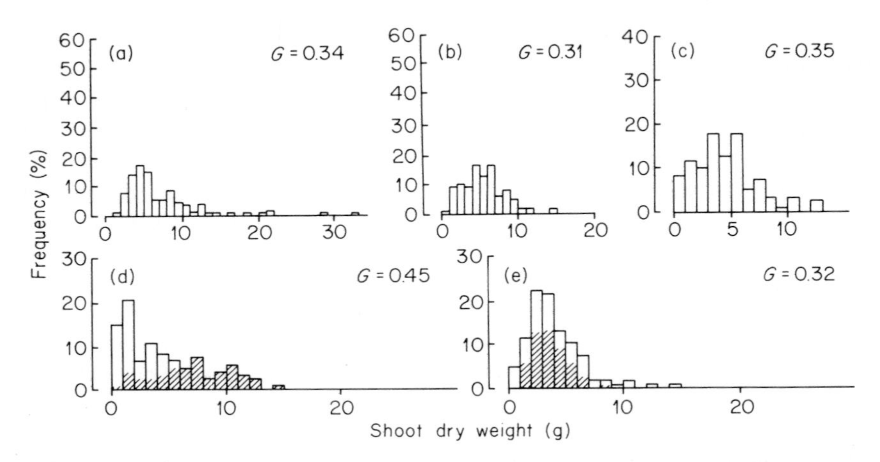

Fig. 2. The distribution of shoot dry weight. a, Monocultures of the conventional pea, Birte. b, Monocultures of the leafless pea, Filby. c, Monocultures of *Avena fatua*. d, A 1:1 mixture of Birte and *Avena*. e, A 1:1 mixture of Filby and *Avena*. All populations were sown at the same density. The contribution made by the pea plants in mixture is shown by the hatched area. *G* is the value of the Gini coefficient. [From Butcher (1983).]

IV. INTERFERENCE IN MIXTURES OF SPECIES

In two-species mixtures the influence of interference on the size hierarchy developed amongst the component species depends on a range of factors such as morphology, relative plant size and the competitive abilities of the plants concerned. Butcher (1983) for example, grew wild oats (*Avena fatua*) in monoculture and 1:1 mixtures with three varieties of leafless pea and three varieties of conventional pea. The peas were also grown in monoculture. An analysis of the size hierarchies in monoculture using the Gini coefficient [see Eq. (6)] showed that there was less inequality in populations of leafless peas than those of conventional peas, which in turn showed less inequality between individuals than in populations of *Avena* (Fig. 2). The lower degree of inequality between individuals in the pea crops may reflect the relative genetic uniformity of the plants with respect to growth rate. On the other hand, the greater hierarchy developed in the leafy than in the leafless pea may reflect the greater growth made by the conventional pea. In mixtures the inequality of peas and wild oats considered together was always greater than in the monocultures. However, the conventional peas had a considerable competitive advantage over the wild oats and this resulted in a more marked hierarchy than in mixtures with the leafless pea.

In a two-species mixture, such as that of peas and wild oats, interference between neighbouring plants may be between either plants of the same species or members of different species. However, although intra- and interspecific competition are different aspects of the same general phenomenon of interplant competition, they have generally been treated in isolation. As outlined earlier the study of interference in monocultures has primarily been through an examination of the reaction of plants to varying density, whilst the study of interference in mixtures has generally involved growing plants at a constant overall density and varying the proportions of the component species (de Wit 1960). Despite the artificiality of the latter experimental design for analysing interference in two-species mixtures, it has generally been accepted (Harper 1977) that competition experiments based on a substitutive design are much easier to analyse than additive experiments, in which both density and frequency are confounded. Rather than avoid the problems of density it would seem preferable, however, to develop mathematical models that can be used to analyse experiments where the density and frequency of both species are varied. Equation (1), which describes the relationship between mean yield per plant (w) and the density of surviving plants (N) in monoculture

$$w = w_m (1 + aN)^{-b} \tag{1}$$

can easily be extended to a two-species mixture (N_1 and N_2) so that the effect of a second species on the yield of the other can be modelled by the equations

$$\begin{aligned} w_1 &= w_{m_1}(1 + a_1(N_1 + \alpha_{12}N_2))^{-b_1} \\ w_2 &= w_{m_2}(1 + a_2(N_2 + \alpha_{21}N_1))^{-b_2} \end{aligned} \tag{7}$$

where α_{12} and α_{21} are competition coefficients that determine the equivalence between species (Watkinson 1981). Equation (2) can similarly be modified so that the density of the surviving plants in a mixture can be related to the initial densities of the two species (Lonsdale 1981)

$$\begin{aligned} N_1 &= N_{i1}(1 + m_1(N_{i1} + \gamma_{12}N_{i2}))^{-1} \\ N_2 &= N_{i2}(1 + m_2(N_{i2} + \gamma_{21}N_{i1}))^{-1} \end{aligned} \tag{8}$$

where γ_{12} and γ_{21} are competition coefficients.

The preceding equations operate in the same way as the well-known Lotka–Volterra competition equations and as such the equations and competition coefficients greatly oversimplify the process of interspecific competition. Indeed the competition coefficients in such equations could be illusory and may often obscure the real mechanisms of competitive interactions (Pianka 1981). They are certainly not constants between pairs of species and may, for example, vary, depending on the relative germination time of the two species and conditions in which the plants find themselves. Consideration of Fig. 2 also makes it abundantly clear that a single individual of one species cannot be defined in terms of

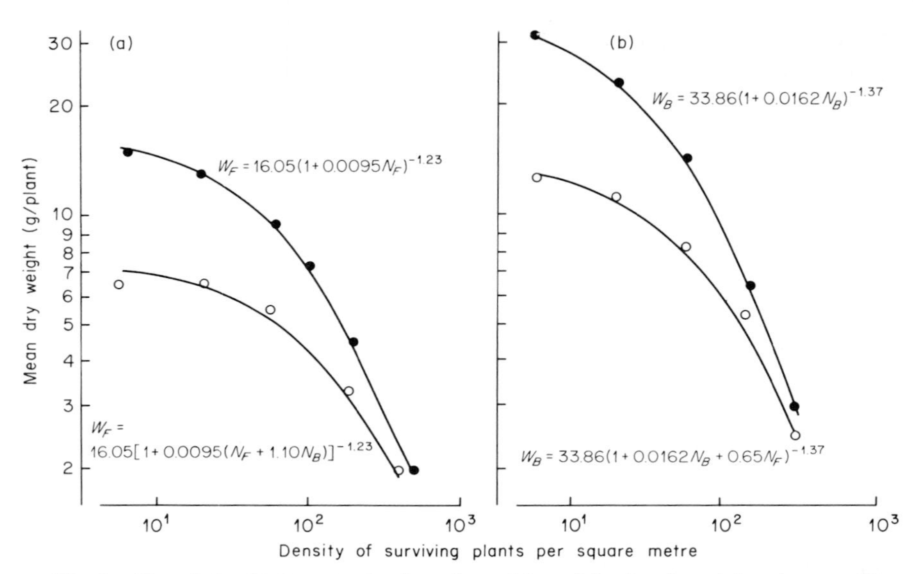

Fig. 3. The relationship between the shoot dry weight and density of surviving plants. a, The leafless pea, Filby. b, The conventional pea, Birte. In pure stand (●) and in the presence of a constant density of *Avena fatua* (○). The curves represent the best fit of Eqs. (1) and (7). [From Butcher (1983).]

so many equivalents of a second species because they both vary so greatly in size. Nevertheless, despite these qualifications, the model described provides a powerful tool for the analysis of the effects of competition. It should be noted that the parameters are estimated over a range of densities and frequencies and do not therefore vary in the same way as the relative yield total and relative crowding coefficient, which are the derived measures of plant performance and competition in the model of de Wit (1960).

Figure 3 shows the density response of two varieties of pea sown over a wide range of densities in pure populations and also in the presence of a constant density of *Avena fatua* to mimic the situation in which a crop is subject to a given level of weed infestation. It can be seen that the competition model provides an extremely good fit to the data (Butcher 1983). For the leafless pea (Filby) grown with *Avena*

$$w_F = 16.05(1 + 0.0095(N_F + 1.10N_B))^{-1.23} \tag{9}$$

and for the conventional pea (Birte)

$$w_B = 33.86(1 + 0.0162(N_B + 0.65N_F))^{-1.37} \tag{10}$$

The parameter estimates indicate that for the leafless pea an isolated plant would yield 16 g (w_m = 16.05) and that the area of resources required to obtain that weight would be approximately 95 cm^2. The parameter *b,* which has a value greater than unity in both varieties, indicates that the biological yield per unit area will decrease at high densities but to a greater extent in the conventional pea, perhaps as a result of the greater lodging in this variety. It is interesting to note that although the ecological neighbourhood area *a* of the leafless pea Filby is smaller than that of the conventional pea Birte, the area required to produce 1 g of tissue is larger. Birte is a larger plant than Filby and, therefore, needs more resources and a larger area, but because of its leaf structure, it has a greater efficiency of light capture and, therefore, requires less area to produce 1 g of tissue.

The use of the competition model [Eq. (7)] need not be restricted to the analysis of experiments involving a simple additive design when the density of one species is kept constant and that of the other is varied. Figure 4 shows the mean yield per plant and the yield per unit area of wheat for a large combination of densities of both wheat (*W*) and *Agrostemma githago* (*A*), once a common annual cornfield weed (Firbank & Watkinson 1985). A least square technique was used to fit Eqs. (1) and (7) to data on the yield of wheat in pure and mixed populations. It was found that the mean yield of wheat (w_W) *at harvest* could best be described by the equation

$$w_W = 46.8(1 + 0.24(N_W + 1.63N_A))^{-0.66} \qquad (11)$$

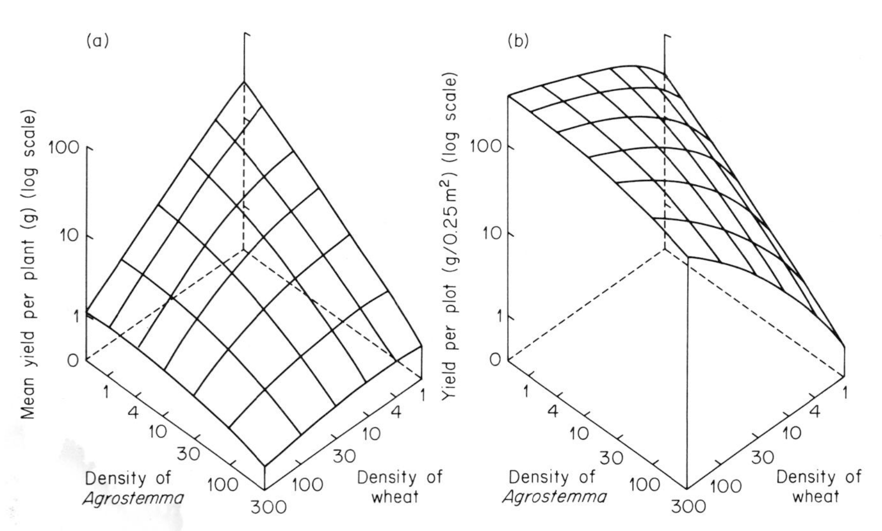

Fig. 4. The influence of weed and crop density on (a) the mean shoot dry weight per plant *w* and (b) the yield per unit area *Y* of wheat in an experiment in which spring wheat (cv Sicco) and *Agrostemma githago* were grown together over a range of densities. The response curves show the best fit of Eq. (7) to the data. a, $w_w = 46.8(1 + 0.24\ (N_w + 1.63\ N_A))^{-0.66}$. b, $Y = w_w N_w$. The density scales are $\log(N + 1)$. [Firbank & Watkinson (1985).]

where N_W and N_A are the densities at harvest of both wheat and *Agrostemma*, respectively. According to Eq. (11) the mean yield per plant of wheat at harvest is reduced by an increase in the density of surviving plants of both species. The competition coefficient indicates that an increase in the density of *Agrostemma* has a greater impact on the yield of wheat than a corresponding increase in the density of wheat. However, although *Agrostemma* had a strongly detrimental effect on wheat yield, the estimate of the competition coefficient in equation (8) indicates that it enhances the survival of wheat since $\gamma = -0.42$.

V. RESPONSES TO CROWDING IN NATURAL POPULATIONS OF ANNUAL PLANTS

Very few competition experiments have been carried out on mixtures containing more than two species (but see Haizel & Harper 1973) and those experiments that have been used to assess the role of competition in determining community structure (e.g. Silander & Antonovics 1982) do not allow the competitive relationships between species to be quantified over a range of densities and frequencies. Is there any reason, therefore, to expect that the equations outlined for one and two species can be used to describe interference in more complex communities?

If it is assumed that the competition coefficients between species are independent of one another, then Eq. (7) can be extended to take account of n species in the following manner:

$$w_i = w_{mi}(1 + a_i(N_i + \sum_{j \neq i}^{n} \alpha_{ij} N_j))^{-b_i} \tag{12}$$

Here n is the number of species (subscripted by i and j), and the other parameters are as previously defined. Rearranging Eq. (12) so that

$$w_i = w_{mi}(1 + a_i \sum_{j \neq i}^{n} \alpha_{ij} N_j)^{-b_i}(1 + a_i(1 + a_i \sum_{j \neq i}^{n} \alpha_{ij} N_j)^{-1} N_i)^{-b_i} \tag{13}$$

and substituting

$$W_i = w_{mi}(1 + a_i \sum_{j \neq i}^{n} \alpha_{ij} N_j)^{-b_i}$$

and

$$A_i = a_i(1 + a_i \sum_{j \neq i}^{n} \alpha_{ij} N_j)^{-1}$$

gives

$$w_i = W_i(1 + A_i N_i)^{-b_i} \tag{14}$$

which is identical in form to Eq. (1). Note that this equation can be used to describe the relationship between surviving plant density and shoot dry weight, the yield of a plant part or reproductive output, whether measured in terms of numbers or mass (Watkinson 1980). Equation (14) can, therefore, be used to describe the relationship between the yield and density of species i in a mixed community, given that the value for the competitive pressure ($\Sigma\ \alpha_{ij}N_j$) remains approximately constant or is very small. It might seem unreasonable to expect either of these conditions to apply, even in populations of annual plants, but the experimental manipulation of plant density in a community alters N_i whilst leaving the term $\Sigma\ \alpha_{ij}N_j$ unaltered. Thus equation (14) has been found to give a good fit to data for *Vulpia fasciculata* in an experiment (Watkinson & Harper 1978) in which natural densities were manipulated on the fixed dunes at Aberffraw in north Wales (Fig. 5). The best fit of equation (14) to data on the number of seeds per plant s and the surviving density of *Vulpia* (Watkinson & Davy 1985) was provided by the equation

$$s = 3.05(1 + 1.93 \cdot 10^{-4}N)^{-0.90} \tag{15}$$

This indicates that an isolated plant of *Vulpia* will produce an average of just over three seeds in a matrix of vegetation containing prostrate perennials such as *Anthyllis vulneraria, Lotus corniculatus* and *Thymus praecox* and annuals such as *Cerastium diffusum, Mibora minima* and *Phleum arenarium*. However, the fact that the negatively density-dependent relationship between reproductive output and density varies little over space or time (A. R. Watkinson, unpublished data) despite considerable changes in the associated vegetation perhaps indicates that there is little interference between *Vulpia* and its associated species in the field. *Cakile edentula* (Keddy 1981), *Diamorpha smallii* (Clay and Shaw 1981) and *Floerkea proserpinacoides* (Smith 1983) have also been shown to exhibit a negatively density-dependent relationship between reproductive output and density when natural densities in the field have been manipulated. Ter Borg (1979) has similarly demonstrated the density-dependent regulation of fecundity in *Rhinanthus angustifolius* (= *R. serotinus*), a hemiparasitic annual, when sown onto experimental plots.

An analysis of data from studies on *Androsace septentrionalis* (Symonides 1979), *Polygonum confertiflorum* (Reynolds 1984) and *Salicornia europaea* agg. (Jefferies, Davy & Rudmik 1981) indicates that equation (14) can be used to describe the relationship between reproductive output and the natural density of plants in certain cases (Fig. 5). For *Salicornia europaea* (Watkinson & Davy 1985) the relationship between the number of seeds per plant s and density could be described by the equation

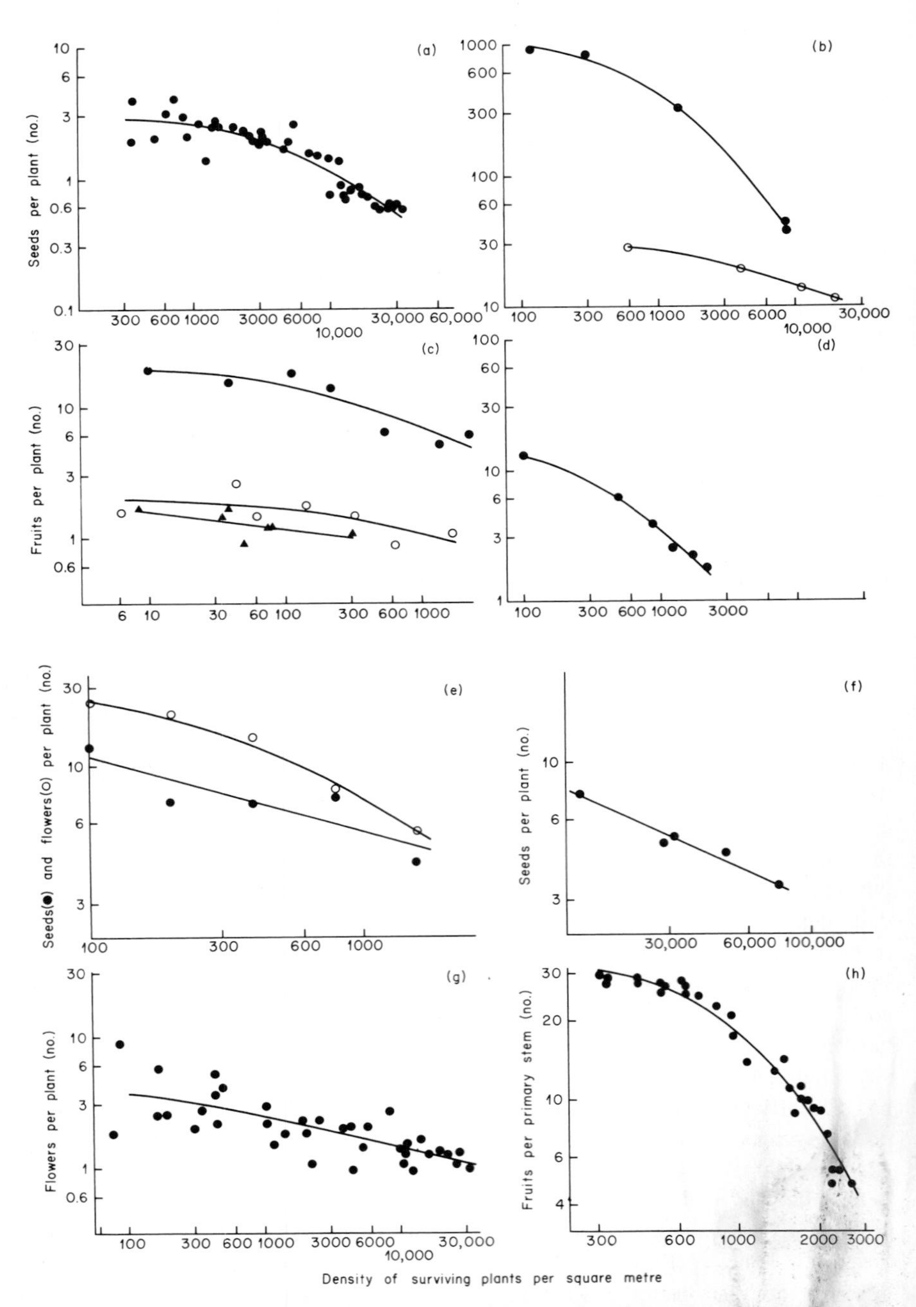

Fig. 5. The relationship between the reproductive output per plant and the density of plants at maturity in populations. a, *Vulpia fasciculata*. [Recalculated from Watkinson and Harper (1978).] b, *Salicornia europaea* on a high (○) and low (●) marsh. [Recalculated from Jefferies *et al.* (1981).] c,

$$s = 1100(1 + 7.57 \cdot 10^{-4}N)^{-1.63} \quad (16)$$

on the low marsh, where *Salicornia* formed almost a monoculture, and

$$s = 33.8(1 + 9.23 \cdot 10^{-4}N)^{-0.36} \quad (17)$$

on the high marsh, where the vegetation cover was much denser and *Salicornia* occurred amongst a matrix of perennials. Interference for nutrients from these perennials presumably limited the growth of *Salicornia* on the high marsh. Of note is that an isolated plant of *Salicornia* requires approximately the same area of ground to produce 34 and 1100 seeds on the upper and lower marshes, respectively.

It is clear from the preceding studies that the negatively density-dependent control of reproduction in annual plants may be a common mechanism for regulating the population size of annual plants. This is not to deny that other regulatory processes may also be important. Amongst the annuals already discussed, Symonides (1979) has, for example, found that the mortality of seedlings in *Androsace* is strongly density-dependent whilst other studies have shown that the activity of pollinators, predators and pathogens may also vary with density (Antonovics & Levin 1980).

Models incorporating information of the type outlined here on density-dependent processes, together with information on the way density-independent factors affect the numbers of births, deaths, immigrants and emigrants in a population, have been valuable in demonstrating the way population size is regulated and abundance determined (e.g. Watkinson & Harper 1978; Watkinson & Davy 1985; Watkinson 1985). Yet they do not contain information on the spatial distribution of individuals or variance in plant size. Few studies on plant population dynamics even contain information on spatial distribution (but see Symonides 1983) and it remains an exciting challenge to integrate the spatial and temporal dynamics of plant populations.

ACKNOWLEDGEMENTS

I thank R. E. Butcher and L. G. Firbank for allowing me to use their unpublished data.

Cakile edentula on the seaward (●), middle (○) and landward (▲) sides of a sand dune. [Recalculated from Keddy (1981).] d, *Rhinanthus angustifolius*. [Recalculated from unpublished data of Piet Drent presented in Ter Borg (1979).] e, *Floerkea proserpinacoides*. [Recalculated from data of Smith (1983).] f, *Polygonum confertiflorum*. [Recalculated from data of Reynolds (1984).] g, *Diamorpha smallii*. [Clay and Shaw (1981).] h, *Androsace septentrionalis*. [Recalculated from data of Symonides (1979).] The lines represent the best fit of Eq. (14) to the data. [a–d from Watkinson and Davy (1985).]

REFERENCES

Aikman, D. P. & Watkinson, A. R. (1980). A model for growth and self-thinning in even-aged monocultures of plants. *Annals of Botany (London),* **45,** 419–427.

Antonovics, J. & Levin, D. A. (1980). The ecological and genetical consequences of density-dependent regulation in plants. *Annual Reveiw of Ecology and Systematics,* **11,** 411–452.

Begon, M. (1984). Density and individual fitness: Asymmetric competition. *Evolutionary Ecology* (Ed. by B. Shorrocks), pp. 175–194. Blackwell Scientific Publications, Oxford.

Burdon, J. J., Marshall, D. R. & Brown, A. H. D. (1983). Demographic and genetic changes in populations of *Echium plantagineum. Journal of Ecology,* **71,** 667–679.

Butcher, R. E. (1983). *Studies on interference between weeds and peas.* Ph.D. thesis, University of East Anglia.

Cannell, M. G. R., Rothery, P. & Ford, E. D. (1984). Competition within stands of *Picea sitchensis* and *Pinus contorta. Annals of Botany (London),* **53,** 349–362.

Clay, K. & Shaw, R. (1981). An experimental demonstration of density-dependent reproduction in a natural population of *Diamorpha smallii,* a rare annual. *Oecologia,* **51,** 1–6.

de Wit, C. T. (1960). On competition. *Verslagen van Landbouwkundige Onderzoekingen,* **66,** 1–82.

Firbank, L. G. (1984). *The population biology of* Agrostemma githago, *L.* Ph.D. thesis, University of East Anglia.

Firbank, L. G., & Watkinson, A. R. (1985). On the analysis of competition within two-species mixtures of plants. *Journal of Applied Ecology,* **22,** 503–517.

Ford, E. D. (1975). Competition and stand structure in some even-aged monocultures. *Journal of Ecology,* **63,** 311–333.

Ford, E. D. & Diggle, P. J. (1981). Competition for light in a plant monoculture modelled as a spatial stochastic process. *Annals of Botany (London),* **48,** 481–500.

Gottlieb, L. D. (1977). Genotypic similarity of large and small individuals in a natural population of the annual plant *Stephanomeria exigua* ssp. *coronaria* (Compositae). *Journal of Ecology,* **65,** 127–134.

Haizel, K. A. & Harper, J. L. (1973). The effects of density and the timing of removal on interference between barley, white mustard and wild oats. *Journal of Applied Ecology,* **10,** 23–31.

Harper, J. L. (1961). Approaches to the study of plant competition. *Mechanisms in Biological Competition* (Ed. by F. L. Milthorpe), pp. 1–39. Academic Press, New York.

Harper, J. L. (1977). *Population Biology of Plants.* Academic Press, London.

Harper, J. L. (1981). The concept of population in modular organisms. *Theoretical Ecology* (Ed. by R. M. May), pp. 53–77. Blackwell Scientific Publications, Oxford.

Jefferies, R. L., Davy, A. J. & Rudmik, T. (1981). Population biology of the salt marsh annual *Salicornia europaea* agg. *Journal of Ecology,* **69,** 17–32.

Keddy, P. A. (1981). Experimental demography of the sand-dune annual *Cakile edentula,* growing along an environmental gradient in Nova Scotia. *Journal of Ecology,* **69,** 615–630.

Koyama, H. & Kira, T. (1956). Intraspecific competition among higher plants. VIII. Frequency distribution of individual plant weight as affected by the interaction between plants. *Journal of the Institute of Polytechnics, Osaka City University, Series D,* **7,** 73–84.

Liddle, M. J., Budd, C. S. J. & Hutchings, M. J. (1982). Population dynamics and neighbourhood effects in establishing swards of *Festuca rubra. Oikos,* **38,** 52–59.

Lonsdale, W. M. (1981). *Self-thinning in pure and mixed populations of plants.* Ph.D. thesis, University of East Anglia.

Mead, R. (1966). A relationship between individual plant spacing and yield. *Annals of Botany (London),* **30,** 301–309.

Obeid, M., Machin, D. & Harper, J. L. (1967). Influence of density on plant to plant variations in fibre flax, *Linum usitatissimum. Crop Science,* **7,** 471–473.

Pianka, E. R. (1981). Competition and niche theory. *Theoretical Ecology* (Ed. by R. M. May), pp. 167–196. Blackwell Scientific Publications, Oxford.

Rabinowitz, D. (1979). Bimodal distributions of seedling weight in relation to density of *Festuca paradoxa* Desv. *Nature (London),* **277,** 297–298.

Reynolds, D. N. (1984). Populational dynamics of three annual species of alpine plants in the Rocky Mountains. *Oecologia,* **62,** 250–255.

Ross, M. A. & Harper, J. L. (1972). Occupation of biological space during seedling establishment. *Journal of Ecology,* **60,** 77–88.

Shinozaki, K. & Kira, T. (1956). Intraspecific competition among higher plants. VII. Logistic theory of the C-D effect. *Journal of the Institute of Polytechnics, Osaka City University, Series D,* **7,** 35–72.

Silander, J. A. & Antonovics, J. (1982). Analysis of interspecific interactions in a coastal plant community—a perturbation approach. *Nature (London),* **298,** 557–560.

Smith, B. H. (1983). Demography of *Floerkea proserpinacoides,* a forest-floor annual II. Density-dependent reproduction. *Journal of Ecology,* **71,** 405–412.

Symonides, E. (1979). The structure and population dynamics of psammophytes on inland dunes. II. Loose-sod populations. *Ekologia Polska,* **27,** 191–234.

Symonides, E. (1983). Population size regulation as a result of intra-population interactions. I. Effect of density on the survival and development of individuals of *Erophila verna* (L.) C.A.M. *Ekologia Polska,* **31,** 839–881.

Ter Borg, S. J. (1979). Some topics in plant population biology. *The Study of Vegetation* (Ed. by M. J. A. Werger), pp. 13–55. Dr. W. Junk, The Hague.

Turner, M. D. & Rabinowitz, D. (1983). Factors affecting frequency distribution of plant mass: The absence of dominance and suppression in competing monocultures of *Festuca paradoxa.* *Ecology,* **64,** 469–475.

Watkinson, A. R. (1980). Density-dependence in single species populations of plants. *Journal of Theoretical Biology,* **83,** 345–357.

Watkinson, A. R. (1981). Interference in pure and mixed populations of *Agrostemma githago.* *Journal of Applied Ecology,* **18,** 967–976.

Watkinson, A. R. (1984). Yield-density relationships: the influence of resource availability on growth and self-thinning in populations of *Vulpia fasciculata.* *Annals of Botany,* **53,** 469–482.

Watkinson, A. R. (1985). On the abundance of plants along an environmental gradient. *Journal of Ecology,* **73,** 569–578.

Watkinson, A. R. & Davy, A. J. (1985). Population biology of salt marsh and sand dune annuals. *Vegetatio,* **62,** 487–497.

Watkinson, A. R. & Harper, J. L. (1978). The demography of a sand dune annual: *Vulpia fasciculata.* 1. The natural regulation of populations. *Journal of Ecology,* **66,** 15–33.

Watkinson, A. R., Lonsdale, W. M. & Firbank, L. G. (1983). A neighbourhood approach to self-thinning. *Oecologia,* **56,** 381–384.

Weiner, J. & Solbrig, O. T. (1984). The meaning and measurement of size hierarchies in plant populations. *Oecologia,* **61,** 334–336.

Yoda, K., Kira, T., Ogawa, H. & Hozumi, K. (1963). Self-thinning in over-crowded pure stands under cultivated and natural conditions. *Journal of Biology, Osaka City University,* **14,** 107–129.

19

The Thinning Rule and Its Application to Mixtures of Plant Populations

James White

Department of Botany
University College
Dublin, Ireland

I. INTRODUCTION

The inverse relationship between the size and the density of plants in closed vegetation has long been recognized and was probably most evident to foresters concerned with the management of trees to maximise timber production. It is difficult to pinpoint its earliest published notice by a plant ecologist, though it was recorded clearly by Weaver (1918), one of Clements's first students. As with so many aspects of early plant ecology, Clements was a seminal source of ideas on plant population dynamics (White 1985), and his various experiments on plant competition (Clements, Weaver & Hanson 1929) reveal his interest in the link between plant size and density. The size–density relationship is one of the most fundamental aspects of the population dynamics of organisms with plastic morphological expression and is particularly relevant to plants with metameric construction (White 1984). Its recognition and analysis by biologists deserve a closer historical study that it has hitherto received: we are still dependent on Clements for the only account of the history of plant competition studies (Clements, Weaver & Hanson 1929).

The first formal statement of the reciprocal relationship between size and numbers in dense single-species populations is apparently that of Reineke (1933), who noted that the mean diameter at breast height (d) of surviving trees in populations undergoing self-thinning was correlated with the number of survivors (N) as a power function, $d \propto N^{-\alpha}$, where $\alpha = 0.62$ or thereabouts. As the forestry literature of the period reveals, this was widely debated, often confirmed, but after about 20 years it became obscured by quantitative models of greater precision and utility to foresters.

STUDIES ON PLANT DEMOGRAPHY:

A similar power function relationship was discussed later by Japanese foresters (e.g. Tadaki & Shidei 1959), but it remained for T. Kira's school at Osaka City University to resuscitate it in the context of a wide-ranging appraisal of the nature of plant competition which they conducted throughout the 1950s and early 1960s, in a notable series of papers on 'intraspecific competition among higher plants'. Following detailed time-series analyses of several plant populations growing at high densities, they proposed the relationship $W \propto N^{-\alpha}$, where W is mean plant weight and α typically had values about 1.5 (Yoda *et al.* 1963). They suggested that this was a very general relationship, even a law, which they termed the '$-\frac{3}{2}$ power law', applicable to many plant populations. As they recognized, the equation was formally similar to Reineke's. There is now copious evidence that W $\propto d^{2.5}$ (White 1981), so they are in fact identical statements, reflecting merely the size parameter employed.

Unlike Reineke's version, the Japanese formulation soon became well known and in the past 20 years has evoked a large and still expanding literature, much of which has been reviewed by Westoby (1984). It has become one of the best founded generalizations of plant population dynamics. Its prominent discussion by Harper (1967) undoubtedly led to its widespread investigation. I still retain an indelible memory of its exposition by John Harper in his celebrated presidential address to the British Ecology Society in 1967, when I first heard it. Within a year the Japanese results were confirmed at Bangor and extended to other species (White & Harper 1970). Later it was realized that the equation could be made more precise by reducing the parameter values to standard, uniform dimensions (White 1975), $W = K \cdot N^{-1.5}$, where K typically has values between 3.2×10^3 and 2.5×10^4 when W was expressed in grams dry weight and N in plants per square metre. The equation in this form held over six orders of magnitude of density, based on thinning trajectories for various trees and herbs (White 1980). Under these conditions, the dimensionality of K is grams per cubic metre, suggesting that it is a mass–density constant which relates plant mass to its occupancy of space (White 1981).

Already in 1967 Harper had asked "whether this formal relationship holds good for a wider range of species" and "how far (it) can be extended to include populations of several species". The first question has been amply answered since, and I propose in this paper to do so more definitively than previously; I shall also modify earlier statements of its range of applicability (White 1980) in the light of new evidence. The second question has remained more open, and I shall attempt to answer it here as comprehensively as present information allows. In order to do so, it is essential to have a clear understanding of the rule as it applies to single-species populations: this provides the basis for a proper and unambiguous answer to the second question.

All discussion in this paper is restricted, without exception, to the aerial structures of plants, for too little information is available on the size of subterranean organs to enable an informed review to be made of the application of the rule to them at present.

II. THINNING IN SINGLE-SPECIES POPULATIONS

There are two types of data which may be used to test the model of Yoda *et al.* (1963). The first consists of time-series size–density combinations for populations undergoing self-induced thinning. Several examples are given by White (1980) and Westoby (1984), not to speak of the primary literature on individual species, where, however, there is seldom the dimensional standardization necessary to appreciate comparative values of the intercept constant K, without preliminary recalculations. Data on forest trees are plentiful, but they usually have the disadvantage that total plant mass is seldom reported: branches and foliage are rarely included except in forest productivity studies of recent times, often conducted as part of the International Biological Programme of the 1970s.

Since the self-thinning lines seem collectively to define an asymptotic thinning boundary zone (White 1980) defined by the values of K, the second type of data useful in testing the rule involves unique combinations of size and density for particular populations for which there are otherwise no time-series records. Such combinations are abundant in the literature of plant productivity and were used instructively by Gorham (1979) to define, in the absence of any thinning lines, an asymptotic boundary of mass and density, $W = 9.67 \times 10^3 N^{-1.49}$, a result remarkably coincident with White's (1980), which held over a density range 10^{-1}–10^4 plants per square metre, from a moss to trees. Such data, being plentiful, are particularly useful in testing the hypothesis that there is a boundary zone of size and density in plant populations.

I have compiled a very wide range of data of both types for seed plants and cryptogams. A summary of those for seed plants only is shown in Fig. 1; details will be published elsewhere. The species range in size from *Lemna* (Ikusima, Shinozaki & Kira 1955; Clatworthy & Harper 1962) to *Sequoia sempervirens* (Westman & Whittaker 1975), spanning a little over 11 orders of magnitude. The size range can by extended downwards by a further five orders of magnitude if cryptogams are included, at least to *Chlorella* and embracing *Polypodium* gametophytes, along the *same* boundary zone (J. White, unpublished).

The density range shown in Fig. 1 spans about seven orders of magnitude, but this is not quite the limit for seed plants. Densities as high as 1.5×10^7 plants per square metre equivalent have been measured by Salisbury (1970) for seedlings of *Juncus bufonius* and 7.5×10^5 for *Anagallis minima* and *Radiola linoides,* all annuals of periodically dried-out muddy or peaty habitats. Annual grass seedlings at densities of 2×10^5 per square metre have been recorded by Biswell and

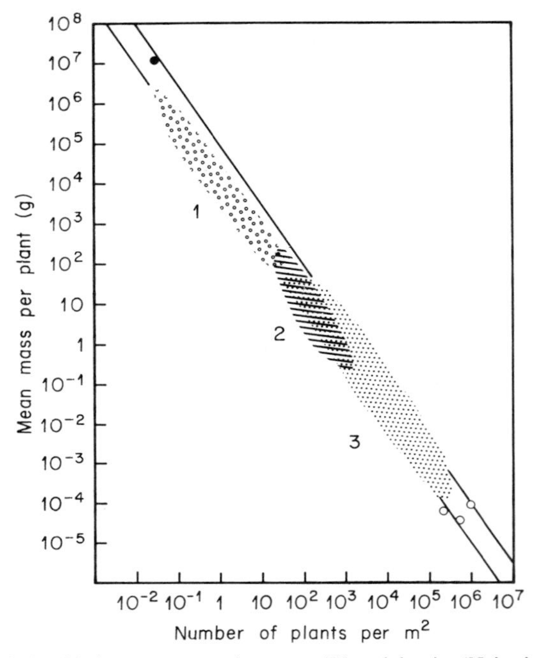

Fig. 1. Relationship between mean plant mass (W) and density (N) in single-species populations of herbaceous plants and trees. Each shaded area encloses the limits of many self-thinning lines or points of maximum conjoint mass and density combinations. 1, Trees. 2, Clonal perennial herbs. 3, Annual herbs and perennial herbs grown at high density in experimental systems. ○ *Lemna* spp.; ● *Sequoia sempervirens*. The positions of boundary lines given by $W = 10^4N^{-3/2}$ and $W = 10^5N^{-3/2}$ are indicated; they are not regression lines.

Graham (1956). Since we may safely assume that seedlings at such densities must thin as some grow in size, the thinning rule probably applies over a range of eight to nine orders of magnitude of density for seed plants.

The shaded areas in Fig. 1 encompass diverse time-series thinning lines and single size–density combinations, chosen in all cases to represent the most extreme (high) values. It is obvious that most of the data lie within a boundary line of $K = 10^5$ in the thinning equation $W = \mathrm{K} \cdot N^{-3/2}$; some of them range down to a line of $K = 3.2 \times 10^3$. That is, most of the data are enclosed by a zone of thinning lines of -1.5 slope, which ranges from log $K = 3.5$ to 5.0; this is an upward extension of the range reported previously from a smaller data set influenced chiefly in its circumscription by examples of trees (White 1980). If one imagines a series of -1.5 slope lines traversing Fig. 1, it is possible to define the position of any thinning line or single size–density point with reference to the intercept value K. For convenience these (imaginary) lines will be referred to by their log K value in the remainder of this paper: the lines actually drawn in Fig. 1

are at log K = 4 and 5. One may characterize the position of the point shown for *Sequoia sempervirens,* for example, as being at a boundary line of log K = 4.72.

A. Trees

The shaded area for trees in Fig. 1 includes most of the data previously given by White (1980) and comprises 30 thinning lines and about 50 points of extreme conjoint mass and density. Except for *Sequoia sempervirens,* shown independently, no tree population exceeds a boundary of log K = 4.4. None of the single-point values for mass and density in the worldwide IBP Woodland Data Set (De Angelis, Gardner & Shugart 1981), from which I have taken a further 52 points for pure stands, exceeds a boundary line of 4.4 and most lie between 3.5 and 4 (J. White, unpublished).

B. Clonal Perennial Herbs

The shaded area for clonal perennial herbs encloses only single-point mass and density combinations, about 60 altogether, without any thinning lines. The genera represented include *Carex, Cyperus, Glyceria, Phragmites, Sasa, Scirpus, Sparganium, Spartina* and *Typha*. The examples include all the published data reported by Gorham (1979) and by Hutchings (1979). No point exceeds a boundary of log K = 5, and only nine exceed 4.5, the most extreme being *Sasa paniculata* and *Typha latifolia.* The failure to find examples of thinning lines with simple size–density relationships comparable to those for trees and non-clonal herbs confirms Hutchings' (1979) experience. Self-thinning in trees and in those herbs without clonal structure involves separate genets. Clonal perennials have, by definition, ramets derived from a single genet. Given the typical scale of sampling for plant productivity studies, from which most of my data here are derived, it is unlikely that more than a few genets, perhaps only one, are responsible for the shoot population sampled. We may further surmise that there is a higher degree of physiological integration among the ramets of a single genet that among separate genets (root grafting in some tree populations notwithstanding); this may inhibit such profligate shoot production in clonal perennial herbs that shoot mortality is inevitable by self-thinning before reproductive maturity is achieved by each shoot. Consequently, we may expect the ramets within a genet of a clonal perennial herb *not* to show the pattern of self-thinning so well known for populations composed of shoots of separate genets. On one interpretation (Hutchings 1979) this may be taken as an illustration that clonal perennial herbs are an exception to the thinning rule. Certainly the time course of size and density during the development of the shoot population in such herbs is not along a uniform $-\frac{3}{2}$ slope (Hutchings 1979; J. White, unpublished data). But in a different sense these shoot populations are bounded by the thinning rule, perhaps even governed by it, in their typical size–density relationships, since they do not transgress an asymptotic upper limit that is characterized by actual thinning lines for genet populations of trees and other herbs.

C. Herbs, Excluding Clonal Perennial Species

The shaded area (3) in Fig. 1 encompasses 21 thinning lines [but excludes five of those in White (1980)] and about 40 single-point mass and density combinations. All but 5 lie within a boundary line of log $K = 5$: these are the most extreme points in populations of *Lolium perenne* (Kays & Harper 1974), *Agrostemma githago* and *Festuca pratensis* (Lonsdale & Watkinson 1983) and thinning lines for *Coix Ma-yuen* (Kawano & Hayashi 1977) and *Vulpia fasciculata* (Watkinson 1984). The thinning line for the annual grass *Coix* is along a boundary line of log $K = 5.46$, which is quite remarkable. The experiment was done by using "extraordinary high nitrogen levels, which have rarely if ever been applied in any previously known cultivation experiments using crop or wild plants", to quote the authors: these produced LA1 values approaching 70, for example. Although the harvested plots were large for such an experiment (1 m^2), there was no explicit statement that edge effects had been discounted, and this may partly explain the rest.

I have excluded from my data, as far as possible, those reports in which small plot or pot size is unaccompanied by a statement that edge effects have been considered. The calculation of mass per unit area is subject to serious error if edge effects are not controlled in experimental plots. Cannell and Smith (1980) suggested that yields would be seriously overestimated if the ratio of the height of measured plants inside the plot to their distance from the edge of the plot (that is, the border width) exceeded 4. I endorse their convention of excluding yield estimates from plots which do not meet this criterion, but I have been unable to apply it rigorously to the mass estimates at a given density for herbs used in the construction of Fig. 1. The high boundary value (log $K = 5.5$) reported for *Vulpia fasciculata* (Watkinson 1984) was obtained from complete populations grown in 10-cm pots, but they were surrounded by plants grown in separate guard pots in an attempt to minimize edge effects which are scarcely avoidable in such small pots. As with the *Coix* experiment, nutrient levels were maintained at high levels in the *Vulpia* experiment.

It is notable that of the five examples which have mass and density combinations over log $K = 5$, four are grasses and the fifth has a canopy of linear leaves. There is, indeed, good evidence that plant geometry influences the position of thinning lines vis-à-vis the log K intercept, as Lonsdale and Watkinson (1983) showed by a reanalysis of White's (1980) data and by their own results. It appears that the position of mass and density combinations within the frame of reference of Fig. 1 is directly related to the ability of plant populations to pack 'biomass' into space. There is, in fact, a monotonic relationship between the position of a particular stand with respect to its thinning line boundary value and its concentration of mass in space (its mass density) expressed in grams per cubic metre (J. White, unpublished).

D. The Position of Self-thinning Lines on the Size–Density Graph

Since Watkinson (1984) reported the trajectory of his *Vulpia* thinning line as $\log W = 5.74 - 1.58 \log N$, and I have stated that the extreme boundary value of his data, in my terms, is 5.5, I should at this point interpolate a justification for this value. I wish also to propose a protocol for judging the position of thinning lines in relation to this 'boundary zone' which I have been discussing so far. Since the suggestion made some years ago that a necessary condition for a comparative survey of thinning lines was a standard frame of reference dimensionally (White 1975, 1980), there seems to have been widespread consensus to use the units of gram dry weight and plants per square metre. This enables values of the intercept constant to be compared quickly if a standard density basis such as one plant per square metre is used. The value of this intercept constant is, however, usually extrapolated from the actual data to which it applies. Generally speaking, trees occur at densities less than one per square metre and herbs at densities higher than one per square metre. Accordingly, one plant per square metre may be a convenient compromise as a baseline density, but for lines of a given slope, quite different intercept constants are possible, depending on whether the line is extrapolated 'downwards' to one plant per square metre or 'upwards' from higher densities, even if the thinning lines occur within a quite narrow boundary zone as defined earlier. Actual thinning lines often occupy only a short extent of the overall pattern revealed in Fig. 1. For example, the actual density range of the *Vulpia fasciculata* thinning line (Watkinson 1984) extended from 10^5 to 2×10^3, and it is all-encompassed on a graph such as Fig. 1 by a boundary line of $\log K = 5.5$. The value given by Watkinson ($\log K = 5.74$) is formally correct for the line he describes, but the population never reached the extrapolated density of one plant per square metre and hence never reached a boundary limit of 5.74.

This problem is a source of potential confusion when comparing the full equation for thinning lines and I suggest that, given the overall generality which Fig. 1 represents, *boundary values* of mass and density in these terms be calculated from the most extreme mass–density points of the thinning line. Since, too, the slope and intercept constants are inextricably interdependent (J. F. White & Gould 1965), small deviations from a slope of -1.5 of a thinning line can greatly enlarge or diminish the intercept constant, the more so the farther the line has to be extrapolated to a base density of one plant per square metre. Another example of this may be noted for illustration (though one could cite many): Malmberg and Smith (1982) reported a thinning line for a mixture of *Medicago sativa* and *Trifolium pratense*, $\log W = 5.58 - 1.75 \log N$. In fact no actual combination of mass and density exceeds a boundary value of $\log K = 4.78$; their intercept is an artifact of extrapolation over 3.3 orders of magnitude of density.

The thinning rule if fully defined by *both* slope and intercept constants in the equation $W = K \cdot N^{-\alpha}$. Exceptions to the rule as a time trajectory of mass and density are known (notably for clonal plants), but several of those reported are invalid, since they fail to recognize that the rule is primarily concerned with the correlation between plant size and the corresponding *maximum* asymptotic density in a population (Yoda *et al.* 1963). This correlation may be difficult to detect in the absence of density-dependent mortality. Other relationships between mass and density apply to less restrictive conditions (White 1981). Rejection of the rule *in its original meaning* (Yoda *et al.* 1963) should take account of the actual position of a data set in the zone defined empirically by those multiple, independent observations for which the rule *is* valid, such as that shown in Fig. 1.

III. THINNING IN MIXED-SPECIES POPULATIONS

Taking the frame of reference shown in Fig. 1 as valid for single-species populations, I shall rephrase Harper's (1967) query: do mixed species populations show departures from the size–density relationships shown in Fig. 1? The data available for this test are relatively meagre, but sufficient to provide, I believe, a fairly unambiguous answer.

A. Trees

In the literature of North American forestry in which I have concentrated my search, there are few examples of published inventories of size and number of multispecies forest stands over time. There are, however, some significant long-term studies which remain unpublished in detail, and I shall describe two of them briefly: both have been conducted on the forests of Harvard University as part of a series of silvicultural researches.

1. Harvard Black Rock Forest, Hudson Highlands, New York

In 1931 two plots, each 0.1 ha, were established in a natural forest which had developed from coppice sprouts after earlier cutting. One plot was used as a control (plot 4a-1c, compartment VII), the other for a thinning treatment. All trees over 4-cm diameter at breast height (dbh) were marked and have been measured at regular intervals since. I shall restrict my remarks here to the control plot. This is a particularly valuable set of data for an analysis of multispecies thinning in a natural forest because there has been almost no recruitment to the

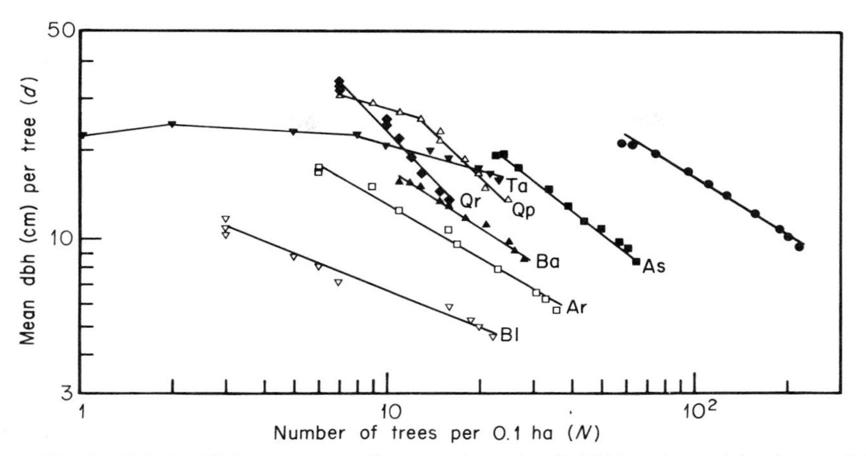

Fig. 2. Relationship between mean diameter at breast height (dbh) per tree and density on a 0.1-ha plot at Harvard Black Rock Forest. The points on the right represent all species in a successive time sequence from 1931 to 1984; the line is fitted by eye as a dbh–density boundary. Ar, *Acer rubrum*. As, *Acer saccharum*. Ba, *Betula alleghaniensis*. Bl, *Betula lenta*. Qp, *Quercus prinus*. Qr, *Quercus rubra*. Ta, *Tilia americana*. Lines are drawn by eye to facilitate distinctions between the overall time trends for each species. (From data supplied by J. J. Karnig.)

stand over 53 years (about 10 trees, which have been omitted in my analysis). The overall depletion curve of trees present in 1931 has remained almost constant in slope, with a half-life of 28 years. Some further details of the patterns of mortality, dominance and suppression of various species have been given by Lorimer (1981).

The size–density dynamics of the trees are illustrated in Fig. 2. There has been a remarkably consistent overall trend in the dbh–density relationship for all species over a long period, with some fall-off in the past few years. The various species behave more individualistically: all show a pattern of depletion in numbers and increasing size of survivors, except *Tilia americana* after 1961. These patterns are readily interpretable in terms of the growth rates and shade tolerance of individual species, though space precludes a discussion of this phenomenon here.

Does the pattern of thinning of all trees collectively conform to the thinning rule? The equation of the line defining the boundary drawn in Fig. 2 is $d = 16.5N^{-0.67}$, where d is dbh in centimetres and N is plants per square metre. This is an expression of Reineke's (1933) rule. A further step is required to convert this equation into a form homologous with those used in Fig. 1, the calculation of tree mass. The derivation of tree volume from stem dbh is widely practised in forestry by means of volume tables. The local Harvard Black Rock Forest volume table shows that $V = 1.09 \times 10^{-4}d^{2.5}$, where V is volume in cubic metres and d is dbh in centimetres. This holds over the range 10–45 cm dbh and gives merchantable wood volume to a top diameter of 10 cm; it is based on the accurate measurement of 180 hardwood trees, consisting of oaks, maples, ash and hickory (J. J. Karnig, personal communication 1973). As such, it can be seen that this underestimates the true volume of plant material probably by 10–20%. This is, unfortunately, a restriction typical of many data derived from forestry sources for testing the thinning rule, but within the relatively coarse scale of the analysis I am outlining in this paper, it makes only a small difference to the position of any thinning line so derived. The power function relationship $V \propto d^{2.5}$ is to be expected and is typical of a large number of trees (White 1981). One further source of error should be noted in deriving volume estimates by such a power function: the problem caused by logarithmic transformation. This method for calculating volume from dbh has a venerable tradition in forestry (e.g. Schumacher & Hall 1933), but the inherent error has only been more recently appreciated. The error is probably of the order of 10% (Whittaker & Marks 1975).

From an analysis of several volume tables appropriate to hardwood trees in the eastern USA, some of which I have employed for examples given later in Fig. 4, the volume equation given is fairly typical: the power exponent varies from 2.35 to 2.52, and the corresponding intercept constants from 1.09 to 1.93×10^{-4}, when volume is expressed as square metres and dbh in centimetres. As always in such equations these values are intertwined and the lowest slopes have the highest intercept constants; within their domain of reference (over dbh of c. 5–40 cm) they are very similar, however.

The final step is the calculation of mass in gram dry weight from volume in cubic metres; this is straightforward from the "Wood Handbook" specific gravity tables (U.S. Forest Products Laboratory 1974). For the species shown in Fig. 2, a median value of 6.41×10^5 g/m^3 was used, being somewhat higher for *Acer saccharum* and *Quercus rubra,* but lower for *Acer rubrum* and *Tilia americana,* for example. Where W is expressed in grams and N in plants per cubic metre, the equation of the boundary line for all species combined in Fig. 2 is $W = 7.7 \times 10^3 N^{-1.675}$, well within the envelope of thinning lines for pure stands in Fig. 1. For the reasons given earlier it is probably an underestimate of the true mass for a given density of about 20%. This is difficult to judge without precise data, which are simply unavailable for mixed stands, to test a relationship of this sort rigorously. Even an error of this size, however, would retain the thinning line within the boundary zone of size and density typical of pure stands, since it only adds a fraction of an order of magnitude to the intercept constant.

2. *Harvard Forest, Petersham, Massachusetts*

Another example of the thinning dynamics of mixed forest stands is derived from two separate, replicate plots about 90 m apart in a second-growth woodlot at Harvard Forest; both are control plots of a silvicultural (thinning) experiment begun in 1956 (Tom Swamp I, Expt. 56-2). Detailed stand maps and measurements of all trees over 5-cm dbh were maintained at regular intervals to 1975, though their recent history has not been recorded. The earlier history of this forest has been summarized by Lutz & Cline (1947). The size–density relationships of the trees on both plots are shown in Fig. 3. The boundary line for all species collectively is not, in this case, a pure thinning line, since there has been recruitment to the stand (about 11% of the number of trees present in 1956), mainly of *Acer rubrum* (nearly 7%). The boundary line common to both plots after 1966/69 is $d = 5.05\,N^{-0.54}$. After transformation in the manner described earlier for the plot at Harvard Black Rock Forest by using a local volume table (Spaeth 1920) and the same estimate of wood specific gravity, the size–density relationship of these stands is $W = 5.3 \times 10^3 N^{-1.36}$, within the range for trees shown in Fig. 1.

In these plots the size–density dynamics of individual trees species show greater differences among themselves than those in Fig. 2. Once again, their behaviours are silviculturally predictable and range from the rapid decline in number without appreciable size increase of some trees (*Betula lenta, Fraxinus americana, Pinus strobus*) to the progressive dominance without much mortality of *Quercus rubra*. Although the overall dynamic relationships conform to the thinning rule, the component species behave individualistically. Some further observations on the diameter distributions and tolerance of these populations have been reported by Lorimer and Krug (1983).

The precise positions of the thinning or boundary lines for all species of these mixed stands are shown in Fig. 4 in the context of a larger-scale size–density arena. Despite its lower slope, ostensibly not in perfect conformity to the thinning rule on that one criterion, the boundary line of Fig. 2 can be seen to be within the range of thinning lines of pure populations. Various other data from mixed-species forests have been assembled in Fig. 4, after the necessary, if perhaps somewhat approximate, transformations to size in terms of mass. There is no evidence of any difference in their thinning trajectories or boundary values from those now well established for pure stands. Forty individual points for mixed stands from the Woodland Data Set (De Angelis, Gardner & Shugart 1981) all lie within a boundary line of 4.4, most of them, as with pure stands, between 3.5 and 4. The most extreme conjoint mass and density combination I have found for mixed forest stands is for an evergreen gallery forest in Thailand reported by Ovington (1964), which lies at a boundary limit of 4.58.

B. Herbs

There are few good thinning trajectories available for experimental populations of herbaceous plants grown in mixtures, and these involve only two-species combinations (Fig. 5). None transgresses the boundary limits already recognized

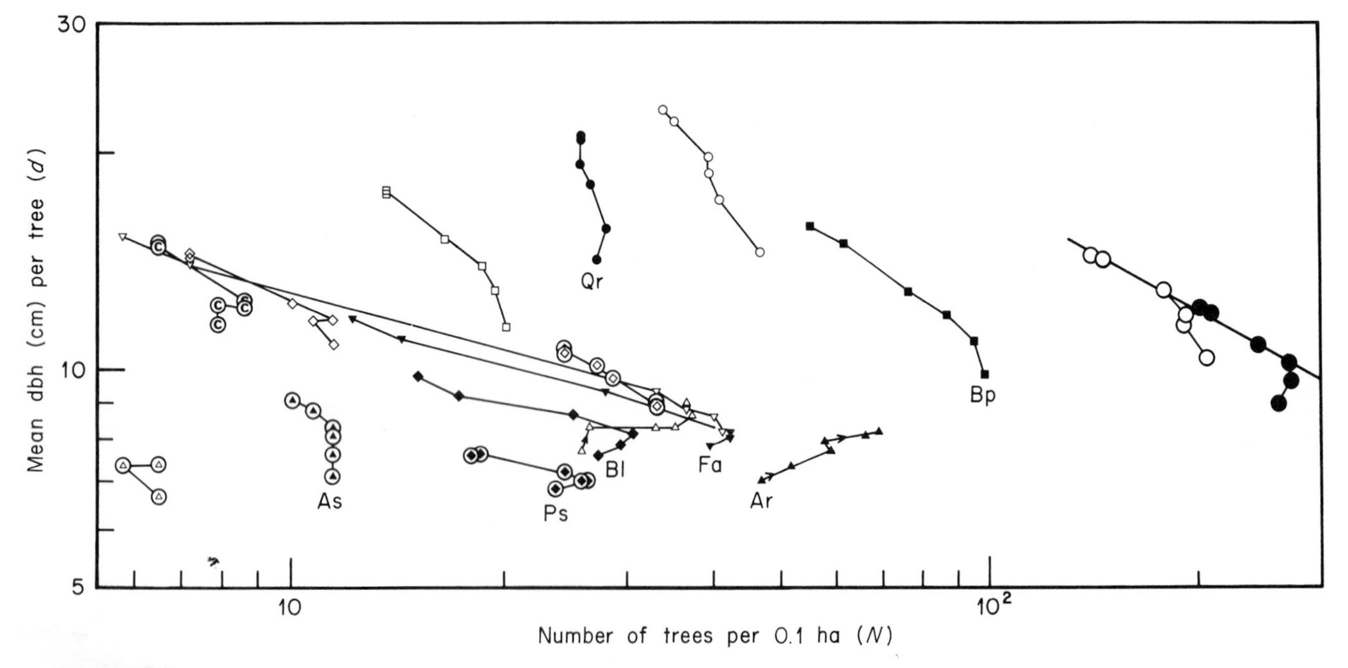

Fig. 3. Relationship between mean diameter at breast height (dbh) per tree and density (per 0.1 ha) on two adjacent 0.14-ha plots at Harvard Forest. The plots are distinguished by closed and open symbols (record units 1–6 and 43–48, respectively) for each species except *Carya* sp. (C), which occurred on only one plot. The points on the right represent all species in a successive time sequence 1956, 1960, 1966, 1969, 1973, 1975; the line is fitted by eye as a dbh–density boundary. Ar, *Acer rubrum*. As, *Acer saccharum*. Bl, *Betula lenta*. Bp, *Betula papyrifera*. Fa, *Fraxinus americana*. Ps, *Pinus strobus*. Qr, *Quercus rubra*. (From data supplied by E. M. Gould and personal records for 1973.)

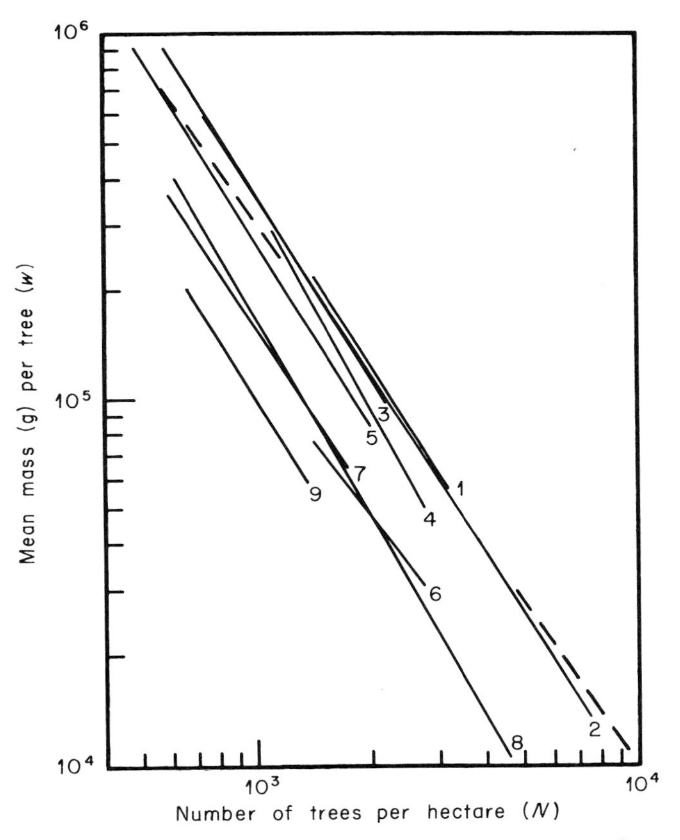

Fig. 4. Relationship between mean plant mass (W) and density (N) for all stems in mixed stands of forest trees. Each line represents either the time course of thinning of all stems over varying time periods or a boundary value of discrete mass and density combinations recorded at different ages. The hatched line shows the position of a boundary line given by $W = 10^4 N^{-3/2}$. 1, *Pseudotsuga menziesii–Alnus rubra,* boundary line for stands aged 29, 41, 53 years. (Binkley [1984].) 2, *Tsuga heterophylla–Picea sitchensis,* predicted thinning line from age 30 to 110 years in stands of site index 110. (Taylor [1934].) 3, Second-growth mixed hardwood forest, Harvard Black Rock Forest, actual thinning line over 53 years (see Fig. 2). 4, Mixed hardwood forest, actual thinning line over 50 years (1927–1977) for all major species on medium moist sites. (Stephens & Waggoner [1980].) 5, *Pinus* spp.–*Abies* spp.–*Pseudotsuga menziesii,* predicted thinning line from age 50 to 150 years in stands of site index 60. (Dunning & Reineke [1933].) 6, Second-growth mixed hardwood forest, Harvard Forest, boundary line for stands from 1956 to 1975 (see Fig. 3). 7, *Castanea dentata–Quercus* spp., predicted thinning line from age 30 to 75 years in stands of site quality I. (Frothingham [1912].) 8, *Quercus* spp. (five predominant species), predicted thinning line from age 20 to 100 years in stands of site index 60. (Schnur [1937].) 9, Mixed hardwood forest, actual thinning line from 1933 to 1978. (Spurr & Barnes [1980], Table 14.1.)

for pure stands. The meagre data may be supplemented by extreme size–density combinations from mixed natural vegetation, and a few examples are also shown in Fig. 5. The size–density dynamic behaviour of individual species in such natural or seminatural vegetation is not known, but we may expect a great diversity, since even in the simpler two-species combinations shown in Fig. 5 there may be striking differences between the component species (Bazzaz & Harper 1976).

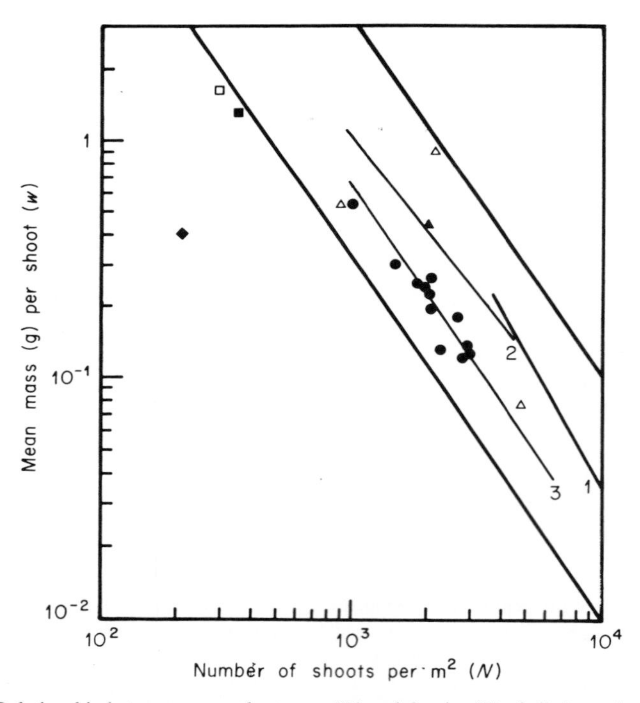

Fig. 5. Relationship between mean plant mass (W) and density (N) of all shoots in mixed stands of herbaceous plants. The numbered lines show the time course of thinning, and individual points indicate discrete mass and density combinations in separate populations. The positions of boundary lines given by $W = 10^4N^{-3/2}$ and $W = 10^5N^{-3/2}$ are indicated. 1, *Medicago sativa–Trifolium pratense,* thinning line over 25 days. (Malmberg & Smith [1982].) 2, *Brassica napus–Raphanus sativus,* thinning line over 140 days. (White & Harper [1970].) 3, *Sinapis alba–Lepidium sativum,* thinning line over 34 days; ▲ most extreme mass and density combination recorded. (Bazzaz & Harper [1976].) ● Plots of seminatural grassland vegetation, ranging in age from 3 to 22 years. (Jukola-Sulonen [1983].) □ Grassland dominated by *Calamagrostis canadensis* and *Epilobium angustifolium* (Mitchell & Evans 1966). ■ Grassland dominated by *Miscanthus sinensis* (Hayashi, Hishinuma & Yamasawa [1981].) ◆ Semi-arid grassland, most extreme mass and density combination (Kumar & Joshi 1972). △ Three tropical grasslands, most extreme mass and density combinations for each. (Singh & Joshi [1979].)

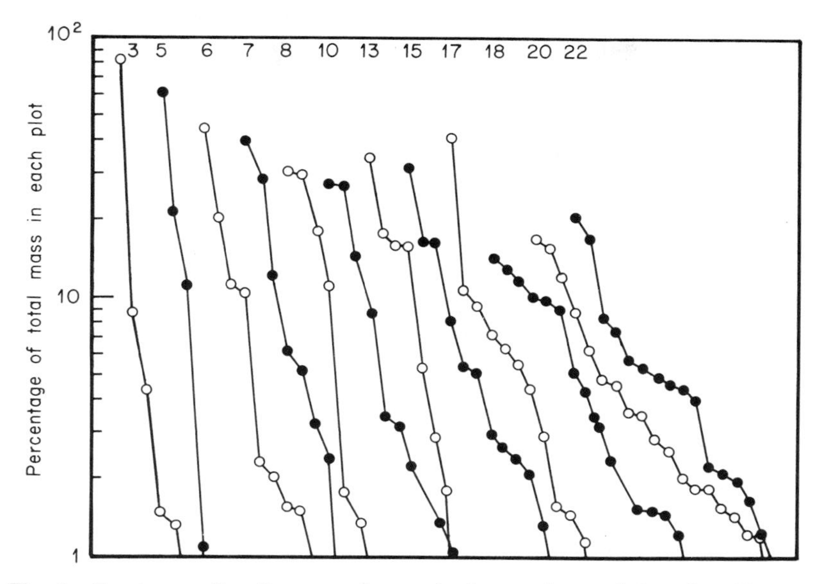

Fig. 6. Dominance–diversity curves for grassland vegetation aged from 3 to 22 years (as indicated). The species are arranged in order of the relative abundances of their biomass within each set of data, down to a threshold of 1%: this includes almost all species with at least a biomass of 4 g/m². Total biomass ranged from 297 to 534 g/m² and was independent of age. (Drawn from data in Jukola-Sulonen [1983].)

One might expect, given the resource differentiation that is possible between species, that complex mixtures might have size–density limits greater than those shown by pure populations. Although this may be so, there is no evidence available to support it and, on the contrary, one good data set refutes it. The data for a group of adjacent, semi-natural sown grasslands shown in Fig. 5 were derived from pastures aged 3–22 years from date of sowing. Samples were taken from 1971 to 1979 at the time of peak standing crop in sites which had been allowed to grow undisturbed for several years. No quadrat was sampled more than once. Species diversity ranged from 12 to 36 species. The boundary line which these points (Fig. 5) define is $W = 8 \times 10^3 N^{-1.38}$ (using measured biomass: Jukola-Sulonen 1983, Appendix 1). There is no simple age sequence among the points shown. Quite diverse size–density relationships hold among the major species individually (Jukola-Sulonen 1983). The dominance–diversity curves for these various grasslands are illustrated in Fig. 6 and show a time-series pattern now well known for natural vegetation. It is clear that neither species dominance nor diversity has any influence on the position of these mixed-species communities on the size–density boundary.

IV. CONCLUSION

The self-thinning rule defined as $W = K \cdot N^{-3/2}$, where K shows values between 3.2×10^3 and 10^5, or $\log K = 3.5 - 5.0$, holds over a size–density space which approaches 8 orders of magnitude of density and 12 of mean plant mass for seed plants. All possible size–density relationships may occur inside this boundary zone, but only rarely outside it. The boundary limits tend to be lower for trees than for herbs, although this may be a sampling artifact, since aerial plant mass is easier to measure in herbs than in trees. It is a coarse-grained rule, insofar as a relatively wide range of mass is achievable per unit area for a given density, but this is relatively small when the large range of the rule's applicability is considered. Much more precise formulations are possible for size–density relationships within parts of this 'global rule' and are indeed highly developed for special parts of it. In the density range 10^{-1}–10 per square metre, for example, the rule has severe limitations for precise predictions in forest management, for which more accurate thinning models have been developed for local circumstances and particular species. The thinning rule, conceived in these terms and on this scale, is in my view analogous to some large-scale astrophysical rule or law: an order of magnitude variation in one scale parameter is negligible if it holds over many orders of magnitude of another related parameter. This is certainly true of the thinning rule, as Fig. 1 shows, and it can be extended by three further orders of magnitude, to almost 10^9 plants per square metre equivalent for some cryptogams (J. White, unpublished). Within this domain of reference, the evidence presently available suggests that the rule holds for complex mixtures of species, too, when all components are collectively considered. Notwithstanding the known exceptions, the thinning rule as originally formulated continues to receive confirmation, and still remains a remarkably general statement of constraint between mass and numbers in plant populations.

ACKNOWLEDGMENTS

I am grateful to Dr. E. M. Gould (Harvard Forest) and Mr. J. J. Karnig (Harvard Black Rock Forest) for permission to use data gathered by them or in their care, and to Dr. E.-L. Jukola-Sulonen for details of information published in her monograph.

REFERENCES

Bazzaz, F. A. & Harper, J. L. (1976). Relationship between plant weight and numbers in mixed populations of *Sinapsis alba* (L.) Rabenh. and *Lepidium sativum* L. *Journal of Applied Ecology,* **13,** 211–216.

Binkley, D. (1984). Importance of size–density relationships in mixed stands of Douglas-fir and red alder. *Forest Ecology and Management,* **9,** 81–85.

Biswell, H. H. & Graham, C. A. (1956). Plant counts and seed production on California annual type ranges. *Journal of Range Management,* **9,** 116–118.

Cannell, M. G. R. & Smith, R. I. (1980). Yields of minirotation closely spaced hardwoods in temperate regions: Review and appraisal. *Forest Science,* **26,** 415–428.

Clatworthy, J. N. & Harper, J. L. (1962). The comparative biology of closely related species living in the same area. V. Inter- and intraspecific interference within cultures of *Lemna* and *Salvinia natans. Journal of Experimental Botany,* **13,** 307–324.

Clements, F. E., Weaver, J. E. & Hanson, H. C. (1929). *Plant Competition: An Analysis of Community Functions.* Carnegie Institution of Washington, Washington, D.C.

De Angelis, D. L., Gardner, R. H. & Shugart, H. H. (1981). Productivity of forest ecosystems studied during the IBP: The woodland data set. *Dynamic Properties of Forest Ecosystems* (Ed. by D. E. Reichle), pp. 567–672. Cambridge University Press, Cambridge.

Dunning, D. & Reineke, L. H. (1933). Preliminary yield tables for second-growth stands in the California pine region. *U.S., Department of Agriculture, Technical Bulletin,* **354,** 1–23.

Frothingham, E. H. (1912). Second-growth hardwoods in Connecticut. *U.S., Department of Agriculture, Forest Service Bulletin,* **96,** 1–70.

Gorham, E. (1979). Shoot height, weight and standing crop in relation to density in monospecific plant stands. *Nature (London),* **279,** 148–150.

Harper, J. L. (1967). A Darwinian approach to plant ecology. *Journal of Ecology,* **55,** 247–270.

Hayashi, I., Hishinuma, Y. & Yamasawa, T. (1981). Structure and functioning of *Miscanthus sinensis* grassland in Sugadaira, Central Japan. *Vegetatio,* **48,** 17–25.

Hutchings, M. J. (1979). Weight-density relationships in ramet populations of clonal perennial herbs, with special reference to the $-\frac{3}{2}$ power law. *Journal of Ecology,* **67,** 21–33.

Ikusima, I., Shinozaki, K. & Kira, T. (1955). Intraspecific competition among higher plants. III. Growth of duckweed with a theoretical consideration on the C-D effect. *Journal of the Institute of Polytechnics, Osaka City University, Series D,* **6,** 107–119.

Jukola-Sulonen, E.-L. (1983). Vegetation succession of abandoned hay fields in central Finland: A quantitative approach. *Communicationes Instituti Forestalis Fenniae,* **112,** 1–85.

Kawano, S. & Hayashi, S. (1977). Plasticity in growth and reproductive energy of *Coix Ma-yuen* Roman. cultivated at varying densities and nitrogen levels. *Journal of the College of Liberal Arts, Toyama University, Natural Sciences,* **10,** 61–92.

Kays, S. & Harper, J. L. (1974). The regulation of plant and tiller density in a grass sward. *Journal of Ecology,* **62,** 97–105.

Kumar, A. & Joshi, M. C. (1972). The effects of grazing on the structure and productivity of the vegetation near Pilani, Rajasthan, India. *Journal of Ecology,* **60,** 665–674.

Lonsdale, W. M. & Watkinson, A. R. (1983). Plant geometry and self-thinning. *Journal of Ecology,* **71,** 285–297.

Lorimer, C. G. (1981). Survival and growth of understory trees in oak forests of the Hudson Highlands, New York. *Canadian Journal of Forest Research,* **11,** 689–695.

Lorimer, C. G. & Krug, A. G. (1983). Diameter distributions in even-aged stands of shade-tolerant and midtolerant tree species. *American Midland Naturalist,* **109,** 331–345.

Lutz, R. J. & Cline, A. C. (1947). Results of the first thirty years of experimentation in silviculture in the Harvard Forest, 1908–1938. Part I. The conversion of stands of old field origin by various methods of cutting and subsequent cultural treatments. *Harvard Forest Bulletin,* **23,** 1–182.

Malmberg, C. & Smith, H. (1982). Relationship between plant weight and density in mixed populations of *Medicago sativa* and *Trifolium pratense. Oikos,* **38,** 365–368.

Mitchell, W. W. & Evans, J. (1966). Composition of two disclimax bluejoint stands in southcentral Alaska. *Journal of Range Management,* **19,** 65–68.

Ovington, J. D. (1964). Quantitative ecology and the woodland ecosystem concept. *Advances in Ecological Research,* **1,** 103–192.

Reineke, L. H. (1933). Perfecting a stand-density index for even-aged forests. *Journal of Agricultural Research (Washington, D.C.),* **46,** 627–638.

Salisbury, E. J. (1970). The pioneer vegetation of exposed muds and its biological features. *Philosophical Transactions of the Royal Society of London, Series B,* **259,** 207–255.

Schnur, G. L. (1937). Yield, stand and volume tables for even-aged upland oak forests. *U.S., Department of Agriculture, Technical Bulletin,* **560,** 1–87.

Schumacher, F. X. & Hall, F. D. S (1933). Logarithmic expression of timber-tree volume. *Journal of Agricultural Research (Washington, D.C.),* **47,** 719–734.

Singh, J. S. & Joshi, M. C. (1979). Primary production. *Grassland Ecosystems of the World: Analysis of Grasslands and their Uses* (Ed. by R. T. Coupland), pp. 197–218. Cambridge University Press, Cambridge.

Spaeth, J. N. (1920). Growth study and normal yield tables for second growth hardwood stands in central New England. *Harvard Forest Bulletin,* **2,** 1–21.

Spurr, S. H. & Barnes, B. V. (1980). *Forest Ecology.* 3rd ed. Wiley, New York.

Stephens, G. R. & Waggoner, P. E. (1980). A half century of natural transitions in mixed hardwood forests. *Bulletin—Connecticut Agricultural Experiment Station, New Haven,* **783,** 1–43.

Tadaki, Y. & Shidei, T. (1959). Studies on the competition of forest trees. II. The thinning experiment on small model stand of sugi (*Cryptomeria japonica*) seedlings. *Journal of the Japanese Forestry Society,* **41,** 341–349.

Taylor, R. F. (1934). Yield of second-growth western hemlock-Sitka spruce stands in southeastern Alaska. *U.S., Department of Agriculture, Technical Bulletin,* **412,** 1–29.

U.S. Forest Products Laboratory (1974). *Wood Handbook.* U.S. Department of Agriculture, Agricultural Handbook, *72* (revised). U.S. Government Printing Office, Washington, D.C.

Watkinson, A. R. (1984). Yield-density relationships: The influence of resource availability on growth and self-thinning in populations of *Vulpia fasciculata. Annals of Botany (London),* **53,** 469–482.

Weaver, J. E. (1918). The quadrat method in teaching ecology. *Plant World,* **21,** 267–283.

Westman, W. E. & Whittaker, R. H. (1975). The pygmy forest region of northern California: Studies on biomass and primary productivity. *Journal of Ecology,* **63,** 493–520.

Westoby, M. (1984). The self-thinning rule. *Advances in Ecological Research,* **14,** 167–225.

White, J. (1975). *Patterns of thinning of plant populations.* Text of lecture circulated at XII International Botanical Congress, Leningrad.

White, J. (1980). Demographic factors in populations of plants. *Demography and Evolution in Plant Populations* (Ed. by O. T. Solbrig), pp. 21–48. Blackwell Scientific Publications, Oxford.

White, J. (1981). The allometric interpretation of the self-thinning rule. *Journal of Theoretical Biology,* **89,** 475–500.

White, J. (1984). Plant metamerism. *Perspectives on Plant Population Ecology* (Ed. by R. Dirzo & J. Sarukhán), pp. 15–47. Sinauer Associates, Sunderland, Massachusetts.

White, J. (1985). The census of plants in vegetation. *The Population Structure of Vegetation* (Ed. by J. White), pp. 33–88. Dr. W. Junk, Dordrecht.

White, J. & Harper, J. L. (1970). Correlated changes in plant size and number in plant populations. *Journal of Ecology,* **58,** 467–485.

White, J. F. & Gould, S. J. (1965). Interpretation of the coefficient in the allometric equation. *American Naturalist,* **99,** 5–18.

Whittaker, R. H. & Marks, P. H. (1975). Methods of assessing terrestrial productivity. *Primary Productivity of the Biosphere* (Ed. by H. Lieth & R. H. Whittaker), pp. 55–118. Springer-Verlag, Berlin.

Yoda, K., Kira, T., Ogawa, H. & Hozumi, H. (1963). Self-thinning in overcrowded pure stands under cultivated and natural conditions. (Intraspecific competition among higher plants XI). *Journal of Biology, Osaka City University, Series D,* **14,** 107–129.

V

The Influence of Pathogens and Predators on Plant Populations

20

Pathogens and the Genetic Structure of Plant Populations

J. J. Burdon

Division of Plant Industry
CSIRO
Canberra City, A.C.T., Australia

I. INTRODUCTION

Over the past decade and a half, demographic analyses have provided quantitative information about the enormous mortality which occurs within plant populations in nature. Studies investigating the genetic structure of such populations have shown that further undercurrents of change may occur as individuals with certain genotypes are favoured (Schaal & Levin 1976; Bazzaz, Levin & Schmierbach 1982; Burdon, Marshall & Brown 1983). Although genetic changes have been observed in these investigations their adaptive significance is often obscure. The genetic structure of agricultural plant populations (which is generally uniform) has long been recognized as an important factor in the vulnerability of crops to pest and pathogen attack. Mixtures of varieties each carrying different resistance genes have frequently been proposed (Suneson 1956; Browning & Frey 1969) and grown (Wolfe & Barrett 1980) as a means by which pathogen populations may be held in check. The implications of these and other, similar observations have aroused considerable interest among theoreticians. It has been suggested that pests and pathogens are largely responsible for the diversity of protein polymorphisms found in plants and animals (Clarke 1976) and even for the evolution and maintenance of sexual reproduction (Levin 1975; Hamilton 1980).

Given these intriguing and far-reaching possibilities it is surprising that relatively little interest has been shown in investigating the role that pathogens play in shaping the genetic structure of plant populations. This essay attempts to address this question by critically assessing the results of empirical studies in

STUDIES ON PLANT DEMOGRAPHY:

light of expectations derived from theoretical models. Although host–pathogen interactions at a population level are mutually reciprocal, emphasis is placed here on the effect that pathogens have on the structure of plant populations. At present, our knowledge of the effect of plants on the structure of pathogen populations is noteworthy only because of the lack of experimental data with which to test a plethora of theoretical models.

II. PATHOGENS AS SELECTIVE AGENTS

Conclusive proof that pathogens actually reduce the size of host populations below that which would occur in their absence has yet to be obtained in non-agricultural systems. Several examples clearly show, however, that a range of different pathogens may reduce the survival or reproductive output of affected individuals. Augspurger (1983) found that damping-off fungi were responsible for the death of more than 70% of seedlings that emerged around adult trees of *Platypodium elegans*; Regehr and Bazzaz (1979) recorded a 53% reduction in the seed production of a population of *Erigeron canadensis* as a result of widespread infection by a mycoplasma; Alexander and Burdon (1985) found a significant negative correlation between the combined incidence of the pathogens *Albugo candida* and *Peronospora parasitica* and seed production in *Capsella bursa-pastoris*. The levels of disease required to reduce the reproductive success of individuals may in fact be quite small because the direct effect of diseases is often exacerbated by competitive interactions between affected and non-affected host individuals. If, in addition to the occurrence of disease, variability exists within a plant population in the relative susceptibility of individuals to attack, then, theoretically at least, the basic requirements necessary to produce disease-related changes in the genetic structure of the host population are present.

A. Difficulties in the Assessment of the Selective Effect of Pathogens

A major problem in studies of genotype-specific survival in plant populations is the identification of different genotypes in the field. Studies of the selective role of pathogens are no different. Superficially, it might appear simple to identify resistant and susceptible phenotypes *in situ*. In practice, however, this is complicated by differences in the phenotypic expressions of resistance genes and variability in the biotic and abiotic environments.

Although resistance genes with major phenotypic effects occur in many plant species [the gene-for-gene hypothesis (Flor 1955) based on this type of resistance has been demonstrated or postulated to occur in over 30 plant–pathogen interactions], in general neither do such genes cause complete suppression of sporulation, nor are the infection-type responses of different resistance genes identical. Rather, resistant reactions may be manifest as the failure to produce macroscopic lesions, as non-sporulating necrotic lesions or as sparsely sporulating pustules surrounded by regions of leaf chlorosis. Moreover, background genomic effects in either host or pathogen may result in differences in the infection-type response educed by the same resistance gene when present in different host lines.

Differences in light, temperature, nutrient status and other aspects of the physical environment also interact with host and pathogen to alter the phenotypic expression of infection types. Interactions which in one environment produce necrotic lesions in others may produce small sporulating pustules or, in more extreme cases, such as the temperature-sensitive *Lr20* and *Sr6* resistance genes in wheat, may result in susceptible infection-type responses even when resistance genes are present. Further complications may arise through the presence of pathogen races of differing virulence.

The interaction of these factors makes accurate field determination of generation-to-generation genetic change almost impossible. For this, the most reliable approaches are detailed controlled environment studies of resistance that complement field assessments of the disease loads borne by individual members of the population. Once the basic phenotypic effect of typical resistance genes is understood, a rapid determination of the genetic structure of populations is possible through the screening of seedlings derived from each flowering individual. Comparisons of the genetic structure of populations over a number of generations or from location to location provide a strong test of the selective nature of pathogen attack.

B. Theoretical Expectations

A number of mathematical models have been devised to investigate numerical interactions between different genotypes in host and pathogen populations (e.g. Jayakar 1970; Gillespie 1975; Leonard 1977). Typically such models have shown that alleles for resistance or susceptibility in plant populations are maintained as complex stable, cyclic or chaotic polymorphisms if, in the absence of the pathogen, there is a fitness penalty associated with resistance. Where this assumption is ignored, polymorphisms are not sustained, and these models are reduced to simple competitive interactions. In the absence of disease both resistant and susceptible individuals are competitively equal, and changes that occur in gene frequencies result from random drift. On the other hand, in the presence of a pathogen, resistant individuals always have a selective advantage and eventually exclude susceptible individuals from the population.

A similar argument can be developed for the pathogen: unless fitness penalties are associated with the possession of virulence alleles superfluous to any particular individual host–pathogen combination, models predict a trend towards populations dominated by a race with an extremely broad range of virulence (e.g. Marshall & Pryor 1978).

If assumptions involving fitness penalties associated with resistance and virulence are accepted, then the fitness of individual host genotypes within host–pathogen systems is dependent on relative gene frequencies in the pathogen population, and that of individual pathogen genotypes is dependent on relative gene frequencies in the host population (Leonard & Czochor 1980). These models have yet to be rigorously extended to include interactions among many resistance and virulence loci, although Person (1966) has argued that such interactions will produce cyclical polymorphisms of great complexity in host populations.

C. Empirical Evidence

To date no complete studies have been carried out to compare the viability and fecundity of susceptible and resistant individuals within populations. A growing body of evidence concerning the role of pathogens is available from studies of agricultural mixtures, biological control programs and genetic analyses of individual wild plant populations.

The importance of primary and secondary centres of diversity of plant species as sources of genes for resistance to a wide range of plant pathogens has been recognized for a considerable time (e.g. Frankel & Bennett 1970). Unfortunately, in traditional resistance surveys, individual plant populations are generally represented by only a few accessions. As a result, although such surveys clearly indicate that some geographic areas are richer in resistance genes than others (e.g. Comeau 1982), little information can be gleaned about the frequency of resistant and susceptible genotypes within individual populations. In a few instances more extensive population-based information is available: representative examples are given in Table I. These results indicate that, in line with the predictions of theoretical models, considerable variability exists both within and among populations of a wide range of plant species (grasses, herbs, trees) for resistance to commonly occurring diseases.

D. Variation in Resistance within Populations

In studies of the occurrence of resistance in wild plant populations the degree of discrimination which can reliably be achieved between different infection-type responses is of major concern. In some cases (e.g. *Avena* sp., *Hordeum* sp.) it has been possible to adapt pre-existing scales for the assessment of infection type. Thus, using the standard 0–4 scale devised by cereal rust pathologists, infection-type responses of individual members of *Avena* spp. populations have

TABLE I.

Representative Examples of Host–Pathogen Interactions for Which Data Concerning Resistance within and between Populations Are Available

Host–pathogen combination	Number of populations	Mean sample size/population	Range of resistance, percentage[c]	Mean level of resistance, percentage	Reference
Pinus taeda–*Cronartium fusiforme*	3	11	12.1– 21.2	16.1	Barber (1966)
Avena sp.–*Puccinia coronata*	11[a]	44[a]	0.0– 95.0[a]	27.3[a]	Burdon, Oates and Marshall (1983b)
	10[b]	46[b]	0.0– 5.0[b]	1.0[b]	Burdon, Oates and Marshall (1983b)
Avena sp.–*Puccinia coronata*	31	20	0.0–?	12.4	Dinoor (1970)
Hordeum spontaneum–*Erysiphe graminis hordei*	15	17	0.0–100.0	51.1	Moseman, Nevo and Zohary (1983)
Helianthus annuus–*Puccinia helianthi*	142	c. 20	0.0–c. 80.0	12.8	Zimmer & Rehder (1976)
Triticum dicoccoides–*Erysiphe graminis tritici*	10	23	16.7–100.0	55.3	Moseman *et al.* (1984)

[a] Environment favourable for the pathogen (see text).
[b] Environment unfavourable for the pathogen.
[c] Resistance measured in a diversity of ways; see original references.

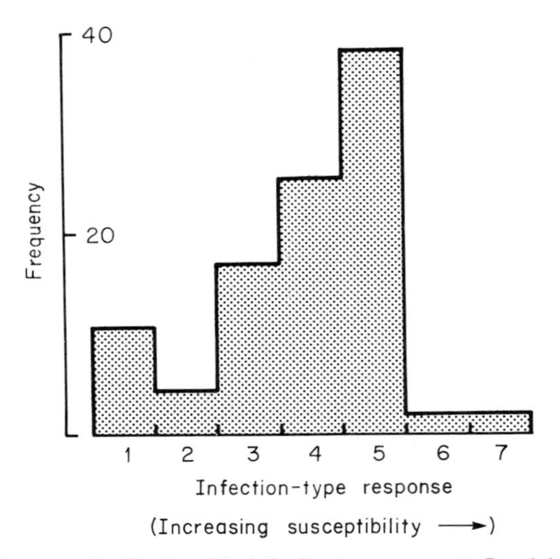

Fig. 1. The frequency distribution of the infection-type response to *Puccinia coronata* race 227 of seedlings derived from 50 *Avena fatua* plants growing in a wild population at Inverell in northern New South Wales. (Burdon, Oates and Marshall [1983] and unpublished data.)

been divided into seven categories ranging from complete resistance to complete susceptibility (Burdon, Oates & Marshall 1983b). In one population, a few individuals (11%) were fully resistant when inoculated with race 227 of *Puccinia coronata* (category one, Fig. 1), a few others (2%) were highly susceptible (category seven), and the majority showed intermediate infection-type responses characterized by pustules of varying size and leaf chlorosis of varying intensity (categories two to six). In a similar study of the resistance of a population of *Hordeum spontaneum* to field strains of the pathogen *Erysiphe graminis,* the division of infection types into three categories showed that 65% of the population was susceptible, 13.5% was resistant and the remaining 21.5% of individuals showed an intermediate response (G. Fischbeck, E. Schwarzbach, A. Segal & I. Wahl, unpublished data cited by Wahl *et al.* 1978).

Unfortunately, these studies *may* provide a very incomplete picture of the resistance structure of plant populations. Because individual hosts can be classified as either resistant or susceptible, depending upon the isolate(s) of pathogen they encounter, any assessment of the resistance structure of a plant population will be strongly influenced by the pathogen isolates used. Isolates with broad ranges of virulence will tend to detect the presence of fewer resistance genes than would isolates with very narrow ranges of virulence. Similarly, studies involving only one race of a pathogen or those involving a mixture of races (many of the traditional resistance surveys fall into this category) will tend to produce a more simplified picture of the extent of variation in resistance than may be obtained through the use of many races separately. In the Australian *Avena–Puccinia* study these problems have been reduced by assessing the seedling response of 50 *Avena* spp. individuals to races 226, 237 and 277 of *P. coronata* (as well as race 227). Each of these additional tests produced resistance profiles which were similar to that obtained in tests involving race 227 (Fig. 1). However, there was no consistent pattern in the infection type produced by particular individuals; some individuals were highly resistant or highly susceptible to all four races, but many showed a differential response.

The complexity that patterns of interaction between host and pathogen population may attain is illustrated by an investigation of an *Avena* population growing at Ein Hashofet in Israel (Dinoor 1977). In this study, seedlings derived from over 100 plants were inoculated separately with six common races of *Puccinia coronata* and the resultant infection-type responses simply categorized as resistant or susceptible. Individual host plants again showed considerable diversity in their responses; however, the extent of this diversity was overwhelming. Although 24% of the population was susceptible to all six races of the pathogen, a total of 17 different combinations of resistance and susceptibility occurred among the remaining 76%. For example, six individuals were resistant to races 264 and 277 alone, 13 to races 264 and 276 alone; a further 25 were protected by genes effective against all three of these races combined. Overall this population presents the pathogen with a two-dimensional mosaic of resistance and susceptibility. Not only do individual host plants react differently to different pathogen races, but the proportion of the population which is susceptible also varies with pathogen phenotype (from 40 to 90%).

E. Variation in Resistance among Populations

Populations from different sites often exhibit markedly different levels of resistance when tested under uniform conditions. Extensive country-wide surveys of the resistance of Israeli populations of *Avena sterilis* to *Puccinia coronata* (Dinoor 1970; Wahl 1970), of *Hordeum spontaneum* to *P. recondita hordei* (Anikster, Moseman & Wahl 1976) and *Erysiphe graminis* (Fischbeck *et al.* 1976; Moseman, Nevo & Zohary 1983), and of *Triticum dicoccoides* to *E. graminis* (Moseman *et al.* 1984) have found distinct regional variation in the frequency of resistance. In general, the geographic distribution of resistance to these pathogens showed an association with rainfall patterns. Populations growing in dry environments (for example, the edge of the Negev desert), where the period suitable for simultaneous growth of both host and pathogen is limited, were mainly susceptible. In contrast, populations growing in more mesic environments (for example, the Golan Heights) were characterized by higher frequencies of resistant individuals.

Similar spatial variation in the frequency of occurrence of resistance was found in the Australian wild oat study. There, both the mean level of resistance of populations of *A. barbata* and *A. fatua* to *P. coronata* and the overall diversity of the resistance profile of populations, were greater in the northern part of New South Wales than in the south (Burdon, Oates & Marshall 1983). These differences could again be correlated with variations in the length of the growing season and general climatic conditions. In addition, however, a 5-year study uncovered complementary differences in the structure of the pathogen population (Oates, Burdon & Brouwer 1983). The pathogen population occurring in the northern part of New South Wales was found to possess greater racial diversity than that in the south. Moreover, the average pathogen isolate in the northern population had a slightly broader range of virulence. These workers suggested that both host and pathogen were evolving together in the north, although in the south there was relatively little interaction between the species.

Even within general climatic regions considerable fluctuations in the frequency of occurrence of resistance are detectable among individual populations. Moseman, Nevo and Zohary (1983) found marked differences between two populations of *Hordeum spontaneum* growing only 6 km apart in northern Israel. In one population 93% of individuals were resistant to *E. graminis;* in the other, 75% of individuals were susceptible.

The extent to which local differentiation in the resistance structure of host populations may occur has not been investigated directly. However, the findings of Snaydon and Davies (1972) indicate that in *Anthoxanthum odoratum,* differential selection for resistance may occur over very short distances. In a study of populations taken from differently treated plots of the Park Grass Experiment at Rothamsted, they found that resistance to *Erysiphe graminis* was greatest in populations growing on plots with high soil nitrogen levels. In addition, populations from plots with tall vegetation were more resistant to *Puccinia poae-nemoralis* than those from short vegetation. High soil nitrogen levels and the occurrence of tall vegetation (through its effect on the humidity of the stand) favour the growth and development of both pathogens, respectively, and should, therefore, have lead to selection in the observed direction. However, these differences in resistance occurred between plots only 30 m apart!

F. Variation in Resistance with Time

The studies reviewed so far provide a picture of the type of variation in resistance which may be found within and between populations of a range of plant species. Unfortunately, however, they all have the major shortcoming of providing only a static picture of the resistance structure of host populations. They give no indication of the speed with which such complexity may develop or the way it is maintained or of how dynamically responsive these systems remain.

To all these questions we have only most incomplete answers.

A number of long-term experiments have shown the importance of physical environmental influences on the success or failure of different plant genotypes growing in mixtures (e.g. Harlan & Martini 1938). In contrast, only one study in which the effect of pathogens has been examined over several generations has been published. In that investigation, seed of five near-isogenic lines of *Avena sativa* was combined together in equal proportions to form a mixture which was then grown in the field for 4 consecutive years (Murphy *et al.* 1982). Half the mixture was inoculated with a composite of five races of *Puccinia coronata*; the other half was protected from this pathogen by periodic fungicide treatment. Over four generations, changes in the frequencies of the five lines in the two treatments were similar. However, in the rust-free environment, despite competitive interactions and year-to-year environmental heterogeneity, there was some evidence that particular genotypes would eventually dominate the mixture and others would be reduced to low frequencies. By contrast, although none of the differences between the two treatments was statistically significant, no individual oat genotype performed as well, or as badly, in the rusted plots as in the rust-free ones.

In the initial mixture constructed by Murphy and his associates, the resistance possessed by the different oat lines overlapped so that each line was susceptible to only one pathogen race. As a consequence, from the point of view of each oat genotype and each pathogen race, the initial mixture was effectively 80% resistant. Seen in this way, it is not surprising that the differences between rusted and non-rusted treatments were minor. Even after four generations, the highest frequency of an individual line in either mixture was only 38%, a level at which the disease-reducing effects of most mixtures are still highly effective (J. J. Burdon, unpublished data). The relative complexity of the resistance profile of this mixture is reminiscent of the diversity of resistance found in many natural populations (e.g. *Avena* spp.) and suggests that such populations may be buffered against rapid directional change.

Simpler mixtures in which initial populations are mainly susceptible may provide a clearer indication of how rapidly the resistance structure of a plant population can respond to differential selection pressure during the early stages of development of a host–pathogen interaction. The only published example is found in changes that occurred in populations of *Chondrilla juncea* in south-eastern Australia after an attempt to control this weedy species through the deliberate release of the rust pathogen *Puccinia chondrillina* (see Cullen & Groves 1977 for full details). *Chondrilla juncea* is an obligate apomict, three distinct morphological forms of which have been identified in Australia (narrow, intermediate and broad-leaved). In 1968, 3 years before the release of a strain of *P. chondrillina* capable of attacking the narrow-leaved form only, seven populations of *C. juncea* in New South Wales were surveyed and the individuals present

TABLE II.

Percentage of Plants of Three Genotypes of *Chondrilla juncea* in Local Roadside Populations in New South Wales, Australia[a]

	C. juncea genotypes in years					
	1968[b]			1980[c]		
Population	N	I	B	N	I	B
1	100	0	0	88	0	12
2	100	0	0	99	1	0
3	98	0	2	5	90	5
4	67	30	3	3	78	19
5	62	24	14	3	80	17
6	51	10	39	8	10	82
7	21	1	78	11	13	76

[a] From Burdon, Groves and Cullen (1981).
[b] 1968: before the release of a rust pathogen capable of attacking the narrow-leaved genotype.
[c] 1980: 9 years after the release of this pathogen. N, I and B are narrow, intermediate and broad-leaved forms, respectively.

classified according to phenotype. In 1980, 9 years after the successful establishment of the rust, these sites were re-examined and the frequency of the different types of *C. juncea* again recorded (Table II) (Burdon, Groves & Cullen 1981). A comparison of the frequency of the three forms at the same sites in 1968 and 1980 shows an increase in the number at which the resistant intermediate and broad-leaved forms occur. Moreover, at sites previously occupied by the narrow-leaved form and either one or both of the other forms, an increase in the frequency of the latter forms has paralleled a decline in the frequency of narrow-leaved individuals.

In long-term studies of changes in the frequency of resistant phenotypes in a population, care must be taken in ascribing observed changes to selection for resistance per se rather than for associated features in the genome. This possibility is of particular concern in strongly inbreeding species, in which populations may rapidly develop highly organized structures featuring coadapted gene complexes (Clegg, Allard & Kahler 1972). In such populations, changes in the frequency of resistance alleles *may* represent no more than a "hitch-hiking" effect in which selection is actually operating on some other aspect of the gene complex. This problem is illustrated by studies of pathogen resistance in barley Composite Cross II (for details of establishment and propagation, see Suneson 1969). Assessments of the frequency of individuals resistant to four different isolates of *Rhynchosporium secalis* in generations 7, 15, 25 and 47 of this cross found a consistent pattern among three of the isolates (Jackson *et al.* 1978). The frequency of phenotypes resistant to these three isolates was less than 30% for the first three sampling dates but rose to more than 75% by the 47th generation. By contrast, the frequency of individuals resistant to the fourth isolate remained low (less than 10%). A similar pattern was found in a later study involving generations 8, 13, 23, and 45 (Muona, Allard & Webster 1982). Initially these results might be thought to reflect selection for certain resistant phenotypes. However, evidence from a variety of sources, including surveys of the racial composition of *Rhynchosporium secalis* and studies of changes in the frequency of blocks of alleles at different loci in this Composite Cross population (Clegg, Allard & Kahler 1972), all make it difficult to assign changes in the frequency of resistant phenotypes solely to the selective advantage that such resistance may confer (Jackson *et al.* 1978). On the other hand, these data do indicate that when selection for disease resistance is sufficient to induce genotypic changes in populations of predominantly selfing plants, the effects of this selection will ramify throughout the genome!

III. CONCLUSION

Because of the confounding effects of environment, field-based measurement of the relationship between the genotype and the overall fitness of individuals remains one of the most difficult problems to solve in plant population biology. In this respect the study of host–pathogen interactions offers the major advantage of being able to relate simply the adaptive value of observed changes to the operation of specific selective forces in the environment. To date, although evidence for the specificity and debilitating effect of pathogens is incontrovertible, experimental data concerning their selective nature within and among populations have yet to be collected. However, considerable circumstantial evidence which supports this hypothesis exists. As a result, host–pathogen interactions offer attractive systems with which to investigate the complexity of changes accompanying the exertion of differential selective pressures on plant populations. Moreover, because of the propensity of pathogens to change, such studies will provide the basis of a more complete understanding of the mechanisms behind the development and maintenance of co-evolutionary interactions.

ACKNOWLEDGMENTS

The helpful comments and suggestions of Dr. S. C. H. Barrett and Dr. A. H. D. Brown are gratefully acknowledged.

REFERENCES

Alexander, H. M. & Burdon, J. J. (1985). The effect of disease induced by *Albugo candida* and *Peronospora parasitica* (downy mildew) on the survival and reproduction of *Capsella bursa-pastoris* (shepherd's purse). *Oecologia,* **64,** 314–318.

Anikster, Y., Moseman, J. G. & Wahl, I. (1976). Parasitic specialization of *Puccinia hordei* Otth. and sources of resistance in *Hordeum spontaneum* C. Koch. *Barley Genetics, Proceedings of* the 3rd International Barley Genetics Symposium, 1975, pp. 468–469.

Augspurger, C. (1983). Seed dispersal of the tropical tree, *Platypodium elegans,* and the escape of its seedlings from fungal pathogens. *Journal of Ecology,* **71,** 759–771.

Barber, J. C. (1966). Variation among half-sib families from three loblolly pine stands in Georgia. *Georgia Forestry Research Paper,* No. 37, pp. 1–5.

Bazzaz, F. A., Levin, D. A. & Schmierbach, M. R. (1982). Differential survival of genetic variants in crowded populations of *Phlox. Journal of Applied Ecology,* **19,** 891–900.

Browning, J. A. & Frey, K. J. (1969). Multiline cultivars as a means of disease control. *Annual Review of Phytopathology,* **7,** 355–382.

Burdon, J. J., Groves, R. H. & Cullen, J. M. (1981). The impact of biological control on the distribution and abundance of *Chondrilla juncea* in southeastern Australia. *Journal of Applied Ecology,* **18,** 957–966.

Burdon, J. J., Marshall, D. R. & Brown, A. H. D. (1983). Demographic and genetic changes in populations of *Echium plantagineum. Journal of Ecology,* **71,** 667–679.

Burdon, J. J., Oates, J. D. & Marshall, D. R. (1983). Interactions between *Avena* and *Puccinia* species. I. The wild hosts: *Avena barbata* Pott. ex. Link, *A. fatua* L. and *A. ludoviciana* Durieu. *Journal of Applied Ecology,* **20,** 571–584.

Clarke, B. C. (1976). The ecological genetics of host-parasite relationships. *Genetic Aspects of Host-Parasite Relationships* (Ed. by A. E. R. Taylor & R. Muller), pp. 87–103. Blackwell Scientific Publications, London.

Clegg, M. T., Allard, R. W. & Kahler, A. L. (1972). Is the gene the unit of selection? Evidence from two experimental plant populations. *Proceedings of the National Academy of Science of the U.S.A.,* **69,** 2474–2478.

Comeau, A. (1982). Geographic distribution of resistance to barley yellow dwarf virus in *Avena sterilis. Canadian Journal of Plant Pathology,* **4,** 147–151.

Cullen, J. M. & Groves, R. H. (1977). The population biology of *Chondrilla juncea* L. in Australia. *Proceedings of the Ecological Society of Australia,* **10,** 121–134.

Dinoor, A. (1970). Sources of oat crown rust resistance in hexaploid and tetraploid wild oats in Israel. *Canadian Journal of Botany,* **48,** 153–161.

Dinoor, A. (1977). Oat crown rust resistance in Israel. *Annals of the New York Academy of Sciences,* **287,** 357–366.

Flor, H. H. (1955). Host-parasite interaction in flax rust—its genetics and other implications. *Phytopathology,* **45,** 680–685.

Fischbeck, G., Schwarzbach, E., Sobel, S. & Wahl, I. (1976). Types of protection against barley powdery mildew in Germany and Israel selected from *Hordeum spontaneum. Barley Genetics, Proceedings of the 3rd International Barley Genetics Symposium, 1975,* pp. 412–417.

Frankel, O. H. & Bennett, E. (Eds) (1970). *Genetic Resources in Plants—Their Exploration and Conservation.* IBP Handbook No. 11. Blackwell Scientific Publications, Oxford.

Gillespie, J. H. (1975). Natural selection for resistance to epidemics. *Ecology,* **56,** 493–495.

Hamilton, W. D. (1980). Sex versus non-sex versus parasite. *Oikos,* **35,** 282–290.

Harlan, H. V. & Martini, M. L. (1938). The effect of natural selection in a mixture of barley varieties. *Journal of Agricultural Research (Washington, D.C.),* **57,** 189–199.

Jackson, L. F., Kahler, A. L., Webster, R. K. & Allard, R. W. (1978). Conservation of scald resistance in barley composite cross populations. *Phytopathology,* **68,** 645–650.

Jayakar, S. C. (1970). A mathematical model for interaction of gene frequencies in a parasite and its host. *Theoretical Population Biology,* **1,** 140–164.

Leonard, K. J. (1977). Selection pressures and plant pathogens. *Annals of the New York Academy of Sciences,* **287,** 207–222.

Leonard, K. J. & Czochor, R. J. (1980). Theory of genetic interactions among populations of plants and their pathogens. *Annual Review of Phytopathology,* **18,** 237–258.

Levin, D. A. (1975). Pest pressure and recombination systems in plants. *American Naturalist,* **109,** 437–451.

Marshall, D. R. & Pryor, A. J. (1978). Multiline varieties and disease control. I. The "dirty crop" approach with each component carrying a unique single resistance gene. *Theoretical and Applied Genetics,* **51,** 177–184.

Moseman, J. G., Nevo, E. & Zohary, D. (1983). Resistance of *Hordeum spontaneum* collected in Israel to infection wtih *Erysiphe graminis hordei.* Crop Science, **23,** 1115–1119.

Moseman, J. G., Nevo, E., El Morshidy, M. A. & Zohary, D. (1984). Resistance of *Triticum dicoccoides* to infection with *Erysiphe graminis tritici. Euphytica,* **33,** 41–47.

Muona, A., Allard, R. W. & Webster, R. K. (1982). Evolution of resistance to *Rhynchosporium secalis* (Orud.) Davis in Barley Composite Cross. II. *Theoretical and Applied Genetics,* **61,** 209–214.

Murphy, J. P., Helsel, D. B., Elliott, A., Thro, A. M. & Frey, F. J. (1982). Compositional stability of an oat multiline. *Euphytica,* **31,** 33–40.

Oates, J. D., Burdon, J. J. & Brouwer, J. B. (1983). Interactions between *Avena* and *Puccinia* species. II. The pathogens: *Puccinia coronata* Cda and *P. graminis* Pers. f. sp. *avenae* Eriks & Henn. *Journal of Applied Ecology,* **20,** 585–596.

Person, C. (1966). Genetic polymorphism in parasitic systems. *Nature (London),* **212,** 266–267.

Regehr, D. L. & Bazzaz, F. A. (1979). The population dynamics of *Erigeron canadensis,* a successional winter annual. *Journal of Ecology,* **67,** 923–933.

Schaal, B. A. & Levin, D. A. (1976). The demographic genetics of *Liatris cylindracea* Michx. (Compositae). *American Naturalist,* **110,** 191–206.

Snaydon, R. W. & Davies, M. S. (1972). Rapid population differentiation in a mosaic environment. II. Morphological variation in *Anthoxanthum odoratum. Evolution,* **26,** 390–405.

Suneson, C. A. (1956). An evolutionary plant breeding method. *Agronomy Journal,* **48,** 188–190.

Suneson, C. A. (1969). Evolutionary plant breeding. *Crop Science* **9,** 395–396.

Wahl, I. (1970). Prevalence and geographic distribution of resistance to crown rust in *Avena sterilis. Phytopathology,* **60,** 746–749.

Wahl, I., Eshed, N., Segel, A. & Sobel, Z. (1978). Significance of wild relatives of small grains and other wild grasses in cereal powdery mildews. *The Powdery Mildews* (Ed. by D. M. Spencer), pp. 83–100. Academic Press, New York.

Wolfe, M. S. & Barrett, J. A. (1980). Can we lead the pathogen astray? *Plant Disease,* **64,** 148–155.

Zimmer, D. E. & Rehder, D. (1976). Rust resistance of wild *Helianthus* species of the north central United States. *Phytopathology,* **66,** 208–211.

21

Sex Ratios, Clonal Growth and Herbivory in Rumex acetosella

Jon Lovett Doust

Department of Biology
Amherst College
Amherst, Massachusetts, USA

Lesley Lovett Doust

Department of Biological Sciences
Mount Holyoke College
South Hadley, Massachusetts, USA

I. INTRODUCTION

In his address to the British Ecological Society, Harper (1967) focused attention on the role of biotic forces in determining the distribution and abundance of species and emphasized the importance of distinguishing the *kinds* of individuals within a population. He postulated that in dioecious species a certain amount of specialization occurs between the niches of males and females (the ''Jack Sprat effect''). Experimental study of sex ratios in two sorrel species, *Rumex acetosella* and *R. acetosa,* had suggested to Putwain and Harper that the sexes showed frequency-dependent interactions, tending to stabilize sex ratios. They concluded this happened because the two sexes occupy at least partly different ecological microhabitats (Putwain & Harper 1970, 1972).

Explanations of separate male and female spatial distributions sometimes involve patchy distribution of resources, possibly acting in conjunction with differential competitive ability of male and female individuals. Measures of light availability, soil moisture, and aspects of soil fertility have all been associated with the spatial or temporal separation of males and females, in a number of dioecious species (Freeman, Klikoff & Harper 1976; Cox 1981; Wade, Armstrong & Woodell 1981; Zimmerman & Lechowicz 1982). Commonly when habitat segregation is observed, females are found in wetter areas than males, although as Willson (1983) has pointed out, this may only reflect the particular species and habitats surveyed rather than any natural trend.

Bawa (1980) drew attention to the fact that dimorphism between the sexes may effectively alter the spatial distribution of resources for pollinators, seed dispersers, and predators. We have found a number of indications that predators and parasites may choose among plants according to the sex of the plants. For example, Keep, Knight, and Parker (1977), studying a sexually dimorphic variety of raspberry, reported an association between plant sex and the plant's response to mildew infection (*Sphaerotheca macularis*). These workers observed that in some cultivars staminate plants were less susceptible to the fungus than were hermaphrodite plants. Sterile plants also seemed to show less susceptibility to mildew than hermaphrodites. Bawa and Opler (1978) reported striking differences in predation by the moth larva *Atteva punctella* upon the inflorescences of staminate and pistillate individuals of the tropical tree *Simarouba glauca.* They noted that female inflorescences were eaten less often than male inflorescences. Similarly, Cox (1982) found that male and hermaphrodite *Freycinetia reineckei* inflorescences suffered significantly greater damage than female inflorescences as a result of the activities of the flying foxes and starlings that pollinate them. In *Silene dioica,* Lee (1981) showed that the rate of infection by the smut *Ustilago violacea* was greater in females than in males and that in some populations over a third of the female reproductive capacity was lost. Lee noted that infection reduced the growth of male plants but induced a large increase in the number of flowers on female plants. Finally, as part of a general survey of gender-related traits in Jack-in-the-pulpit (Lovett Doust & Cavers 1982) we found that females were distinct from males in several respects, and in particular were 10 times more likely to be infected with the host-specific rust fungus *Uromyces ariestriphylli* than were male shoots. Individual genets of Jack-in-the-pulpit may change sex from one year to the next (possibly as a function of their stored reserves) so the differential susceptibility of females cannot be attributed simply to sex .

These observations led us to speculate further about differential herbivory on male and female plants. Here we examine the proposition that insect herbivory may be an important factor affecting population sex ratios and size of individuals in a clonal dioecious plant. Male and female plants are considered as sources of forage for a grazing animal [tobacco wireworm, *Conoderus vespertinus* (Elateridae)]. We describe a preliminary field study of sex ratios in red sorrel, *Rumex acetosella,* in which we were interested in the following questions: (a) Is there evidence of sex-specific herbivory? (b) How does clonal growth of a genet affect the frequency of shoots, in particular when superimposed upon patterns of herbivory?

II. Methods

Rumex acetosella is a short-lived perennial species, capable of clonal growth. It is one of three dioecious sorrels in the northeastern American flora (Gleason 1963). Eight populations were studied in the area of Amherst–South Hadley, Massachusetts. Sites were sampled by using one to five line transects, 50 m in length.

Along every transect, all *Rumex* shoots touching or immediately beneath the measuring tape were recorded as to their location along the transect and their sex (male, female, or nonflowering). Then 200 male and 200 female shoots were randomly collected from each site for further analysis. Leaves of *R. acetosella* commonly contained holes which we believed to be the result of grazing by tobacco wireworms. For each site the 400 shoots were scored for the following variables: shoot height, length of the flowering spike (i.e., the length from the first reproductive lateral shoot to the tip of the main stem), total number of leaves, number of "unscorable" leaves (i.e., broken or otherwise unmeasurable leaves), number of grazed leaves (i.e., leaves having at least one complete hole), number of ungrazed leaves, and total number of grazed holes on the shoot (i.e., sum of all holes on all scorable leaves).

The leaf area of plants was determined as follows. After random sampling, 100 leaves were selected from plants of each sex at every site, and their leaf areas measured (i.e., total $N = 1600$ leaves). Finally, a subsample of 50 leaves of each sex was collected, and the areas were individually determined for all holes in the leaves.

An index of the extent of grazing at each site was calculated as the product of the average number of grazed holes per plant of each gender and the average area of holes on male and female plants. Ratios of leaf area grazed to leaf area available were calculated for males and females at all sites.

Putwain, Machin, and Harper (1968) showed that in permanent grasslands red sorrel only rarely reproduces by seed, and the main means of spread is by root buds. Such clonal growth in *R. acetosella* may allow shifts in the sex ratio of ramets from that established by genets at the seedling stage of the life cycle. As a very coarse estimate of the size of a genet (albeit in a linear dimension only) we performed a nearest-neighbor analysis. For this purpose we assumed that ramets of the same sex immediately adjacent to each other were members of the same genet. (This cannot account for the possibility that two genets of the same sex may be contiguous or that genets may cross the transect, interrupting a series of shoots of the same clone.) Vegetative shoots were interpreted as the end of a clone.

TABLE I.

Population Density and Sex Composition of *Rumex acetosella* at Eight Sites in Central Massachusetts

Site	Transect	Density, shoots per meter	Total individuals	Number of males (and percentage)	Number of females (and percentage)	M/F	Significance of bias[a]
Lower Atkins	1	7.7	384	339 (88.3)	45 (11.7)	7.53	***
	2	7.0	352	194 (55.1)	158 (44.9)	1.23	NS
Site total		7.4	736	533 (72.0)	203 (28.0)	2.63	***
Upper Atkins	1	4.4	220	126 (57.3)	94 (42.7)	1.34	*
	2	4.1	204	127 (62.3)	77 (37.7)	1.65	***
	3	6.2	308	176 (57.1)	132 (42.9)	1.33	*
	4	6.0	300	139 (46.3)	161 (53.7)	0.86	NS
Site total		5.2	1032	568 (55.0)	464 (45.0)	1.22	**
Chicopee	1	7.8	390	103 (26.4)	287 (73.6)	0.36	***
Cemetery	2	11.0	552	271 (49.1)	281 (50.9)	0.96	NS
	3	9.6	481	266 (55.3)	215 (44.7)	1.24	*
	4	3.2	159	142 (89.3)	17 (10.7)	8.35	***
	5	12.5	625	200 (32.0)	425 (68.0)	0.47	***
Site total		8.8	2207	982 (44.0)	1225 (56.0)	0.80	***
Amherst,	1	9.7	485	220 (45.4)	265 (54.6)	0.83	*
Drake Road	2	9.3	467	196 (42.0)	271 (58.0)	0.72	***
Site total		9.5	952	416 (44.0)	536 (56.0)	0.78	***
Amherst Landfill	1	5.0	252	17 (6.8)	235 (93.2)	0.07	***
	2	10.6	532	54 (10.2)	478 (89.8)	0.11	***
Site total		7.8	784	71 (9.1)	713 (90.9)	0.10	***
Chicopee	1	16.9	843	134 (15.9)	709 (84.1)	0.19	***
McDonald's	2	16.1	806	288 (35.7)	518 (64.3)	0.56	***
Site total		16.5	1649	422 (25.6)	1227 (74.1)	0.34	***
Chicopee	1	27.2	1362	524 (38.5)	838 (61.5)	0.63	***
Supermarket	2	7.7	387	306 (79.1)	81 (20.9)	3.78	***
	3	6.9	346	74 (21.4)	272 (78.6)	0.27	***
Site total		14.0	2095	904 (43.2)	1191 (56.8)	0.76	***
Granby	1	24.2	1208	712 (58.9)	496 (41.1)	1.44	***
All sites		10.2	10,663	4,608 (43.2)	6,055 (56.8)	0.76	***
Results of Putwain & Harper (1972)[b]			9,776	4,625 (47.3)	5,151 (52.7)	0.89	***

[a] According to χ^2 tests: ***, $p < .001$; **, $p < .01$; *, $p < .05$; NS, not significant.

III. RESULTS

A. Population Characteristics

A total of 10,663 shoots from eight sites was examined. Overall, there was a significant female bias in the sex ratio of shoots ($p < .001$). At three sites the sex ratio was significantly male-biased,.and at the other five sites a significant female bias was detected (see Table I). The pattern of shoot sexuality within each population was calculated as the ratio of male to female shoot frequency and ranged from 72% male at the lower Atkins site to 9.1% male at the Amherst landfill site. The value of the heterogeneity χ^2 was significant, indicating inconsistency in the bias of the male–female sex ratio in transects within most sites and between sites.

An estimate of the number of genets of each sex present in a transect and at a site was obtained as mentioned previously, by assuming that contiguous shoots of the same sex were members of the same clone and then analyzing sequences of shoots of the same sex. Overall, the estimated genet sex ratio did not differ significantly from unity (Table II), despite the shoot sex ratio being significantly female-biased. The estimated genet sex ratio did not depart significantly from unity at any site except one (Granby), which was male-biased in terms of both shoot and clone frequency. This site with an excess of male clones also had the highest density of both clones and shoots. There was no apparent relationship between the number of clones per transect and either the sex ratio itself or the degree to which the population sex ratio departs from unity. The Granby site seems an interesting case in that it contained many, comparatively small, male genets. The clone having greatest size at that site, in terms of linear dimension,

TABLE II.

Sex Ratio of Shoots and of Estimated Genets for *Rumex acetosella* at Each Site[a]

Site	Shoot sex ratio Male/Female	Significance of χ^2 test	Genet sex ratio Male/Female	Significance of χ^2 test
Lower Atkins	533–203	***	43–43	NS
Upper Atkins	568–464	**	70–58	NS
Chicopee Cemetery	982–1225	***	107–117	NS
Amherst, Drake Road	416–536	***	72–75	NS
Amherst Landfill	71–713	***	19–26	NS
Chicopee McDonald's	422–1227	***	39–38	NS
Chicopee Supermarket	904–1191	***	43–50	NS
Granby	712–496	***	109–54	***
Total	4608–6055	***	502–461	NS

[a] Significance values: ***, $p < .001$; **, $p < .01$; NS, not significant.

TABLE III.

The Length and Number of Shoots in the Largest Estimated Genet at Each Site

Site	Linear dimension, m	Number of shoots
Lower Atkins	13.08	113
Upper Atkins	27.60	80
Chicopee Cemetery	16.14	84
Amherst, Drake Road	11.16	80
Amherst Landfill	13.86	160
Chicopee McDonald's	10.20	218
Chicopee Supermarket	19.30	313
Granby	8.35	166

was 8.35 m long and as such was shorter than the longest clone at any other site (see Table III). This clone contained 166 shoots, however, and was one of the larger clones in the study in terms of number of ramets. *Rumex* clones at that site were therefore dense, but restricted in area.

B. Shoot Characteristics

The male and female *Rumex* shoots were collected when females were in full fruit. Overall there were no significant differences between male and female shoots in terms of height or the length of the flowering spike, though there were differences at individual sites (Table IV). When males and females from all sites were ranked in terms of average height, there was high agreement in the ranking between sexes ($r_s = 0.83$, $p < .05$, Spearman's rank correlation coefficient). Similarly, when shoot height and spike length (a rough measure of reproductive output) were ranked according to site, there was, again, a significant correlation ($r_s = 0.93, p < .01$). There was no significant relationship between shoot density and size of individual shoots. The total number of leaves per shoot did not differ overall between males and females (Table IV), though there were differences across sites, with some having significantly greater numbers of leaves on female shoots, and others having more leaves on male shoots. The average area of a leaf on male shoots was significantly greater than that of a leaf on a female shoot ($p < .001$) (Table V).

C. Leaves and Herbivory

The total number of holes and the number of grazed leaves per shoot did not differ overall between the sexes but did differ significantly on a site-by-site basis

TABLE IV.

A Summary, by Sites, of Male and Female Shoot Characteristics in *Rumex acetosella*[a]

Site		Shoot height (cm)			Length of flowering spike (cm)			Total number of leaves		
		Male	Significance of difference	Female	Male	Significance of difference	Female	Male	Significance of difference	Female
Lower Atkins	$\bar{x}$	26.7	NS	25.7	11.8	NS	12.0	11.5	**	13.1
	sd	6.2		6.4	5.0		7.0	4.6		7.1
Upper Atkins	$\bar{x}$	30.7	NS	32.1	14.7	NS	14.0	16.0	**	13.0
	sd	7.8		8.4	6.6		4.9	11.1		9.1
Chicopee Cemetery	$\bar{x}$	29.9	**	32.6	14.8	NS	15.0	12.2	NS	12.8
	sd	8.2		8.6	6.0		5.1	6.1		6.9
Amherst, Drake Road	$\bar{x}$	33.3	**	31.2	14.5	NS	14.8	11.9	*	13.2
	sd	6.5		6.8	4.9		4.8	5.3		6.0
Amherst Landfill	$\bar{x}$	35.5+	NS	34.8	18.3+	***	15.0	22.8+	**	15.7
	sd	8.2		8.4	8.3		5.5	13.3		8.6
Chicopee McDonald's	$\bar{x}$	27.5	NS	28.0	11.9	NS	12.3	10.6	NS	10.1
	sd	6.0		7.5	4.6		4.9	5.2		4.2
Chicopee Supermarket	$\bar{x}$	31.9	NS	32.7	15.1	NS	15.7	10.8	*	12.1
	sd	8.2		8.4	6.2		5.5	5.4		5.7
Granby	$\bar{x}$	46.7	***	51.1	19.8	***	23.7	13.9	**	11.8
	sd	10.5		9.9	7.8		6.6	8.1		4.7
Total	$\bar{x}$	32.8	NS	33.5	15.1	NS	15.3	12.8	NS	12.7
	sd	6.3		7.7	2.8		3.6	0.6		1.6

[a] *N* in all cases except one was 200; +, sample size of 49. Student's *t*-test between male and female means was carried out. Significance values: *, $p < .05$; **, $p < .01$; ***, $p < .001$; NS, not significant.

TABLE V.

A Summary, by Sites, of Male and Female Leaf Characteristics of *Rumex acetosella*[a]

Site		Number of grazed leaves			Number of holes per shoot			Mean area of a leaf (mm^2)		
		Male	Significance of difference	Female	Male	Significance of difference	Female	Male	Significance of difference	Female
Lower Atkins	$\bar{x}$	3.9	***	8.8	12.3	***	29.7	35.5	*	28.1
	se	0.7		0.9	1.6		2.4	2.8		2.5
Upper Atkins	$\bar{x}$	3.4	***	2.4	8.2	NS	6.2	48.9	NS	42.4
	se	0.2		0.2	0.9		0.6	5.1		4.1
Chicopee Cemetery	$\bar{x}$	2.4	***	3.5	7.5	NS	9.6	47.4	*	34.7
	se	0.2		0.2	0.9		0.8	5.1		3.0
Amherst, Drake Road	$\bar{x}$	1.2	NS	1.5	3.7	NS	5.4	29.6	NS	29.6
	se	0.1		0.2	0.8		1.2	3.2		3.1
Amherst Landfill	$\bar{x}$	7.8+	***	5.5	20.3+	NS	20.3	53.4	**	33.6
	se	0.6		0.3	2.0		1.7	5.7		3.3
Chicopee McDonald's	$\bar{x}$	3.5	*	4.1	9.6	***	16.9	35.6	NS	30.7
	se	0.2		0.2	0.8		1.6	3.1		2.6
Chicopee Supermarket	$\bar{x}$	2.9	*	2.3	11.5	***	5.2	44.4	NS	42.8
	se	0.2		0.2	1.7		0.7	4.4		3.7
Granby	$\bar{x}$	2.1	***	2.9	4.1	**	6.4	62.8	NS	49.1
	se	0.1		0.2	0.5		0.5	5.7		4.5
Total	$\bar{x}$	3.4	NS	3.9	9.7	NS	12.5	44.7	***	36.4
(N = 8 sites)	se	0.7		0.8	1.9		3.2	1.1		0.8

[a] In all cases $N = 200$, except one where noted with + ($n = 49$), and for leaf area determination, where $N = 100$ leaves per sex per site. Significance values: *, $p < .05$; **, $p < .01$; ***, $p < .001$, NS, not significant.

(Table V). As might be expected, within a site the gender which had more grazed leaves had more holes per shoot than did the other gender. Herbivory was therefore not focused on a few leaves. When the number of grazed leaves was examined as a percentage of the number of leaves present, males and females from a site ranked similarly between sites ($r_s = 0.82, p < .05$, Spearman's rank correlation). In other words, at a site where grazing intensity is, say, intense upon males it is also heavy on females, although the actual percentage of leaves grazed may differ between the sexes. When we ranked sites in terms of the extent of herbivory (*herbivory* in this case being assessed as the number of holes found on the average shoot), this ranking did not differ significantly from ranking of the sites according to the extent to which their sex ratio departs from unity. The departure from 1:1 was calculated as the arc–sine transformation of the percentage of males in each population. The more biased shoot populations (in whichever direction) tend to be the more heavily grazed ($r_s = 0.67, p < .10$, Spearman's rank correlation). Another way of looking at the relative degree of herbivory at each site is to examine the number of grazed leaves as a percentage of the total number of leaves per shoot. When this estimate of herbivory is ranked according to size, it corresponds to the degree of male bias in the shoot ratio ($r_s = 0.70, p < .10$, Spearman's rank correlation).

The extent of male and female foliage at each site is shown in Table VI. Foliage available to herbivores should be a function of average area of a leaf, number of leaves, and relative abundance of shoots of each gender. Those sites which are male-biased in their shoot sex ratio are, consequently, male-biased in terms of the surface area of foliage presented at that site. The pattern in Table VI

TABLE VI.

The Availability of Male and Female Foliage of *Rumex acetosella* at Each Site[a]

Site	Male foliage, cm^2	Female foliage, cm^2	M/F foliage, ratio	Grazed male leaves, %	Grazed female leaves, %
Lower Atkins	2175.8	741.0	2.94	34	68
Upper Atkins	4444.8	2557.1	1.74	21	18
Chicopee Cemetery	5679.1	5441.3	1.04	20	27
Amherst, Drake Road	1464.9	2094.8	0.70	10	11
Amherst Landfill	864.1	3761.7	0.23	34	35
Chicopee McDonald's	1592.1	3804.7	0.42	33	41
Chicopee Supermarket	5579.5	6167.4	0.90	27	19
Granby	6214.7	2873.3	2.16	15	25
All sites	3501.9	3430.2	1.27	24.3	30.5

[a] Leaf area × number of leaves per shoot × number of shoots. The number of grazed leaves is presented as a percentage of total number of leaves per shoot.

TABLE VII.

Area of Foliage Grazed as a Percentage of Foliage Available on Shoots of Each Sex of *Rumex acetosella*

	Males			Females		
Site	Leaf area grazed per shoot	Total leaf area per shoot	Percentage	Leaf area grazed per shoot	Total leaf area per shoot	Percentage
Lower Atkins	9.84	408.25	2.4	24.95	324.35	7.7
Upper Atkins	6.56	782.40	0.8	5.21	551.20	0.9
Chicopee Cemetery	6.00	578.28	1.0	8.06	444.16	1.8
Amherst, Drake Road	2.96	352.24	0.8	4.54	390.72	1.2
Amherst Landfill	16.24	1217.52	1.3	17.05	527.52	3.2
Chicopee McDonald's	7.68	377.36	2.0	14.20	310.07	4.6
Chicopee Supermarket	9.20	479.52	1.9	4.37	517.88	0.8
Granby	3.28	872.92	0.4	5.41	579.38	0.9
All sites	7.76	570.37	1.4	10.50	465.19	2.3

shows perfect correspondence with the ranking of sites according to shoot sex ratio.

We calculated the area of foliage grazed on a shoot as a function of the total leaf area available on male or female shoots at a site and found that female leaves at all sites except one (Chicopee supermarket) were more heavily grazed (Table VII). Also, the two sites which show greatest excess of herbivory on females are those that were most strongly male-biased (see Table I).

IV. DISCUSSION

Charnov (1982) has described plant breeding systems as the consequence of selection pressures acting upon separate male and female sexual strategies. Separation of the sexes in part should reflect dimorphic patterns of resource allocation for male and female functions, moulded as a consequence of differential selection pressures (Lovett Doust & Lovett Doust 1983). One such pressure might be differential herbivory and parasitism on males and females of the same species. We had considered that in a clonal species such as *Rumex acetosella* skewed shoot sex ratios may be produced by the differential action of selective herbivores as well as by differential vegetative vigour of male and female plants. Differential herbivory could cause reduced clonal growth and thus fewer aerial flowering spikes of one sex, or it might be sufficient, especially in the early stages of population establishment, to cause differential mortality of male and female genets.

In our comparisons of male and female plants there were no overall differences between the sexes in terms of shoot height, length of the flowering spike, or number of leaves per shoot. The sexes did, however, differ in terms of total leaf area per shoot. Males tended to have significantly more foliage, a finding that supports earlier inferences of greater male vegetative vigour in *R. acetosella*. In the present study the number of grazed leaves per shoot and number of holes grazed per shoot did not differ overall between the sexes but did differ between sites; where grazing was intense on one sex at a given site it was likely to be relatively intense on the other sex at that site. Grazing therefore probably reflects the abundance of herbivores rather than food preference. Sites with evidence of greater herbivory (whether measured as the number of holes, grazed leaves, or area grazed) had relatively more skewed sex ratios. The two most strongly male-biased sites were also those which suffered the greatest excess of herbivory on females compared with that on males, and it is possible that selective grazing may have helped to establish more males there.

Another way of describing the herbivory is to consider the proportion of available male and female leaf area that is grazed. We found that females on average were more heavily grazed, inasmuch as a greater fraction of their total foliage was removed by herbivores (Table VII). The area of foliage that was grazed, expressed as the percentage of the total leaf area of a particular sex, available at a site, was greater for females at seven of the eight sites we studied. Percentage values ranged from 7.7 to 0.8 for females and from 2.4 to 0.4 for males. However, since the percentage of female shoots did not correlate with intensity of grazing, we conclude that differences in shoot sex ratios seem not to be due to significant herbivore influence.

In the eight sites studied here there was no consistent bias in the shoot sex ratio although, as in Putwain and Harper's (1972) study of nine populations in North Wales, females were more common overall, and there was a preponderance of sites which were female-biased (see Table I). Lloyd (1974) reviewed female-predominant sex ratios in higher plants and noted that this situation is generally less common than male predominance. He cited evidence that a preponderance of females in species of four genera (including *Rumex*) may be due to gamete selection. There may be more frequent fertilization by female-determining pollen nuclei than by male-determining nuclei. However, this explanation is not supported by the data of Putwain and Harper (1972), which show that populations of *Rumex acetosella* grown from seed occur at sex ratios of 1 : 1. Similarly, after estimating genet sex ratios at each site, we infer drift from a founder population in which the two sexes were equally represented, toward biased ramet sex ratios. The Granby site was the single exception, where the genet sex ratio appeared male-biased. At this site the subsequent production of ramets by each clone may accentuate the initial bias and create a shoot population with a very deviant sex ratio. The Granby site seemed to be the oldest of all the sites studied, as evidenced by the presence of woody species, herbaceous perennial weeds and tussock grasses among the crowded *Rumex* shoots.

In New Zealand, Harris (1968) observed populations of *Rumex acetosella* which departed significantly from a 1 : 1 ratio. He found that males seemed to predominate in older, better-established plant communities where the vegetation formed closed tussocks. This condition was in contrast to the situation in newly sown pastures and disturbed areas such as roadsides, where he found no significant deviation from a 1 : 1 sex ratio. It may be that our Granby site represents an old-established colony, like those examined by Harris, and that the male-biased clone ratio has been brought about by the gradual elimination of female genets since the population was founded.

Harris (1968) argued that the sex ratio of populations of *Rumex acetosella* is significantly influenced by interspecific competition, and he postulated that male-biased sex ratios were due to differential vegetative vigor, a result of the smaller energy demands made on them for sexual reproduction, which could improve their ability to compete. This phenomenon would explain the differential survivorship of males and females. However, Harris was looking only at ramet sex ratios, and these biased ratios may actually be a consequence of differential clonal growth rates of males and females, rather than differential mortality of male and female genets. The shift between our genet and ramet sex ratios would support this idea. Grant and Mitton (1979) have described differential growth rates between the sexes of the clonal tree *Populus tremuloides*. Female aspen clones consistently showed a greater radial growth increment than males. This result is at odds with reports that attribute higher vegetative growth rates to males, because of the presumably higher cost of sexual reproduction for females. We are at present making an experimental study of the vegetative and associated clonal growth rates of males and females, and the influence of crowding in *R. acetosella*. Preliminary data suggest that females retain leaves longer than males and can produce more shoots than males, by the end of the growing season.

Putwain and Harper (1972) pursued the question of whether in *Rumex* there exists a greater reproductive resource burden on females (through flower and seed production) than that imposed on males by production of male flowers and pollen. Like Harris, they reasoned that if this were the case then males might have comparatively more resource available for clonal growth, causing the sex ratio (especially in older populations) to become male-biased. Putwain and Harper indeed found that male shoots of *R. acetosella* allocated proportionately more of their dry matter to clonal growth, whereas females allocated proportionately more to sexual reproduction. A number of workers have questioned the adequacy of dry weight (or caloric) comparisons as the currency of total metabolic cost (e.g. Watson 1984; Lovett Doust 1986). It could be that cost involves, for example, different limiting nutrients for each sex. Further, in dioecious plants there may be differences between the sexes in such components of evolutionary fitness as, for example, relative root and shoot growth rates, overwintering ability, lifespan, and response to resource limitation.

It seems to us interesting that in our study some sites are presently *female*-biased in terms of reproductive shoots, despite the potentially greater costs of female reproduction and the greater herbivory they suffer, bearing in mind the lower leaf area of the average female plant. One hypothesis is that females are in some way better able to withstand environmental buffeting than males and thus show greater survivorship. Melampy (1981) suggested that survivorship data of dioecious *Thalictrum* spp. indicated that females may be more stress-tolerant than males. He postulated that in *Thalictrum* females may have a greater niche breadth with respect to particular environmental aspects. In their laboratory study of *Rumex acetosella,* Zimmerman and Lechowicz (1982) concluded that, although males and females did show different tolerances to drought, the higher tolerance of males seemed to be more a consequence of greater biomass allocation to roots and leaves (possibly enabling more effective exploitation of available water) than of greater physiological efficiency of water use.

Consideration of the temporal partitioning of the reproductive effort of males and females may further explain populations with a female bias (see also Lloyd & Webb 1977). David Lloyd (personal communication) suggests there is a need to explain why male flowering shoots do not persist and has proposed an explanation based on seasonal phenology and resource allocation, as follows: Male reproductive effort is lower than that for females, but it occurs earlier in the growing season and therefore is drawn from smaller supplies of resources, gathered over a shorter period and when conditions for growth (such as insolation) are not at their peak. Female reproductive effort may be a greater drain on resources, but it comes later in the growing season, when the female plant has presumably accumulated more photosynthate than flowering males. Therefore, although female reproductive effort is, in absolute terms, considerable, it may represent an equal or lesser proportionate drain on resources than does male reproductive effort, especially if the resources of a shoot increase markedly as the season progresses. Thus males do not persist or outgrow females because they lack adequate reserves at the time they are flowering. In the present study male foliage senesced around anthesis. Meanwhile females continued to produce new leaves and photosynthesize until seeds matured. There is surely a greater reproductive drain on females than on males, but this may be compensated by the long growing season occupied by female shoots.

ACKNOWLEDGMENTS

We are grateful to David Lloyd, Spencer Barrett, Paul Cox, Bob Cruden, and Mark Schlessman for comments on an earlier draft.

REFERENCES

Bawa, K. S. (1980). Evolution of dioecy in flowering plants. *Annual Review of Ecology and Systematics,* **11,** 15–39.

Bawa, K. S. & Opler, P. A. (1978). Why are pistillate inflorescences of *Simarouba glauca* eaten less than staminate inflorescences? *Evolution* **32,** 673–676.

Charnov, E. L. (1982). *The Theory of Sex Allocation.* Princeton University Press, Princeton, New Jersey.

Cox, P. A. (1981). Niche partitioning between sexes of dioecious plants. *American Naturalist,* **117,** 295–307.

Cox, P. A. (1982). Vertebrate pollination and the maintenance of dioecism in *Freycinetia. American Naturalist,* **120,** 65–80.

Freeman, D. C., Klikoff, L. G. & Harper, K. T. (1976). Differential resource utilization by the sexes of dioecious plants. *Science,* **193,** 597–599.

Gleason, H. A. (1963). *The New Britton and Brown Illustrated Flora of the North-eastern United States and Adjacent Canada.* Hafner, New York.

Grant, M. C. & Mitton, J. B. (1979). Elevational gradients in adult sex ratios and sexual differentiation in vegetative growth rates of *Populus tremuloides Michx. Evolution,* **33,** 914–918.

Harper, J. L. (1967). A Darwinian approach to plant ecology. *Journal of Ecology,* **55,** 247–270.

Harris, W. (1968). Environmental effects of the sex ratio of *Rumex acetosella. Proceedings of the New Zealand Ecological Society,* **15,** 51–54.

Keep, E., Knight, V. H. & Parker, J. H. (1977). An association between response to mildew (*Sphaerotheca macularis* (Fr.) Jaczewski), sex and spine colour in the raspberry. *Journal of Horticultural Science,* **52,** 193–198.

Lee, J. A. (1981). Variation in the infection of *Silene dioica* (L) Clairv. by *Ustilago violacea* (Pers.) Fuckel in northwest England. *New Phytologist,* **87,** 81–89.

Lloyd, D. G. (1974). Female-predominant sex ratios in Angiosperms. *Heredity,* **32,** 35–44.

Lloyd, D. G. & Webb, C. J. (1977). Secondary sex characteristics in seed plants. *Botanical Review,* **43,** 177–216.

Lovett Doust, J. (1986). Plant reproduction and problems of resource allocation. *Reproductive Strategies of Plants* (Ed. by J. Lovett Doust). CRC Press, Boca Raton, Florida. (In press.)

Lovett Doust, J. & Cavers, P. B. (1982). Sex and gender dynamics in jack-in-the-pulpit, *Arisaema triphyllum* (Araceae). *Ecology,* **63,** 797–808.

Lovett Doust, J. & Lovett Doust, L. (1983). Parental strategy: Gender and maternity in higher plants. *BioScience,* **33,** 180–186.

Melampy, M. N. (1981). Sex-linked niche differentiation in two species of *Thalictrum. American Midland Naturalist,* **106,** 325–334.

Putwain, P. D. & Harper, J. L. (1970). Studies in the dynamics of plant populations. III. The influence of associated species on populations of *Rumex acetosa* L. and *R. acetosella* L. in grassland. *Journal of Ecology,* **58,** 251–264.

Putwain, P. D. & Harper, J. L. (1972). Studies in the dynamics of plant populations. V. Mechanisms governing the sex ratio in *Rumex acetosa* and *R. acetosella. Journal of Ecology,* **60,** 113–129.

Putwain, P. D., Machin, D. & Harper, J. L. (1968). Studies in the dynamics of plant populations. II. Components and regulation of a natural population of *Rumex acetosella* L. *Journal of Ecology,* **56,** 421–431.

Wade, K. M., Armstrong, R. A. & Woodell, S. R. J. (1981). Experimental studies on the distribution of the sexes of *Mercurialis perennis* L. I. Field observations and canopy removal experiments. *New Phytologist,* **87,** 431–438.

Watson, M. A. (1984). Developmental constraints: Effect on population growth and patterns of resource allocation in a clonal plant. *American Naturalist,* **123,** 411–426.

Willson, M. F. (1983). *Plant Reproductive Ecology*. Wiley, New York.

Zimmerman, J. K. & Lechowicz, M. J. (1982). Responses to moisture stress in male and female plants of *Rumex acetosella* L. (Polygonaceae). *Oecologia,* **52,** 305–309.

22

The Role of the Grazing Animal

Rodolfo Dirzo

Departamento de Ecología
Instituto de Biología
Universidad Nacional Autónoma de México
Mexico City, Mexico

I. INTRODUCTION

The particular set of physical conditions prevailing in a given area of the earth constitutes the scenario in which plants play a drama characterized by two major dichotomies: to survive or die, and to leave many or few descendants. In this scenario some individual plants (and animals) leave more descendants than others, and the attributes that determine these differences between plants (or animals) are heritable. A crucial issue for a plant, however, is that this evolutionary drama is played within the context of a very complex matrix of biotic 'threats', which are defined by neighbouring plants (conspecific or otherwise), pathogens and predators and of biotic 'alliances', which are defined by pollinating and dispersal agents. Harper (1977) emphasised the distinction between the physical and biotic forces of the environment [Wallacian and Darwinian forces, respectively, in his terminology (Harper 1977 pp. 749–750)] acting upon plants and suggested a more profound effect of the biotic forces in shaping the ecology and evolution of plants in populations. Among the biotic forces he assigned a preponderant place to predators, and as such, he dedicated 6 of the 24 chapters of his book (Harper 1977) to discussing the role of predators (including pathogens) on the population biology of plants.

From a phytocentric point of view, two major lines of research characterise the study of plant–herbivore interactions. One deals with the structural and functional response of plants to a presumed strong and omnipresent selective pressure exerted by herbivory. This leads to the evolution of a vast array of physical, chemical and phenological plant traits such as vestures of defensive spines and trichomes (Levin 1973); complex metabolic machineries to synthesise, transport and sequester defensive secondary metabolites such as tannins, alkaloids, glucosides, non-protein aminoacids, etc. (Rosenthal & Janzen 1979); evolution

STUDIES ON PLANT DEMOGRAPHY:

of mast-seeding behaviour (Janzen 1976; Silvertown 1980) and specific seed and seedling shadows (Janzen 1970, 1971).

The second line concerns the study of the way herbivores affect, in ecological time, the survival, growth, competitive ability and reproduction of plants in populations (e.g. Bentley, Whittaker & Malloch 1980; Bruce 1956; Harper 1977; Kulman 1971; Varley & Gradwell 1962; Whittaker 1979) with the aim, ideally, of understanding when and how herbivores are important in influencing the distribution and abundance of plants (Harper 1969, 1977). Quite clearly, John Harper's incursions into the field of plant–herbivore interactions belong to this second line of research, though the first line was not absolutely deprived of his participation (e.g. Harper 1969, 1979; Dirzo & Harper 1982). He has rescued a number of fundamental and yet overlooked pieces of work, mainly from the agronomic literature (e.g. Jones 1933a–c; Milton 1940, 1947), and used them together with more recent work in an attempt to provide a coherent image of the way herbivores interact with plants in populations (Harper 1977 Chapters 12–15). He has continuously insisted on the use of experimental procedures as *the only way* to understand fully the role that the grazing animal may play in the population dynamics of plants. "If an animal plays a critical role in determining the distribution and abundance of a plant the activity is not likely to be obvious. A quite new type of experimental approach to plant ecology is required if the role of animals is to be exposed" (Harper 1977 p. 433). In this essay I would like to present a series of examples from the ecological literature in which the experimental approach suggested by Harper has been used and to discuss the extent to which it has helped us in gaining knowledge of the ways the grazing kingdom shapes the dynamics of populations in the plant kingdom.

A variety of experimental procedures are available to the ecologist: the spraying of biocides directed at individual or guilds of herbivores (insecticides, molluscicides), exclusion (or enclosure) experiments (cages, fences), manipulation of density and diversity of both seeds and growing plants, and several types of manipulative, selective predation (seed damage, pruning, defoliation). Besides this range of techniques, some accidental and deliberate introductions of herbivores (e.g. in biological control programmes) can be taken as experiments, although the precision of results and their realism are often unsatisfactory (e.g. an *exotic* animal feeding upon and regulating the plants of a *native* population). Harper (1969) suggested the types of experiments needed to disentangle the role animals may play on vegetational diversity; here I shall follow his example.

II. THE EFFECTS OF GRAZING ON PLANT PERFORMANCE IN A LOCAL POPULATION

Leaving aside the more ecophysiological studies of defoliation in relation to individual plant performance (e.g. Sweet & Wareing 1966), here I include ex-

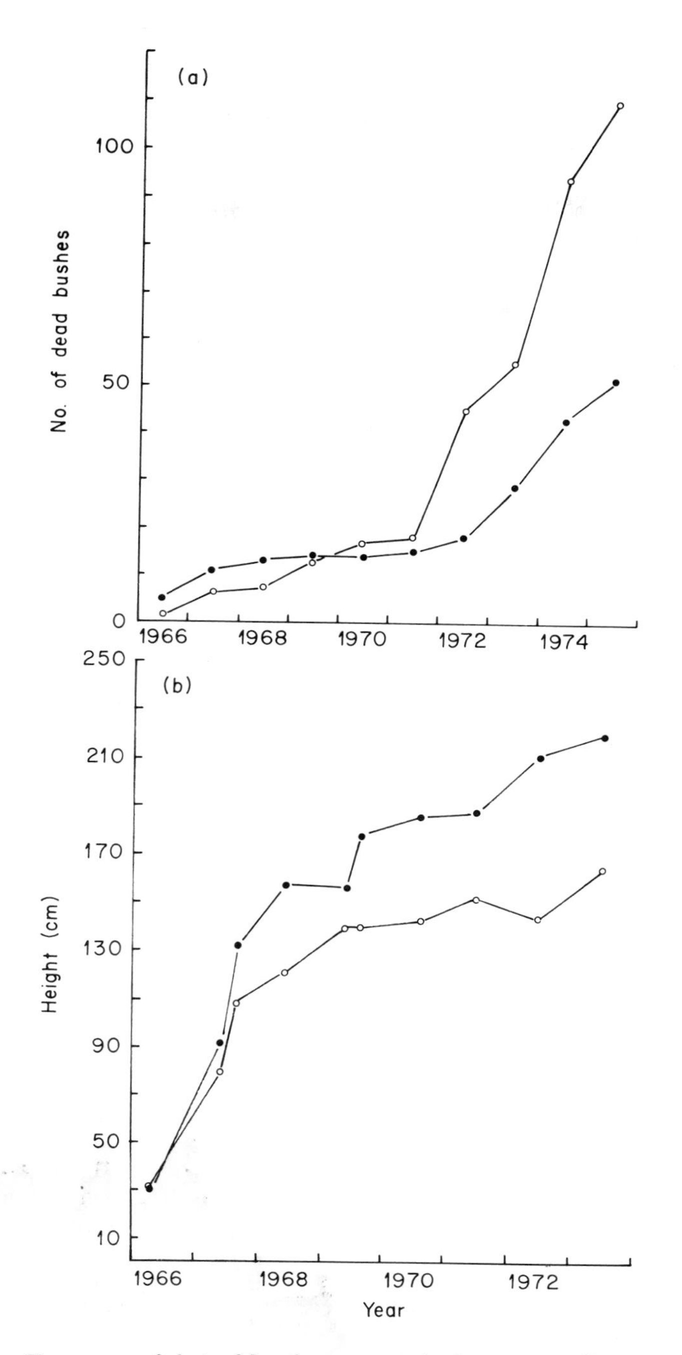

Fig. 1. The response of plants of *Sarothamnus scoparius* from sprayed (●) and unsprayed (○) plots. a, Plant mortality. b, Plant growth (height). (From Waloff & Richards [1977].)

periments in which the main purpose seems to be to describe by how much herbivores are able to reduce plant survival, growth or reproduction within a given natural or artificial population. The experimental procedures include artificial herbivory (e.g. Becker 1983; Maun & Cavers 1971; Rockwood 1973), fencing of herbivores (e.g. Hatto & Harper 1969 in Harper 1977; Dirzo & Harper 1980), and insecticide spraying (e.g. Cantlon 1969; Louda 1985; Waloff & Richards 1977). A common denominator of most of these studies is a high degree of precision at the expense of realism; experiments range from a rather restricted physiological level, to the use of neatly spaced plants in the field, to the use of carefully even-aged monocultures under controlled greenhouse conditions. The common outcome of these experiments is that, if most of the other interacting variables are eliminated, there is an enormous *potential* for herbivores to reduce plant performance, sometimes quite spectacularly. A good example of this is the long-term study of Waloff & Richards (1977) in which growth, survival (Fig. 1) and reproduction (Table I) of broom plants (*Sarothamnus scoparius*) from a plot sprayed with insecticide were dramatically higher than those of the unsprayed (control) plants.

A few of the experiments quoted previously attempt a higher degree of realism, and in these, the results are not so spectacular (but see Louda 1985). For example, Becker (1983) defoliated the seedlings from naturally established populations of two Dipterocarps (*Shorea leprosula* and *S. maxwelliana*) in Malaysia, after a mast fruiting had occurred in 1976. Defoliation treatments were 25 and 100% leaf area removal, together with complete apical bud removal. Prior to the defoliation he had found that natural leaf damage levels were 13 and 56% of the surface area for *S. leprosula* and *S. maxwelliana,* respectively. Strikingly, seedling mortality was affected only by the 100% defoliation treatment, a level

TABLE I.

The Number of Seeds per Pod in Sprayed and Unsprayed Plots of *Sarothamnus scoparius*[a,b]

Year	Sprayed	Unsprayed	p
1970	8.92 ± 0.73	5.15 ± 0.42	<0.01
1971	9.81 ± 0.83	8.04 ± 9.67	<0.001
1972	9.67 ± 0.78	4.54 ± 0.42	<0.001
1973	8.64 ± 0.59	6.30 ± 0.45	<0.001
1974	9.84 ± 0.69	5.61 ± 0.40	<0.001
1975	6.90 ± 0.49	5.34 ± 0.38	<0.001
$\bar{x}$ of 6 years	8.96 ± 1.13	5.83 ± 1.23	<0.01[c]

[a] Data are means (±S.E.) for 6 consecutive years.
[b] Modified from Waloff & Richards (1977).
[c] Two-tailed t-test, $t_{(10)} = 4.17$.

several times the natural average, suggesting that natural herbivory is inconsequential to the survival of both species' seedlings, except for those rare occasions in which insects consume all or most of a seedling's foliage. Becker argues that one or several ecophysiological responses may have compensated for the loss in photosynthetic capacity due to defoliation, such as enhancement or photosynthetic rates of the remaining foliage, or the presence of high levels of energy reserves remaining and being gradually used after defoliation, thus averting seedling death. There are other possible explanations, however: (a) A *sustained* level of defoliation, even though small (say in the order of the average found in the field), may show the true impact of herbivory. Becker defoliated his trees only once; I have found levels of seedling herbivory in a tropical rain forest in Mexico of the same magnitude as Becker's, but the damage is continuously present throughout most of the seedling's life time. (b) As Becker states, defoliation may have affected other fitness traits, such as competitive ability, the consequences of which may not have been obvious during the study period. (c) Other complicating factors, such as subtle variations in plant density or the specific distribution of defoliation regimes among neighbouring plants, may have been overlooked.

In this context, it is instructive to describe a defoliation experiment carried out in Mexico (B. Zagorin & R. Dirzo, unpublished) involving seedlings of *Omphalea oleifera,* a common canopy tree in the Los Tuxtlas tropical rain forest. Plants were grown in trays within enclosures and treatments included the following: (a) two light conditions, sunshine and shade (to simulate a natural light gap and a mature-forest condition: both quite realistic environments experienced by *O. oleifera* seedlings); (b) two densities, low and high (to simulate the average and maximum, respectively, found in the field); (c) four defoliation levels, 0, 5, 25 and 75% leaf area removal (average natural levels range from 5 to 25%; 75% occurs occasionally). The results are shown in Fig. 2. Even the highest levels of defoliation had no effect on plant survival under conditions of low density; the full impact of herbivory was only expressed when plants were *also* experiencing interference, and this was accentuated under conditions of shading. Had the experiment not included the density treatment, it would have been wrongly concluded that average and extreme levels of herbivory, even in the shade, are inconsequential to plant survival.

An enclosure experiment which (besides describing the extent to which herbivores affect growth, survival and reproduction) attempted to explore a more specific question, is that of Dirzo and Harper (1980). They sowed monospecific populations of *Capsella bursa-pastoris* and *Poa annua* at high densities in the presence and absence of the slug *Agriolimax caruanae,* to investigate the way grazing might influence the well-established relationship between the numbers of plants that survive to maturity and the size to which they grow: the $-\frac{3}{2}$ thinning law (Yoda *et al.* 1963; White 1980). The effect of grazing by the slugs was to kill

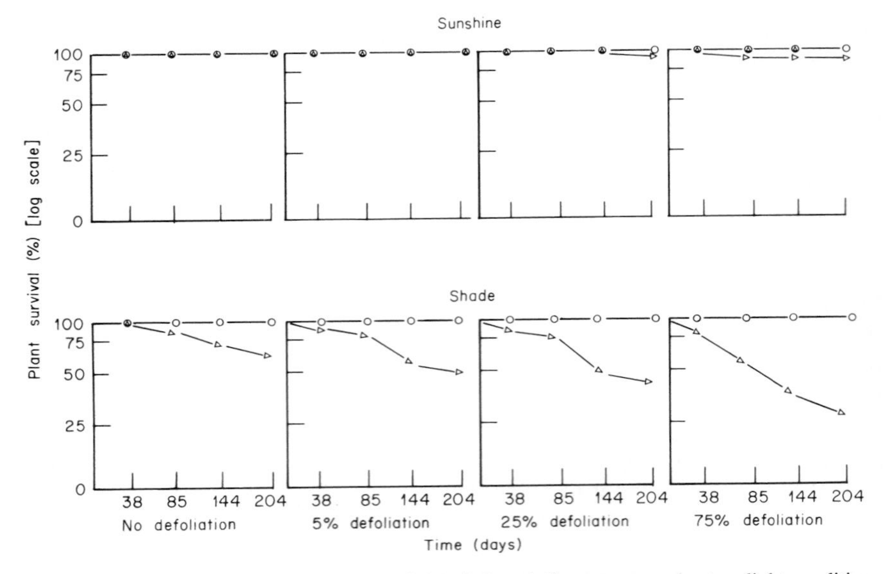

Fig. 2. The survivorship of seedlings of *Omphalea oleifera* grown under two light conditions (sunshine and shade), two plant densities [low (○) and high (△)] and four defoliation treatments (0, 5, 25 and 75% leaf area removed). (From Dirzo [1984].)

many young seedlings of *Poa annua,* thus reducing the density of survivors, which then grew larger than plants in control populations (without slugs). In populations of *Capsella bursa-pastoris,* in contrast, slugs heavily consumed the rosette leaves, thus reducing the size of the grazed plants, which in turn had a better survivorship than the plants in the ungrazed plots: these were bigger but fewer. Despite the contrasting modes of grazing on the two species, the populations of both species adjusted to density by reasonable conformity to the $-\frac{3}{2}$ thinning law. Since there were not significant differences between the thinning rates of the control and grazed populations of both species, it appears that grazing simply removed, or reduced the size of, individuals that would have been equally affected by density in the absence of slugs. Interestingly, in the context of the present discussion, the authors argue that "The substitution of one cause of plant death or plastic change by another . . . has much relevance to interpreting field experiments in which the activity of herbivores is measured by using exclosures". "Changes that occur in exclosure plots will represent not only the results of preventing grazing but also the consequent increased level of interference between the ungrazed plants" (Dirzo & Harper 1980).

From these examples it appears that the next logical step in the development of experiments to understand the impact of grazing on plant performance is to include other interacting variables of which the omnipresent effect of interfering neighbouring plants is one. Several reports from the ecological literature (e.g. Cromartie 1975; Rausher 1981; Thompson 1978; Thompson & Price 1977), as well as from the more agronomically-oriented literature (e.g. Risch 1981), provide evidence of this three-way interaction (herbivore–plant–plant) as a fundamental aspect of the plant–grazer interface. From these reports, it appears that experimental manipulation is the most fruitful approach of study.

III. GRAZING AND PLANT DIVERSITY

Harper (1969) envisaged four major types of experiments to assess the role of grazing on diversity: (a) withdrawal of a herbivore from vegetation, (b) addition of a herbivore to vegetation, (c) control of the intensity or periodicity of herbivory on vegetation and (d) breakdown of the diversity of vegetation to create monotony. By 1985 it appeared that only a few experiments had been made, particularly of types (a) and (b), and experiments of type (d), at least for natural systems are very scarce. Exceptional in this context are a number of studies in which experimental manipulations of plant diversity have been made, but with a major zoocentric emphasis; in some of these studies collateral data on the subsequent effects on plants are also included (e.g. Bach 1980; Rausher 1981; Rausher & Feeny 1980; Solomon 1981; Thompson 1978). Some of the most outstanding experiments with a major phytocentric emphasis seem to have been made with marine systems (e.g. Hay 1981; Lubchenco 1978; Lubchenco & Cubit 1980; Menge 1976). Many of these experimental studies have addressed and explored in detail a point that had been raised by Harper (1969), that in some instances grazing appeared to increase diversity, although in others it appeared to reduce it; this, he believed depended on the relationship between the degree of acceptability (i.e. the feeding preferences of the herbivore) and competitive ability of the plants. The experimental studies [in particular that of Lubchenco (1978) with the herbivorous marine snail *Littorina littorea*] seem to confirm that (a) when the better competitor is also the preferred food, grazing prevents competitive monopoly and plant diversity is high, and (b) when the competitive dominant is not the preferred food, competitive exclusion is enhanced and diversity decreases. When the competitive dominant is preferred, changes in diversity are also the result of the intensity of grazing, a result also hinted by Harper from his type (c) experiments: low intensities of grazing permit competitive dominants to outcompete other species, thus reducing diversity; moderate grazing permits inferior competitors to persist and diversity is high; whereas very intense grazing eliminates most individuals and species, leaving only the most inedible plants in a low-diversity environment.

Lubchenco and Gaines (1981) indicated that it remains to be defined to what extent these feeding preferences are a consequence of trade-offs between com-

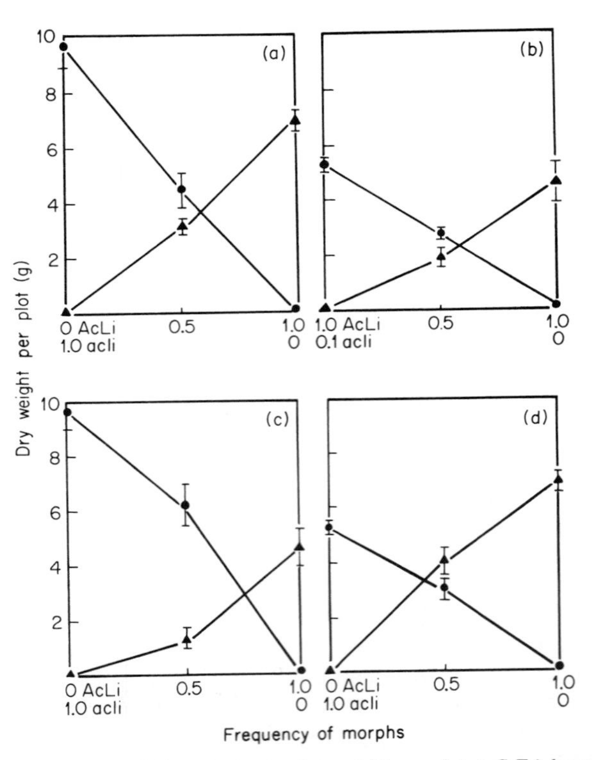

Fig. 3. The production of biomass (mean dry weight per plot ± S.E.) by cyanogenic, AcLi (▲), and acyanogenic, acli (●), morphs of *Trifolium repens* grown as monocultures and 50 : 50 mixtures under four treatments of simulated herbivory. a, Both undefoliated. b, Both morphs defoliated. c, Only AcLi defoliated. d, Only acli defoliated. (From Dirzo [1984].)

petitive abilities and anti-herbivore defences in the plant. Dirzo (1984) presented some evidence in support of such trade-offs, from competition experiments coupled with simulated defoliation in plants of *Trifolium repens,* polymorphic for cyanogenesis, a clearly demonstated anti-herbivore defensive trait (Dirzo & Harper 1982). The experiments (Fig. 3) indicated that the protected forms of *T. repens* appear to be poorer competitors and that a competitive equilibrium appeared only when the unprotected forms were selectively defoliated. The inclusion of this experiment in the present discussion may be germane for it poses the question of investigating to what extent and under what circumstances intraspecific diversity is affected (perhaps maintained?) by grazing. I envisage that the sort of experiments suggested by Harper, applied to a wide variety of, for example, biochemically polymorphic or dioecious plant populations, might give us quite exciting and significant results in plant population biology.

IV. HERBIVORY AND PLANT SUCCESSION

Several authors have suggested that biotic factors may play an important role in plant succession (e.g. Connell & Slayter 1977; Hartshorn 1980; van Hulst 1978; White 1979); very little work has been done, however, at least with terrestrial (natural) plant communities, to test this experimentally. The few experiments that have been described involve insecticide spraying and animal exclusion and enclosure. An outstanding experimental study is that of Lubchenco (1983) in a marine successional system at Massachusetts, on a wave-protected site characterised by a simple successional sequence, in which various ephemeral algae appear first; these early colonists are usually replaced (within a year) by *Fucus vesiculosus,* which in turn is replaced, eventually, by *Ascophyllum nodosum. Fucus vesiculosus* is the dominant species in semi-protected sites, and *A. nodosum* is dominant in protected sites. It appears that these two fucoid species may remain dominant for about 8 years, and Lubchenco (1983) considers them the climax species for the degree of wave exposure of the area. The herbivores of the area are gastropods, mainly *Littorina littorea*. Her experiments were as follows: (a) gastropod exclusions (cages), (b) gastropod enclosures (cages), (c) roof controls (roofs without sides) and (d) unmanipulated controls (no roof or cage). Additionally, in some enclosures, ephemeral algae were removed as well. The experiments attempted to assess the nature of the interaction between early successional algae and *Fucus* and the effect of gastropod grazing on the interaction. The results are shown in Fig. 4. The different experimental and control plots were colonised differently, probably as a result of patchiness in spore and zygote settlement. However the outcome of the colonisation process depended on the experimental manipulation of the herbivores. The controls (Fig. 4a and b) show the natural sequence: ephemerals are abundant and *Fucus* is rare at first, but later on the situation is inverted. When *Littorina* and the ephemerals are excluded, *Fucus* is dominant throughout (Fig. 4c); exactly the same situation is given when colonisation is *only* by *Fucus* and herbivores are excluded (Fig. 4d); thus, quite clearly, *Fucus* grows initially much faster when both ephemerals and gastropods are absent (Fig. 4c and d) than in their presence (Fig. 4a and b). When only ephemerals colonised in the exclusions (Fig. 4e), these remained dominant throughout, demonstrating that when a single type of alga colonised it took over the plot, provided that herbivores were absent (compare Figs. 4c and d to Fig. 4e); however, when both ephemerals and *Fucus* colonised, the outcome depended on the gastropods: in their absence, ephemeral algae quickly took advantage and retained dominance (Fig. 4f); however, in their presence, *Fucus* attained final dominance (Fig. 4a and b). Thus, the outcome of the competitive interactions and therefore the successional status of the plots appears to depend on the herbivores; in their absence, early colonists outcompete the middle-stage successionals, arresting succession; when they are present the opposite is true.

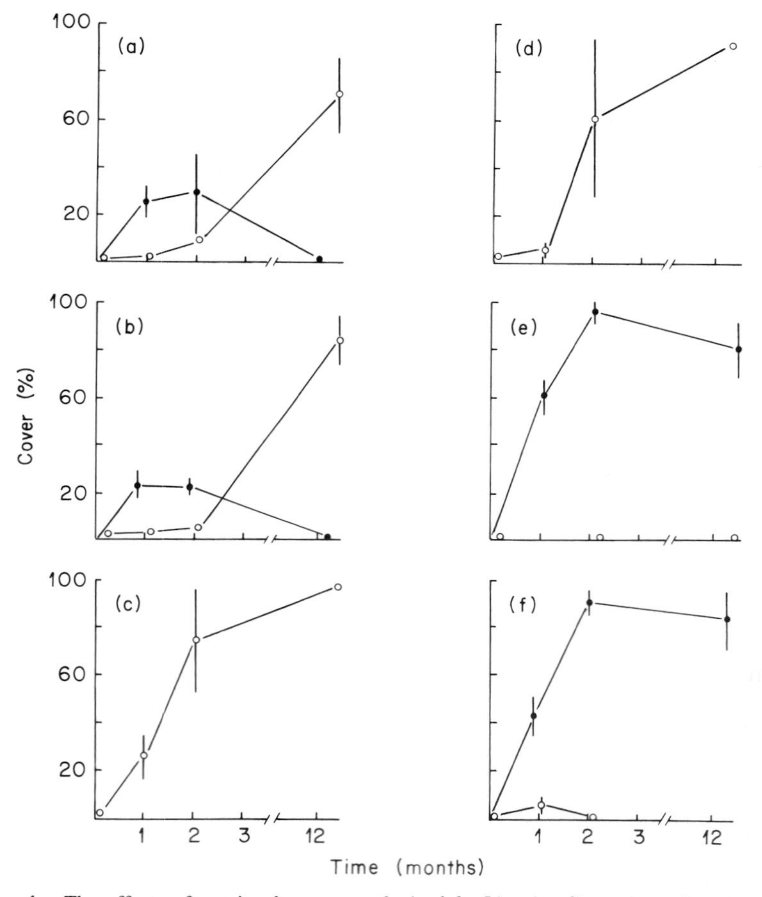

Fig. 4. The effects of grazing by gastropods (mainly *Littorina littorea*) on the competitive interactions between *Fucus vesiculosus* (○) and early successional (ephemeral) algae (●). Percentage cover values are means ±95% confidence limits. a, Unmanipulated control. b, Roof control. c, Herbivore exclusion with ephemeral algae removal. d, Herbivore exclusion (*Fucus* only). e, Herbivore exclusion (ephemeral algae only). f, Herbivore exclusion (ephemeral algae plus *Fucus*). For more details, see text. (After Lubchenco [1983].)

This leads to the general conclusion that herbivores, for this sort of system, *accelerate* succession. (Lubchenco notes that under natural conditions, over a longer time period, *Fucus* eventually establishes, even without gastropods; thus grazing does accelerate succession).

In one of the few experiments reported for terrestrial communities, McBrien, Harmsen & Crowder (1983) found a rather contrasting situation: in abandoned hayfields in southeastern Ontario, an early successional sequence is given by a period of dominance by perennial grasses and other short-lived perennials, followed by a comparatively longer period of dominance by herbaceous perennials, in particular *Solidago canadensis* (which effectively outcompete the former). In this locality, *S. canadensis* is heavily eaten by *Trirhabda* spp. (Chrysomelidae). A comparison of insecticide-sprayed and unsprayed plots showed that herbivores cause a significant reduction of *S. canadensis* cover (from 40–70% to c. 1%) and a consequent increase in cover of the early successionals. Thus, the role of herbivores here appears to *arrest* succession.

Quite clearly, many more studies of the sort described are needed before any generalisations about the role of herbivores on succession can be attempted. Clearly, also, it would have been extremely difficult to arrive at Lubchenco's convincing conclusions in the absence of experimental manipulation.

V. CONCLUSION

From the examples presented here, it appears that most of what we have learned about the role the grazing animal plays in plant populations was discovered through the application of experimental manipulations of the sorts suggested by Harper. As in any other aspects of experimental plant ecology, these experiments are set up in the context of a compromising trilogy of precision, realism and generality (J. L. Harper, personal communication). The crucial issue is to find a combined optimum for the specific problem one is investigating. It seems to me that the greatest challenge lies in seeking for generality. In the study of plant–herbivore interactions the scissors, the insecticide, the cage and the enclosure are all there for us to use with whatever degree of ingenuity may be necessary to arrive at an optimum of precision, realism and, ideally, generality, for each particular plant–herbivore problem we might want to investigate.

ACKNOWLEDGMENTS

I am grateful to Juan Núñez for drawing my attention to one important reference I had overlooked. Fernando Chiang made several important corrections to the manuscript.

REFERENCES

Bach, C. E. (1980). Effects of plant diversity and time of colonization on an herbivore-plant interaction. *Oecologia*, **44**, 319–326.

Becker, P. (1983). Effects of insect herbivory and artificial defoliation on survival of *Shorea* seedlings. *Tropical Rain Forest: Ecology and Management* (Ed. by S. L. Sutton, T. C. Whitmore & A. C. Chadwick), pp. 241–252. Blackwell Scientific Publications, Oxford.

Bentley, S., Whittaker, J. B. & Malloch, A. J. C. (1980). Field experiments on the effects of grazing by a chrysomelid beetle (*Gastrophysa viridula*) on seed production and quality in *Rumex obtusifolius* and *Rumex crispus*. *Journal of Ecology*, **68,** 671–674.

Bruce, D. (1956). Effect of defoliation on growth of long leaf pine seedlings. *Forest Science*, **2,** 31–35.

Cantlon, J. E. (1969). Stability of natural populations and their sensitivity to technology. *Brookhaven Symposia in Biology*, **22,** 197–205.

Connell, J. H. & Slayter, R. D. (1977). Mechanisms of succession in natural communities and their role in community stability and organization. *American Naturalist*, **111,** 1119–1144.

Cromartie, W. J. (1975). The effect of stand size and vegetational background on the colonization of cruciferous plants by herbivorous insects. *Journal of Applied Ecology*, **12,** 517–533.

Dirzo, R. (1984). Herbivory: A phytocentric overview. *Perspectives on Plant Population Ecology* (Ed. by R. Dirzo & J. Sarukhán), pp. 141–165. Sinauer Associates, Sunderland, Massachusetts.

Dirzo, R. & Harper, J. L. (1980). Experimental studies on slug-plant interactions. II. The effect of grazing by slugs on high density monocultures of *Capsella bursa-pastoris* and *Poa Annua*. *Journal of Ecology*, **68,** 999–1011.

Dirzo, R. & Harper, J. L. (1982). Experimental studies on slug-plant interactions. III. Differences in the acceptability of individual plants of *Trifolium repens* to slugs and snails. *Journal of Ecology*, **70,** 101–117.

Harper, J. L. (1969). The role of predation in vegetational diversity. *Brookhaven Symposia in Biology*, **22,** 48–62.

Harper, J. L. (1977). *The Population Biology of Plants*. Academic Press, London.

Harper, J. L. (1979). Review of *Biochemical Aspects of Plant and Animal Coevolution* (Ed. by J. B. Harborne). *New Phytologist*, **83,** 601–602.

Hartshorn, G. (1980). Neotropical forest dynamics. *Biotropica*, **12,** Suppl., 23–30.

Hay, M. E. (1981). Herbivory, algal distribution, and the maintenance of between-habitat diversity on a tropical fringing reef. *American Naturalist*, **118,** 520–540.

Janzen, D. H. (1970). Herbivores and the number of tree species in tropical forests. *American Naturalist* **104,** 501–528.

Janzen, D. H. (1971). Seed predation by animals. *Annual Review of Ecology and Systematics*, **2,** 465–492.

Janzen, D. H. (1976). Why bamboos wait so long to flower. *Annual Review of Ecology and Systematics*, **7,** 347–391.

Jones, M. G. (1933a). Grassland management and its influence on the sward. *Empire Journal of Experimental Agriculture*, **1,** 43–57.

Jones, M. G. (1933b). Grassland management and its influence on the sward. II. The management of a clovery sward and its effects. *Empire Journal of Experimental Agriculture*, **1,** 122–28.

Jones, M. G. (1933c). Grassland management and its influence on the sward. III. The management of a grassy sward and its effects. *Empire Journal of Experimental Agriculture*, **1,** 224–234.

Kulman, H. M. (1971). Effect of insect defoliation on growth and mortality of trees. *Annual Review of Entomology*, **16,** 289–324.

Levin, D. A. (1973). The role of trichomes in plant defense. *Quarterly Review of Biology*, **48,** 3–15.

Louda, S. M. (1985). Herbivore effect on stature, fruiting and leaf dynamics of a native crucifer. *Ecology*, **66** (in press).

Lubchenco, J. (1978). Plant species diversity in a marine intertidal community: Importance of herbivore food preference and algal competitive abilities. *American Naturalist*, **112,** 23–39.

Lubchenco, J. (1983). *Littorina* and *Fucus*: Effects of herbivores, substratum heterogeneity, and plant escapes during succession. *Ecology*, **64,** 1116–1123.

Lubchenco, J. & Cubit, J. (1980). Heteromorphic life histories of certain marine algae as adaptations to variations in herbivory. *Ecology*, **6,** 676–687.

Lubchenco, J. & Gaines, S. D. (1981). A unified approach to marine plant-herbivore interactions. I. Populations and communities. *Annual Review of Ecology and Systematics,* **12,** 405–437.

Maun, M. A. & Cavers, P. B. (1971). Seed production in *Rumex crispus.* I. The effects of removal of cauline leaves at anthesis. *Canadian Journal of Botany,* **49,** 1123–1130.

McBrien, H., Harmsen, R. & Crowder, A. (1983). A case of insect grazing affecting plant succession. *Ecology,* **64,** 1035–1039.

Menge, B. A. (1976). Organization of the New England rocky intertidal community: Role of predation, competition and heterogeneity. *Ecological Monographs,* **46,** 355–364.

Milton, W. E. J. (1940). The effect of manuring, grazing and liming on the yield, botanical and chemical composition of natural hill pastures. *Journal of Ecology,* **28,** 326–356.

Milton, W. E. J. (1947). The composition of natural hill pasture under controlled and free grazing, cutting and manuring. *Welsh Journal of Agriculture,* **14,** 182–195.

Rausher, M. D. (1981). The effect of native vegetation on the susceptibility of *Aristolochia reticulata* (Aristolochiaceae) to herbivore attack. *Ecology,* **62,** 1187–1195.

Rausher, M. C. & Feeny, P. (1980). Herbivory, plant density, and plant reproductive success: The effect of *Battus philenor* on *Aristolochia reticulata. Ecology,* **61,** 905–917.

Risch, S. J. (1981). Insect herbivore abundance in tropical monocultures and polycultures: An experimental test of two hypotheses. *Ecology,* **62,** 1325–1340.

Rockwood, L. L. (1973). The effect of defoliation on seed production of six Costa Rican tree species. *Ecology,* **54,** 1363–1369.

Rosenthal, G. A. & Janzen, D. H. (1979). *Herbivores: Their Interaction with Secondary Plant Metabolites.* Academic Press, New York.

Silvertown, J. W. (1980). The evolutionary ecology of mast seeding in trees. *Biological Journal of the Linnean Society* **14,** 235–250.

Solomon, B. P. (1981). Response of a host-specific herbivore to resource density, relative abundance, and phenology. *Ecology,* **62,** 1205–1214.

Sweet, G. B. & Wareing, P. F. (1966). Role of plant growth in regulating photosynthesis. *Nature (London),* **210,** 77–79.

Thompson, J. N. (1978). Within-patch structure and dynamics in *Pastinaca sativa* and resource availability to a specialized herbivore. *Ecology,* **59,** 443–448.

Thompson, J. N. & Price, P. W. (1977). Plant plasticity, phenology, and herbivore dispersion: Wild parsnip and parsnip webworm. *Ecology,* **58,** 1112–1119.

Van Hulst, R. (1978). On the dynamics of vegetation: Patterns of environmental and vegetational change. *Vegetatio,* **38,** 65–75.

Varley, G. C. & Gradwell, G. R. (1962). The effect of partial defoliation by caterpillars of the timber production of oak trees in England. *Proceedings of the 11th International Congress of Entomology, 1960,* **2,** 211–214.

Waloff, N. & Richards, O. W. (1977). The effect of insect fauna on growth, mortality and natality of broom, *Sarothamnus scoparius. Journal of Applied Ecology,* **14,** 787–789.

White, J. (1980). Demographic factors in population of plants. *Demography and Evolution in Plant Populations* (Ed. by O. T. Solbrig), pp. 21–48. Blackwell Scientific Publications, Oxford.

White, P. S. (1979). Pattern, process and natural disturbance in vegetation. *Botanical Review,* **45,** 229–299.

Whittaker, J. B. (1979). Invertebrate grazing: Competition and plant dynamics. *Population Dynamics* (Ed. by R. M. Anderson B. D. Turner & L. R. Taylor), pp. 207–222. Blackwell Scientific Publications, Oxford.

Yoda, K., Kira, T., Ogawa, H. & Hozumi, K. (1963). Self-thinning in overcrowded pure stands under cultivated and natural conditions. *Journal of Biology, Osaka City University,* **14,** 107–129.

VI

Plant Reproductive Biology

23

Islands and Dioecism: Insights from the Reproductive Ecology of Pandanus tectorius *in Polynesia*

Paul Alan Cox

Department of Botany and Range Science
Brigham Young University
Provo, Utah, USA

> Many of the enigmas of insular floras would be solved if we could interpret aright the 156 species of *Pandanus*
>
> Guppy 1906 p. 157

I. THE COLONIZATION OF ISLANDS BY DIOECIOUS PLANTS

The establishment of a breeding population of a dioecious plant species on an island requires the arrival of at least one male and one female propagule. For pollination to occur, the pollen vector must be found on the island, and, depending upon the nature of the vector, the two propagules must become established within relative proximity to each other. In addition, the colonization events must not be so separated in time as to exceed the lifespans of the plants involved. None of these problems is faced by self-fertile hermaphroditic species, which at best require only the establishment of a single propagule for the beginning of a reproductive population. The relative disadvantages of dioecious species in island colonization have been stated by Baker (1955, 1967) in what Stebbins (1957) has called *Baker's law*: self-fertile taxa will be favored in long-distance dispersal. Although the general aspects of Baker's law are relatively intuitive, a quantitative appreciation of the obstacles faced by dioecious species in long-distance dispersal can be gained through numerical experiments.

Through use of simple algorithms, long-distance dispersal to islands can be simulated with the aid of a computer. Using 100 by 100 matrices as islands onto which male or female propagules land at random, a lifespan is assigned to each propagule, at the end of which lifespan the propagule disappears from the island.

A mean annual immigration flux indicative of the number of propagules landing on the island through time is chosen with a prescribed variance fitting a binomial distribution. If during an iteration of the program a propagule arrives on the island, each of the cells adjacent and diagonal to its position on the island is examined. If a neighboring propagule of the opposite sex is found to occur in an adjacent or diagonal cell, a breeding population is said to be founded, and the number of years taken to establish the population is calculated. If no neighboring propagule of the opposite sex is found, the program is continued to the next iteration.

Through such numerical experiments, I examined the effects of various values for mean annual immigration flux and plant lifespans on the number of years required for establishment of breeding populations of dioecious plants on islands of fixed area. Each numerical experiment was repeated 500 times, and means of the trials were calculated. The results of experiments for 72 different combinations of plant lifespan and immigration flux are represented as a three-dimensional surface in Fig. 1. As can be seen, the number of years required for population establishment is inversely proportional to the number of seeds arriving per year. If the mean annual immigration flux is only 0.2 seeds, then approximately 322 years are required for population establishment. If, however, the mean annual immigration flux is 1.8 seeds, then only 35 years are required.

The effect of various lifespans on the number of years required for population establishment is even more profound. Given a mean immigration rate of 1.8 seeds per year, about 35 years are required for population establishment for dioecious species with lifespans ranging from 50 to 500 years. However, 179 years are required for establishment of dioecious species with lifespans of 5 years, and dioecious annuals require an average of 1918 years for population establishment. Given such enormous disadvantages of dioecious species in long-distance dispersal, we could therefore predict a relative paucity of dioecious plants on islands (unless dioecism evolved after arrival on the island) and a perennial habit in those dioecious plants that have colonized new islands.

Analysis of island floras lends partial support to one of these predictions; very few dioecious annuals occur on oceanic islands. Such analyses, however, indicate that dioecious taxa fare far better than predicted in long-distance dispersal since dioecious taxa do not appear to be significantly underrepresented in island floras when these floras are compared with source floras or mainland floras of similar latitude and ecological character (Baker & Cox 1984). Why dioecious species are less badly impaired than expected in long-distance dispersal has been something of a mystery. It has previously been conjectured that many dioecious taxa on islands have evolved from originally self-fertile colonists (Baker 1967; Carlquist 1965, 1966, 1974; Gilmartin 1968). However, the list of taxa which have probably undergone such autochthonous evolution of dioecism has recently

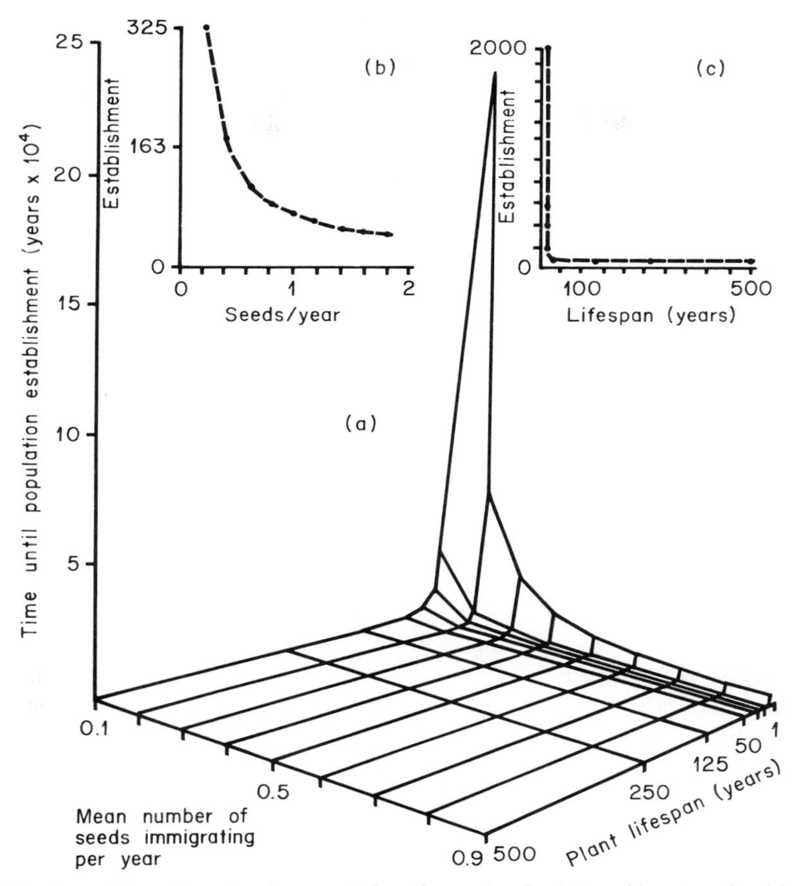

Fig. 1. a, Three-dimensional representation of computer simulation of invasion of an island by dioecious plants. Each point represents the mean of 500 trials. Note that annual plants with a low immigration flux require the greatest number of years for establishment. b, Detail of a single rib of surface for plant lifespan = 500 years. c, Detail of a single rib of surface for mean annual immigration flux = 1.8 seeds.

been reduced by those who point to the existence of the same or closely related dioecious taxa in other island groups (Bawa 1982; Godley 1979).

Another approach to explaining the unexpected success of dioecious plants in long-distance dispersal is to examine features of the breeding systems or dispersal mechanisms that might mitigate the otherwise deleterious effects of dioecism of dispersal ability (Baker & Cox 1984). Thus studies of dioecious plants at the population level occasionally reveal the existence of "leaky dioecy," or aberrations from strict dioecism. For example, populations of the dioecious liana *Freycinetia reineckei* (Pandanaceae) in Samoa produce a consistent but low percentage of hermaphroditic individuals (Cox 1981), even though such individuals are at a selective disadvantage compared to unisexual individuals as a result of the feeding behaviors of the large flying foxes and birds that pollinate them (Cox 1982, 1984). Similar hermaphroditism has been found in *F. negrosensis* in the Philippines and *F. strobilaceae* in Indonesia (Stone as quoted in Cox 1981; Cammerloher 1923). In *F. scandens* in Australia, monoecious individuals have been shown to be self-fertile (Cox, Wallace & Baker 1984). Reports of leaky dioecy in other island genera have been summarized by Baker and Cox (1984).

Another possible mechanism that might mitigate the deleterious effects of dioecism on dispersal ability is the dispersal of seeds as groups rather than singly. Frugivorous birds and bats may defecate entire seed populations at a single time. Thus the evolution of dioecism may be less disadvantageous in plants with fleshy fruits than in plants without endozootic dispersal, a reversal of the arguments of Bawa (1980, 1982) and Givnish (1980) concerning the apparent correlation of fleshy fruits and dioecism. Multipropagule dispersal units occur in dioecious taxa with abiotic dispersal as well. Thus monoecious species of *Cotula* in New Zealand have the achenes dispersed singly, whereas in dioecious species the head is dispersed as a unit (Lloyd 1972). Similar adaptations can be found in wild species of *Spinacia*.

These and other mechanisms by which dioecious plants have overcome their inherent disadvantages in long-distance dispersal can perhaps best be understood by close examination of the reproductive ecology of a dioecious species that has been extraordinarily successful in island colonization: *Pandanus tectorius*.

II. THE REPRODUCTIVE ECOLOGY OF *PANDANUS TECTORIUS*

Pandanus tectorius (Pandanaceae) sensu latissimo is particularly appropriate for a discussion of islands and dioecious plants since it has probably the largest range of any dioecious taxon in Oceania, ranging from French Polynesia to Africa. These arborescent monocotyledons are common on most vegetated islands and atolls in the tropical Pacific and Indian oceans and play an important role in the structure of insular plant communities (Stone 1976, 1982). *Pandanus tectorius* trees are regarded as second only to the coconut, *Cocus nucifera,* in usefulness by Pacific islanders. In nearly all Polynesian dialects the common name of *P. tectorius* is a cognate of *lau fala* or *lau hala* (literally: 'mat leaf') since the long linear leaves are used in the construction of mats, baskets, and other useful articles (Rickard & Cox 1984). The sweet flesh surrounding the woody syncarps of the large multiple fruits is eaten by children throughout the Pacific. In Samoa, the aerial roots are used in medicinal compounds, while in many archipelagos the fragrant male inflorescences are used in perfumery. In Tahiti, the local beer is named 'Hinano' after the male *P. tectorius* inflorescence; in Hawaii hinano was used as a love potion.

The usefulness of *Pandanus tectorius* trees is complemented by their beautiful, albeit unusual, appearance. Their architecture is uncommon and most closely conforms to Stone's model: the growth of the main orthotropic axis is continuous and produces orthotropic branches, which themselves branch sympodially (Hallé, Oldeman & Tomlinson 1978). However, despite their potential biological interest and local economic value, little has been known concerning their reproductive biology.

In 1982 and 1983 field studies on the reproductive ecology of *Pandanus tectorius* were initiated in populations on the islands of Moorea and Tahiti in French Polynesia and on the islands of Kauai, Maui, and Hawaii in Hawaii. The results permit preliminary conclusions to be made on certain aspects of *P. tectorius* population structure and reproductive ecology as they relate to colonization ability.

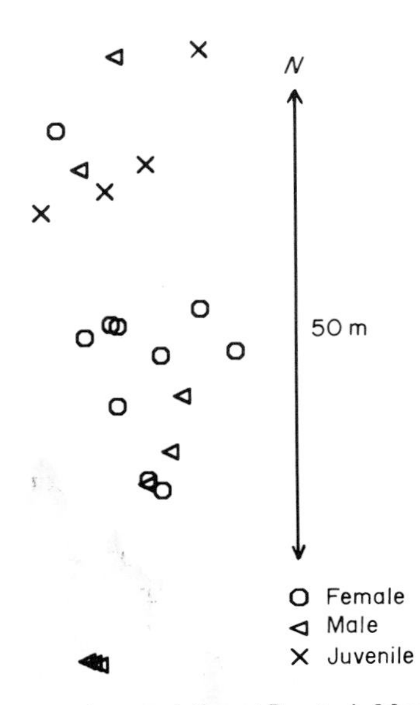

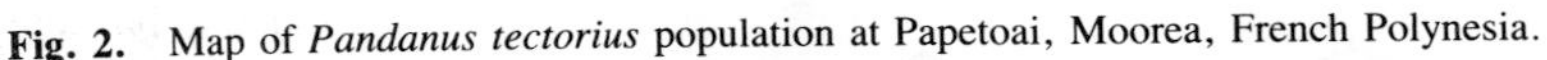

Fig. 2. Map of *Pandanus tectorius* population at Papetoai, Moorea, French Polynesia.

A. Sex Ratio and Population Structure

1. Methods

The sex ratio and population structure of an isolated *Pandanus tectorius* population at Papetoai, Moorea, French Polynesia, was determined by completely mapping the population. The sex ratio of a large *P. tectorius* population at Hana, Maui, Hawaii, was also estimated by sampling the population with 15 separate 30-m transects; at each 10-m interval, the sex of the nearest *P. tectorius* individual was recorded.

2. Results

A map of the Papetoai population (Fig. 2) reveals a roughly equal sex ratio of flowering individuals with no statistically significant deviation from unity (chi-square = 0.222, N.S.). A sex ratio near unity for flowering individuals was also found in the Hana population; 25 females and 22 males were recorded in the transects (chi-square = 0.19, N.S.). In the Papetoai population, no nonflowering mature plants were present; in the Hana population, however, six nonflowering mature plants were recorded. Thus there is a possibility that the sex ratio of flowering individuals may not represent the true sex ratio of all adults in the Hana population. Vernet (1971) has shown the same populations of *Asparagus acutifolius* sampled at different times yield dramatically different estimates of sex ratio as a result of the presence of large numbers of nonflowering individuals. However, assuming that the greatest deviation of the estimated sex ratio in the Hana population from the true adult sex ratio would occur if the sterile adults proved to be all females, the calculated chi-square statistic (1.528, N.S.) for such a situation still indicates no significant deviation from unity. In both populations, juvenile plants were present; in the Papetoai population four juveniles were observed, and in the Hana population five juveniles were recorded in the transects. Alhtough the sex ratios of these juvenile cohorts are unknowable given current technology, they are of importance to the estimation of the adult sex ratio only if these juveniles do not represent true genetic individuals but are instead ramets produced by rhizomatous growth from nearby adults. Rhizomatous growth has never been noted in *P. tectorius* and is unlikely to occur since all vegetative axes are orthotropic. However, the possibility of rhizomatous growth was checked by completely excavating a large individual of *P. tectorius* at a population in Ha' apiti, on the island of Moorea. The excavation revealed no subterranean or plagiotropic shoots. The original establishment growth of the primary axis was found to assume a "saxophone" shape common in many palms and other arboreceous monocotyledons. As can be seen from Fig. 2, the Papetoai *P. tectorius* population is dense; this is characteristic of coastal *P. tectorius* populations throughout the Pacific.

B. Pollination and Reproduction

1. Methods

Floral visitors were observed and recorded during diurnal and nocturnal observations at Papetoai, Moorea; Hana, Maui; and Kailua, Hawaii. Electronic observations using an infrared beam apparatus attached to an automatic camera flash system were also made. Ultraviolet dyes were dusted on staminate inflorescences. Insects captured on pistillate inflorescences were examined with a portable UV light in the field for the presence of these dyes. Preserved specimens of these insects were later examined with scanning electron microscopy for the presence of *Pandanus* pollen. Periods of stigmatic receptivity of pistillate inflorescences at different stages of development were measured by using two different peroxidase assays: 30% hydrogen peroxide solution was placed on stigmatic surfaces which were then observed with a hand lens for oxygen elution, and a paper assay specific to peroxidase was used. Sticky slides were placed at periodic distances from staminate individuals to measure pollen flux. Pistillate inflorescences in the Temai, Moorea, and Hana, Maui, populations were bagged in either cloth or mesh bags. *Pandanus* pistillate inflorescences from Maui and Moorea were preserved in FAA in large fluid-tight bags, with measures taken to avoid crushing or altering in any manner their three-dimensional geometry during transport. Laminar flow characteristics around these *Pandanus* inflorescences were studied in a wind tunnel, using speeds ranging between 0.5 and 3 m/sec, utilizing techniques patterned after those of Niklas (1982, 1984). Flow patterns around pistillate inflorescences were also studied in the field at the Papetoai and Temai populations in Moorea, and at the Hana, Maui, populations by using a portable Draeger airflow test apparatus coupled with an automatic camera and high-speed winder.

2. Results

Although insect visits to staminate inflorescences were common in both French Polynesia and Hawaii (Fig. 3a), few insect visits were made to pistillate inflorescences, and no insects were observed traveling between staminate and pistillate inflorescences. Ultraviolet dye was not found on any insects captured on pistillate inflorescences. Scanning electron microscopy also did not reveal any *Pandanus* pollen on insects captured on pistillate inflorescences, with the exception of a single individual of the Forficulidae from Moorea that had *Pandanus* pollen in its mouth parts. Paper chromatography of fluids found within pistillate inflorescences did not reveal any significant concentration of sugars. It thus appears that *Pandanus* inflorescences neither produce nectar nor are pollinated primarily by insects, although insects rob pollen from staminate inflorescences. Visual and electronic surveillance of *Pandanus* inflorescences in Tahiti and

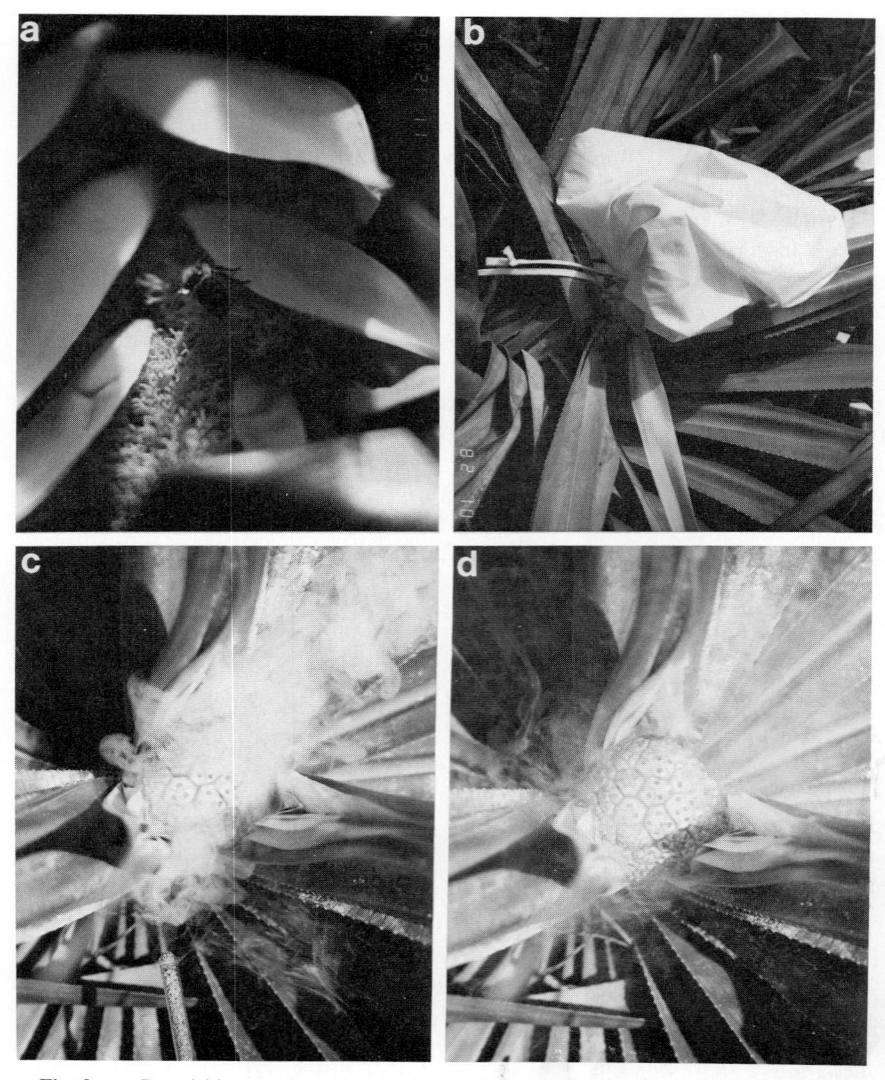

Fig. 3. a, Bee visiting staminate *P. tectorius* population at Papetoai, Moorea, French Polynesia. b, Cloth-bagged *P. tectorius* individual at Hana, Maui. c,d, *In situ* flow experiment with Draeger flow equipment at Moorea. Smoke injected by nozzle visible at bottom of photo (c) begins to curl around bracts and 5 sec later (d) has begun to spiral down to stigmatic surfaces. Wind speed was approximately 0.5 m/sec.

Hawaii failed to reveal any vertebrate visitors, although a few lizards were photographed lurking on pistillate (but never staminate) inflorescences.

The peroxidase assays indicated stigmatic receptivity to begin at a very early stage of inflorescence development and to extend well after fruit formation, an unusually long period of stigmatic receptivity for an entomophilous plant. It therefore appears that (a) pistillate *Pandanus* inflorescences do not offer any edible rewards to potential pollinators, (b) the visitation of insects to *Pandanus* inflorescences results in a negligible amount of pollen transfer from staminate to pistillate inflorescences, (c) insect visits to staminate inflorescences are probably associated with pollen robbing, and the infrequent visits to pistillate inflorescences are probably due to mistakes (Baker 1976). Experiments were therefore performed to investigate the possibility of anemophily in *Pandanus tectorius*.

Flow patterns around *Pandanus tectorius* inflorescences were investigated in the field and in a laboratory wind tunnel. The pistillate inflorescence is radially symmetrical as a result of its position on the terminal end of a tristichous axis. I found that regardless of inflorescence orientation, laminar flows (in the laboratory) or slightly turbulent flows (in the field) in the range of 1 m/sec result in back eddies that decelerate in the region of the stigmatic surfaces (Fig. 3c, d). A spiral flow pattern in both the *xy* and *xz* planes is also apparent, with a result that particles carried in the flow undergo a dramatic deceleration near the stigmatic surfaces. Thus *Pandanus* inflorescences appear to function as highly efficient pollen receivers and are hydrodynamically analogous to some filter-feeding marine invertebrates (Koehl 1982; Vogel 1981). The form of the *Pandanus* pistillate inflorescence probably cannot, however, be strictly interpreted as an adaptation, since I found inflorescences from the outgroup (as determined by a preliminary cladistic analysis of the Pandanaceae by me and M. J. Donoghue; see also Stone 1972) *Freycinetia* to exhibit nearly identical flow patterns in the laboratory wind tunnel. Since *Freycinetia* species are vertebrate-pollinated but not wind-pollinated (Cox 1982, 1983a–c, 1984), the observed flow patterns around the inflorescence probably are a simple consequence of the tristichous vegetative axis. Since none of the likely outgroups of the Pandanaceae is primarily anemophilous, these flow patterns around *Freycinetia* inflorescences could be interpreted as a preadaptation to wind pollination. The *Pandanus* inflorescence may not therefore represent the result of natural selection for an efficient filtering mechanism, but rather the inheritance of a preadaptation from a non-anemophilous ancestor.

The conclusion that *Pandanus tectorius* is primarily anemophilous was confirmed by the bagging studies. In Hana, Maui, a total of 18 pistillate inflorescences were bagged prior to anthesis in either net or cloth bags during October 1982. Unbagged control inflorescences, which were manipulated in the same manner as bagged inflorescences (removing and cutting leaves beneath the inflorescences), were also selected. The bagging experiments were checked in

January and May 1983. All inflorescences, regardless of bag type, were found to have produced fruit. As a test against the possibility that bagged inflorescences were inadvertently contaminated with pollen prior to or subsequent to bagging, samples of stigmatic surfaces were examined with scanning electron microscopy for *Pandanus* pollen grains; none was found. The bagging experiments therefore not only indicate anemophily in *Pandanus,* since net bagged inflorescences produced fruit, but also suggest that *Pandanus tectorius* is apomictic since cloth-bagged inflorescences produce fruit. To determine whether the apomictic system of *Pandanus* is obligate or facultative, isozyme variation of endosperm samples from either cloth-bagged or unbagged inflorescences was studied by using starch gel electrophoresis. If apomixis were obligate, then all endosperm samples from different syncarps within the same cephalium should be genetically identical, regardless of bagging treatment. However, if apomixis is facultative, then endosperm samples from syncarps within an unbagged cephalium should show genetic variability if pollen from different paternal genotypes had resulted in fertilization. The five loci studied were esterase, malate dehydrogenase, acid phosphatase, fluorescent esterase, and alcohol dehydrogenase. Endosperm samples from syncarps from bagged and unbagged cephalia were assigned code numbers and run on the same gels in blind tests. The starch gel electrophoresis revealed significant isozyme variability within unbagged cephalia, but no variability in endosperm samples from cloth-bagged cephalia. It thus appears that *Pandanus tectorius* is facultatively apomictic, being capable of asexual reproduction if pollination does not occur.

The mechanism of the apomictic system may be unusual. In an early embryological study Campbell (1911) noted migration of somatic nuclei into the embryo sac; thus adventitious embryogeny may be the apomictic mechanism, with competition occurring between sexual and asexual embryos. This would explain the observations of Kurz (1867), who found isolated *Pandanus dubius* females to produce fruit; Fagerlind (1940), who found parthenocarpy in *Pandanus columnaeformis*; and Cheah and Stone (1975), who found supernumerary nucellar nuclei in the embryo sac of *Pandanus parvus* from the chalazal region. The indication of facultative rather than obligate apomixis in *Pandanus tectorius* is consistent with the determination of sex ratios approaching unity in both Tahitian and Hawaiian populations, since obligate apomixis would likely result in strongly female-biased sex ratios.

C. Dispersal

The primary dispersal unit of *P. tectorius* is the syncarp. The syncarps are brightly colored; the red outer flesh is sweet and as mentioned previously is eaten by children; the fibrous mesocarps fuse into a hard, impermeable structure that is buoyant. The fibrous parts of the syncarps can often be found in beach-drift

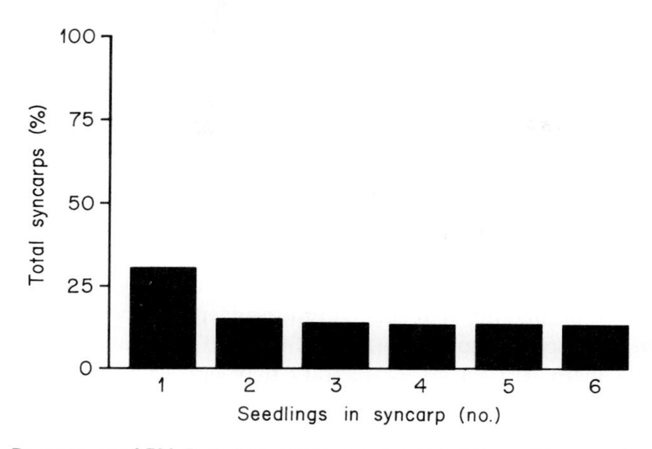

Fig. 4. Percentages of 782 *P. tectorius* syncarps sampled at Hana, Maui, producing one, two, three, four, five, or six seedlings.

throughout the Pacific (Gunn & Dennis 1976). Both Guppy (1906) and Ridley (1930) consider *P. tectorius* to be primarily dispersed by ocean currents. The syncarps have been found on Cocos-Keeling Island, where the species does not occur (Ridley 1930 p. 320). The color and sweetness of the syncarps indicate that local dispersal of *P. tectorius* may be effected by birds and other animals; recently local dispersal of *P. tectorius* by land crabs (*Cardisoma carniflex*) has been reported from Fanning Island (Mary Ann B. Lee, personal communication). Thus local dispersal of *P. tectorius* syncarps may be by biotic vectors, and long-distance dispersal may be primarily effected by ocean currents. Guppy (1906), however, conjectured that now-extinct large birds of the Columbae and Megapodidae may have once carried the syncarps of various *Pandanus* species among islands. Regardless of the vectors responsible for transportation, *P. tectorius* syncarps seem well adapted as dispersal units since each syncarp contains several seeds. A founder population of both male and female plants could thus potentially arise from the establishment of a single syncarp. In order to assess this possibility, four square plots were established in the Hana population, each plot being 0.5 m on a side. Every *P. tectorius* syncarp within the plot was censused for the number of seedlings emerging from it. In the four plots a total of 782 syncarps were found to have produced one or more seedlings. As can be seen in Fig. 4, about 31% of the syncarps produced only one seedling, and 69% of the syncarps produced between two and six seedlings. The relative percentages of syncarps producing one, two, three, four, five, or six seedlings as shown in Fig. 4 can be used to estimate the probability that any given established syncarp will produce at least one male and one female plant. We can assume an equal ratio of sexually produced male and female seedlings within a given syncarp; although the mechanism of sex determination is unknown in *Pandanus,* this assumption is consistent with the nearly equal ratio of the sexes found in both the Hawaiian and French Polynesian populations. The probability that any established syncarp will produce at least one male and one female seedling is equal to

$$p(x) = \sum_{i=1}^{6} p(i)(.5)^{(i-1)}$$

where *p(i)* is the probability that the synacarp will produce *i* seedlings. Using data from the Hana population for *p(i)* (Fig. 4), we find the value of *p(x)* equal to .551. Thus, there is a 55.1% probability that any established syncarp will produce at least one male and one female seedling.

III. DISCUSSION

Although dioecious plants appear to be at a significant disadvantage in the colonization of islands, *Pandanus tectorius* has been tremendously successful throughout the Pacific and Indian oceans. This success can undoubtedly be attributed in part to a number of reproductive mechanisms that mitigate the otherwise deleterious effects of a dioecious breeding system. First, a single *Pandanus tectorius* propagule is likely to produce more than one seedling, and a sexually produced propagule has a better than 50% chance of producing at least one male and one female seedling. Thus a single immigration event may result in the establishment of an entire breeding population. Pollination is not constrained by the faunal composition pollination of the island since wind is the predominant pollen vector. The pollination mechanism itself is very efficient for a wind-pollinated system, since the female inflorescence functions hydrodynamically to filter pollen out of the air. If, however, only a single male plant is established, the perennial habit of *Pandanus* may allow persistence until a female seedling arrives. If only a single female plant is established, at reproductive maturity, apomictic seed can be produced, and perhaps in time some of the syncarps of this entirely apomictic population may colonize new islands. In any case the apomictic population can persist until a male plant or pollen arrives, allowing sexual reproduction to resume.

Pandanus tectorius thus incorporates several different mechanisms for reducing the deleterious effects of dioecism. In addition, the existence of staminodia and pistillodia in individuals of some species of *Pandanus* (Stone 1968) raises the possibility that further exploration will one day reveal the existence of bisexual individuals, as have recently been discovered in *Freycinetia* (Cox 1981; Cox, Wallace & Baker 1984). It is possible that other dioecious island species have similar reproductive systems. However, the possibility of dioecious plants' possessing apomictic or facultatively apomictic breeding systems has not been carefully investigated, although Barlow *et al.* (1978) suggested apomixis as a possible explanation for the highly female-skewed sex ratios in *Viscum album* (Loranthaceae) in Romania. Studies on other dioecious taxa are needed to determine whether apomixis is common or rare.

ACKNOWLEDGMENTS

I thank Herbert and Irene Baker for advice, and Barbara Cox, David Cox, Hugues Hanere, Charles Germain, Dan Omer, and Patty Omer for assistance in data collection. I thank also the Pacific Tropical Botanical Garden for access to *P. tectorius* populations, and Paul Groff for bringing the *Cotula* story to my attention. Joel Kingsolver assisted in the wind tunnel studies. This study was funded by grants from the American Philosophical Study, the National Geographic Society, and Mr. John Elliott and a fellowship from the Miller Institute for Basic Research in Science, University of California, Berkeley.

REFERENCES

Baker, H. G. (1955). Self-compatability and establishment after "long-distance" dispersal. *Evolution (Lawrence, Kans.),* **9,** 347–348.

Baker, H. G. (1967). Support for Baker's law- as a rule. *Evolution (Lawrence, Kans.),* **21,** 853–856.

Baker, H. G. (1976). "Mistake" pollination as a reproductive system, with special reference to the Caricaceae. *Tropical Trees: Variation, Breeding and Conservation* (Ed. by J. Burley & B. T. Styles), pp. 161–170. Academic Press, London.

Baker, H. G. & Cox, P. A. (1984). Further thoughts on islands and dioecism. *Annals of the Missouri Botanical Garden,* **71,** 230–239.

Barlow, B. A., Weins, D., Weins, C., Busby, W. H. & Brighton, C. (1978). Permanent translocation heterozygosity in *Viscum album* and *V. cruciatum,* Sex association, balanced lethals, sex ratios. *Heredity,* **40,** 33–38.

Bawa, K. S. (1980). Evolution of dioecy in flowering plants. *Annual Review of Ecology and Systematics,* **11,** 15–40.

Bawa, K. S. (1982). Outcrossing and the incidence of dioecism in island floras. *American Naturalist,* **119,** 866–871.

Campbell, D. H. (1911). The embryo-sac of *Pandanus. Annals of Botany (London),* **25,** 773–789.

Cammerloher, H. (1923). Over eenige Minder bekende Lukmiddelen van Bluemen. *Tropische Natur,* **12,** 147–151.

Carlquist, S. J. (1965). *Island Life.* Natural History Press, Garden Grove, California.

Carlquist, S. J. (1966). The biota of long-distance dispersal. IV. Genetic systems in the flora of oceanic islands. *Evolution (Lawrence, Kans.),* **28,** 433–455.

Carlquist, S. J. (1974). *Island Biology.* Columbia University Press, New York.

Cheah, C. H. & Stone, B. C. (1975). Embryo sac and microsporangium development in *Pandanus* (Pandanaceae). *Phytomorphology,* **25,** 228–238.

Cox, P. A. (1981). Bisexuality in the Pandanaceae: new findings in the genus *Freycinetia. Biotropica,* **13,** 195–198.

Cox, P. A. (1982). Vertebrate pollination and the maintenance of dioecism in *Freycinetia*. *American Naturalist,* **120,** 65–80.

Cox, P. A. (1983a). Sex and the single flower. *Natural History (N.Y.),* **92,** 42–50.

Cox, P. A. (1983b). Extinction of the Hawaiian avifauna led to a change of pollinators for the ieie *Freycinetia arborea. Oikos,* **41,** 195–199.

Cox, P. A. (1983c). Natural history observations on Samoan bats. *Mammalia,* **47,** 519–523.

Cox, P. A. (1984). Chiropterophily and ornithophily in *Freycinetia* (Pandanaceae) in Samoa. *Plant Evolution and Systematics,* **144,** 277–290.

Cox, P. A., Wallace, B. & Baker, I. (1984). Monoecism in the genus *Freycinetia* (Pandanaceae). *Biotropica* (in press).

Fagerlind, F. (1940). Stemplebau und embryosackentwicklung bei einigen Pandanazeen. *Annales du Jardin Botanique de Buitenzorg,* **49,** 55–78.

Gilmartin, A. J. (1968). Baker's law and dioecism in the Hawaiian flora; an apparent contradiction. *Pacific Science,* **22,** 285–292.

Givnish, T. J. (1980). Ecological constraints on the evolution of breeding systems in seed plants: Dioecy and dispersal in gymnosperms. *Evolution (Lancaster, Pa.),* **34,** 959–972.

Godley, E. J. (1979). Flower biology in New Zealand. *New Zealand Journal of Botany,* **17,** 441–446.

Gunn, C. R. & Dennis, J. V. (1976). *World Guide to Tropical Drift Seeds and Fruits*. Quadrangle, New York Times Press, New York.

Guppy, H. B. (1906). *Observations of a Naturalist in the Pacific between 1896 and 1899. Vol. II. Plant Dispersal*. Macmillan, London.

Hallé, F., Oldeman, R. A. A. & Tomlinson, P. B. (1978). *Tropical Trees and Forests*. Springer-Verlag, Berlin.

Koehl, M. A. R. (1982). The interaction of moving water and sessile organisms. *Scientific American,* **287,** 124–136.

Kurz, S. (1867). Revision of the Indian screwpines and their allies. *Journal of Botany, British and Foreign,* **5,** 93–106, 125–136.

Lloyd, D. G. (1972). A revision of the New Zealand subantarctic and South American species of *Cotula* section Leptinella. *New Zealand Journal of Botany,* **10,** 277–372.

Niklas, K. J. (1982). Pollination and airflow patterns around conifer ovulate cones. *Science,* **217,** 442–444.

Niklas, K. J. (1984). The motion of windborne pollen grains around conifer ovulate cones, implications on wind pollination. *American Journal of Botany,* **71,** 356–374.

Rickard, P. P. & Cox, P. A. (1984). Custom umbrellas (Poro) from *Pandanus* in Solomon Islands. *Economic Botany,* **38,** 314–321.

Ridley, H. N. (1930). *Dispersal of Plants throughout the World*. Reeve, London.

Stebbins, G. L. (1957). Self-fertilization and population variability in the higher plants. *American Naturalist,* **41,** 337–354.

Stone, B. C. (1968). Morphological studies in Pandanaceae. I. Staminodia and Pistillodia of *Pandanus* and their hypothetical significance. *Phytomorphology,* **18,** 498–509.

Stone, B. C. (1972). A reconsideration of the evolutionary status of the family Pandanaceae and its significance in monocotyledon phylogeny. *Quarterly Review of Biology,* **47,** 34–45.

Stone, B. C. (1976). On the biogeography of *Pandanus* (pandanaceae). *Comptes Rendus des Seances de la Societe de Biogeographie (Paris),* **458,** 69–90.

Stone, B. C. (1982). *Pandanus tectorius* Parkins in Australia: a conservative view. *Botanical Journal of the Linnean Society,* **85,** 133–146.

Vernet, P. (1971). La proportion des sexes chez *Asparagus acutifolius* L. *Bulletin de la Société Botanique de France,* **118,** 345–358.

Vogel, S. (1981). *Life in Moving Fluids*. Willard Grant Press, Boston, Massachusetts.

24

The Meaning and Measurement of Reproductive Effort in Plants

F. A. Bazzaz and E. G. Reekie

Department of Organismic and Evolutionary Biology
The Biological Laboratories
Harvard University
Cambridge, Massachusetts, USA

I. INTRODUCTION

The measurement of reproductive effort (RE) in contrasting species and in different environments has been the focus of many recent studies. The impetus for these studies has been a desire to test some of the theoretical predictions about the way plants should allocate their resources between vegetative and reproductive activities.

Harper (1967) has suggested that colonizing plants should devote more of their resources to reproduction than plants from more mature habitats. Individuals in open environments will face little interference from neighbours; consequently, the chance of leaving descendants will be closely linked to fecundity. In a crowded community where resources are more limiting, however, individual success is much more dependent upon the ability to capture a share of the resources. The individual which sacrifices competitive ability for fecundity may not survive to reproduce. Gadgil and Solbrig (1972) make similar arguments in terms of r and K selection, predicting that under conditions of high density-independent mortality (in open, disturbed environments, for example) RE should be high, whereas under conditions of density-dependent regulation (in closed communities, for example), RE should be lower.

Abrahamson (1975) and Williams (1975) have both suggested that in organisms that propagate both vegetatively and sexually, sexual RE will increase as

STUDIES ON PLANT DEMOGRAPHY:

density increases. At low densities, increased allocation to clonal growth facilitates rapid local spread with low risk. At higher densities, local conditions for growth are more unfavourable, and sexual reproduction allows dispersal and colonization of new sites.

Life history models that consider the effects of juvenile versus adult mortality predict that in environments in which adult mortality is high in relation to juvenile mortality, plants should have a high RE; in environments where juvenile mortality is high, RE should be lower. These predictions are qualitatively the same as those made by Harper. High resource availability in open, disturbed environments would favour juvenile survivorship, although adult survivorship would be reduced by disturbance. In less disturbed environments, adult mortality would be reduced, thereby increasing competition and making it more difficult for juveniles to establish. Life history models also predict that annuals should have a higher RE than perennials and that semelparous species should have a higher RE than iteroparous species. It is only advantageous to sacrifice future growth and reproduction if present reproduction is increased (see reviews of Hirshfield & Tinkle 1975; Schaffer & Gadgil 1975; Horn 1978; Michod 1979; Bell 1980; Stearns 1980). More recently, models have been constructed for populations classified in terms of variables other than age, such as size, developmental stage, or spatial location (Caswell 1982a) and for non-equilibrium populations (Caswell 1982b). Models have also been constructed to consider the optimal switching pattern between vegetative and reproductive growth (King & Roughgarden 1982a, b, 1983) and the importance of resource storage to allocation patterns (Schaffer, Inouye & Whittam 1982; Chiariello & Roughgarden 1984).

Although many studies have compared the RE of a variety of plants in different environments, it is still unclear whether or not the theoretical predictions are borne out. Some studies have confirmed the predictions, although others have produced conflicting results (see discussion in Pitelka, Stanton & Peckenham 1980; Soule & Werner 1981; Willson 1983). We see three possible explanations for this conflict: (a) present methods of assessing RE may in some cases distort the results; (b) many of the tests described in the literature are not appropriate tests of the theories; and (c) the theories are either incorrect or not general enough to apply to all these specific situations. The primary focus of this essay will be on the first of these proposed explanations.

There are two major procedural difficulties involved in assessing RE in plants: (a) deciding what currency to use (i.e. in terms of what resource or combination of resources should RE be calculated?) and (b) deciding what structures and processes should be considered part of RE. Most studies calculate RE as the proportion of the plant's biomass allocated to seeds and other 'obvious' reproductive structures. The questions of whether or not biomass allocation is a good measure of total carbon allocation (i.e. biomass carbon plus respiratory carbon) and of whether or not carbon allocation necessarily approximates that of other resources have not been adequately addressed. Available evidence, however, suggests that neither of these commonly made assumptions is necessarily correct. It has been shown that biomass allocation may not be a good indicator of the way that some mineral nutrients are allocated (van Andel & Vera 1977; Lovett Doust 1980; Abrahamson & Caswell 1982). Jurik (1983) has shown that in *Fragaria,* RE calculated in terms of biomass and in terms of total carbon differ substantially, and that there is no straightforward relationship between the two measures.

A third problem with the measurement of RE concerns one of the underlying assumptions of the various theories generated to explain resource allocation patterns. As pointed out by Harper (1977), these concepts depend absolutely on the idea that different structures or activities are alternatives, that a gain in one results in a loss in another. Ideally then, from the point of view of the various theories, any measure of the 'effort' involved in reproduction would be based not on the allocation of resources to reproduction but rather on the degree to which vegetative processes were decreased by reproduction.

It has been demonstrated that many plants produce reproductive structures that are capable of supplying a large proportion of their own carbon needs through photosynthesis. Consequently, neither biomass nor total carbon allocation to these structures will reflect the net carbon costs of producing them (Bazzaz, Carlson & Harper 1979).

II. CARBON ALLOCATION TO REPRODUCTION

A. Reproductive versus Vegetative Structures

Reproduction involves not only the production of propagules but also the production of male flower parts and other ancillary and support structures such as sepals, petals, nectar, etc. Because the proportion of resources invested in these structures can be substantial and these structures are part of the 'effort' involved in reproduction, they should be included in any measure of RE (Thompson & Stewart 1981). When this approach is examined carefully, however, difficulties are encountered in deciding which structures are reproductive. Flower and fruit parts are obviously reproductive, but many plants produce fruiting structures on specialized stems that may or may not bear leaves. The status of these structures is less clear. Thompson and Stewart (1981) have suggested that all structures not possessed by the vegetative plant be considered reproductive. Even this definition, however, may omit some reproductive structures. In many grasses, for example, the stem internodes elongate when the plant flowers. Presumably this facilitates the dispersal of pollen and seeds, or perhaps it assists in the receipt of pollen. These structures are indistinguishable from normal vegetative stems, the only difference being that the proportion of stem material has increased in relation to that in the vegetative plant. Furthermore, it can be argued that since the reproductive parts require water and mineral nutrients, and these resources are supplied by the roots, a portion of the root material should be considered part of the RE. Similar arguments can be made for other plant parts normally considered part of vegetative growth.

B. Reproductive Respiration

Resource allocation measured at maturity is not necessarily a good indication of total allocation to a particular function. Total carbon allocation to some function involves not only allocation to the biomass of various structures, but also the carbon involved in the respiration associated with the growth and maintenance of these structures. There is a wide range in the conversion efficiencies of different plant constituents, and these are reflected in the almost twofold difference in the sucrose required for the synthesis of various plant tissues (Penning de Vries 1975). In order to get an accurate picture of either carbon or energy allocation it will be necessary to measure the respiration associated with reproductive and vegetative growth.

Several studies (Harper & Ogden 1970; Hickman & Pitelka 1975; Abrahamson & Caswell 1982) calculate energy allocation to reproduction by determining the energy content of the biomass. The results of these studies indicate that biomass allocation and energy allocation are more or less equivalent in cases in which lipids are not a significant storage product. Because of the differences in energy requirement to produce various plant constituents, reproductive effort calculated in this way, however, does not necessarily reflect the energy required to produce these structures.

Furthermore, direct measures of the respiration of flowers and fruits will reveal only a portion of the respiratory cost of reproduction. The respiration associated with the uptake and transport of the mineral nutrients required for reproductive growth, for instance, would not be reflected in the respiration of the flowers and fruits. Reproductive respiration, however, can be defined in the same manner as reproductive structures, as the respiration associated with the production and maintenance of propagules and any ancillary and support structures, regardless of where this respiration takes place.

Similar problems are encountered when the allocation of other resources is examined closely. The water content of various structures tells us little about the relative amounts of water required by different structures since the vast majority of the water used by a plant is transpired over the course of its growth. Furthermore, many mineral nutrients are very mobile within plants and can be reallocated from one structure to another over the course of the growing season.

C. Reproductive Photosynthesis

A number of flowers and fruits are capable of photosynthesis and can supply a significant proportion of their own carbon needs (Bazzaz & Carlson 1979; Werk & Ehleringer 1983). Consequently, total carbon allocation to these structures will not accurately reflect the net carbon cost to the vegetative plant (Bazzaz, Carlson & Harper 1979). The act of reproduction may also enhance the photosynthetic rate of the leaves (e.g. Hansen 1970). The quantitative significance of this latter process in reducing the cost of reproduction, however, is not yet known. Regardless of the means, any increase in photosynthesis due to reproduction complicates the definition of RE, because photosynthesis is a vegetative function. Ideally, some proportion of these 'reproductive' structures should be assigned to vegetative growth. There are obvious practical difficulties in following such an approach. Thompson and Stewart (1981) have suggested that this problem could be avoided by using some resource other than carbon as the currency of allocation. This solution is not very satisfactory, however, because of the important structural and energetic role of carbon in plants.

There are two possible ways of dealing with this problem. The first is to assume that the cost of reproductive photosynthesis is minimal. It can be argued that the only costs associated with reproductive photosynthesis are the production and maintenance of the chlorophyll and photosynthetic enzymes in the reproductive structures. The cost of producing the structures themselves can be attributed entirely to reproduction since they would be produced regardless of whether there were reproductive photosynthesis or not. The cost of reproductive photosynthesis is probably further reduced by the breakdown of the photosynthetic apparatus and the probable re-allocation of the resulting products to the developing seeds as the reproductive structures mature. If the cost of reproductive photosynthesis is minimal in relation to the cost of the entire structure, then RE can simply be calculated as the total carbon allocated to reproduction divided by total photosynthesis. The second approach is to subtract reproductive photosynthesis from the cost of producing the reproductive structures to get a net reproductive cost (Bazzaz & Carlson 1979; Bazzaz, Carlson & Harper 1979). This net cost can then be expressed in relation to the total photosynthesis of the vegetative parts. Which of these two approaches is taken will depend upon the objectives of the study. If the objective is to determine the carbon allocated to reproduction, the first approach is the closer approximation. On the other hand, if the objective is to determine the potential effect of reproduction on vegetative activity (i.e. what is the carbon cost to the vegetative plant), then the second approach is better. From the point of view of life history theory, which is concerned with the tradeoff between vegetative and reproductive functions, the latter approach is the more relevant.

D. The Experimental Separation of Vegetative and Reproductive Functions

Using carbon as the currency of allocation, vegetative allocation can be taken to include all those structures directly involved in the capture of carbon plus any necessary support structures and activities.

In practice it is relatively simple to determine the structures and activities involved in the uptake of carbon but much more difficult to measure the structures and activities involved in the production of propagules directly. Many plants go through a prolonged vegetative state before reproducing; therefore, the amount of root and stem material required to support the leaves and the respiration required to support these structures can be determined directly. On the other hand, no plant exhibits only reproductive growth. Even in species which exhibit a sharp transition from vegetative to reproductive growth, some vegetative activity continues in the vegetative structures produced before the change in allocation pattern, making it impossible to determine directly the structures and activities required for the production of propagules. It is possible, however, to determine reproductive allocation indirectly as the difference between total carbon use and the allocation of carbon to vegetative growth.

In plants exhibiting both vegetative and reproductive growth, the structures and activities required for the support of the leaves can be determined by comparison with vegetative plants growing in similar environments. The allometric ratios between leaves and roots and between leaves and stems for vegetative plants could be used to determine the proportion of stem and root material necessary for leaf growth in reproductive plants. The stem and root material above that needed for leaf support would be classified as reproductive structures. Vegetative and reproductive plants could be produced by controlling the photoperiod in photoperiod-sensitive plants, or, alternatively, the ratios between plant parts could be examined at different stages in the natural life cycle.

Reproductive respiration and photosynthesis could be determined in the same fashion. The respiration and photosynthetic rates of vegetative plants are then used to estimate the respiration and photosynthesis of the vegetative parts of the reproductive plants. The difference between total respiration or photosynthesis of the reproductive plants and the estimated vegetative activity is attributed to

TABLE I.

Summary of the Various Direct and Indirect Measured of Reproductive Effort[a]

Direct measure		Indirect measure	
RE1	(Inflorescence biomass)/(total above-ground biomass)	RE6	(Total biomass of vegetative plant − vegetative biomass of reproductive plant)/(total biomass of vegetative plant)
RE2	(Inflorescence biomass)/(total plant biomass)	RE7	(Total biomass and respiration of vegetative plant − vegetative biomass and respiration of reproductive plant)/(total biomass and respiration of vegetative plant)
RE3	(Reproductive biomass)/(total biomass of reproductive plant)		
RE4	(Reproductive biomass + reproductive respiration)/(total biomass and respiration of reproductive plant)		
RE5	(Reproductive biomass + reproductive respiration − reproductive photosynthesis)/(total biomass and respiration of reproductive plant − reproductive photosynthesis)		

[a] Respiration and biomass carbon were expressed in the same terms by assuming equivalency between dry weight and carbohydrate. Reproductive efforts, although calculated in terms of carbon here, can be calculated in terms of desired currency, e.g. nitrogen or phosphorous.

reproduction. This approach would not only attribute respiration and photosynthesis of the reproductive structures themselves to reproduction, but also any increase in these activities in the vegetative parts of the plants.

The primary motivation for ecological studies of RE in plants is to compare patterns of resource allocation among plants growing in different environments and among plants with different growth habits. It could be argued, therefore, that the various refinements discussed previously, although necessary to obtain a true measure of RE, are not necessary to test the predictions of ecological theory. It is possible that measures of RE based upon biomass allocation to 'obvious' reproductive structures are adequate for comparative purposes. Such measures will bias the comparisons, however, unless the proportion of allocation to ancillary and support structures and any associated activities is constant over the comparisons being made. Because the amount of root and stem material, the number of seeds per fruit or seed head, the amount of respiration per unit of growth, and photosynthetic rates are all likely to vary with environment and species, this assumption is at least suspect and should be tested rigorously.

Reproductive effort has been calculated in five different ways (Table I). The first two which are most commonly used represent measures of RE based upon biomass allocation to 'obvious' reproductive structures. The third measure is also based upon biomass allocation but includes allocation to all reproductive structures (as we have defined them). The fourth measure is perhaps the best measure of the total carbon allocated to reproduction, because it takes into account both the biomass and respiration associated with reproduction. The last measure, in addition, takes into account reproductive photosynthesis, expressing RE as the net carbon cost of reproduction in relation to the total photosynthesis of the vegetative parts of the plant. It is, therefore, a measure of the carbon cost to the vegetative parts of the plant. Indirect measures of reproductive efforts will be discussed later.

E. Currency of Allocation: Past Approaches

One approach to the currency problem is to assume that the allocation of one resource will reflect the allocation pattern of the others. Studies that have compared the allocation patterns of biomass and mineral nutrients have found that the proportion of the total resource allocated to reproductive structures differs significantly among resources (Lovett Doust 1980; van Andel & Vera 1977; Abrahamson & Caswell 1982). From the point of view of life history analysis, however, the absolute magnitude of RE is less important than qualitative differences in RE among populations or environments. Therefore, it can be argued that, providing the relative order of different populations remains the same, it does not matter which resource is used to calculate RE. Abrahamson and Caswell (1982) found that the qualitative ranking of three populations of *Verbascum thapsus* remained more or less the same regardless of whether RE was calculated in terms of biomass or various mineral nutrients. However, distinct differences were found when the allocation patterns of individual nutrients were compared; for example, the qualitative rankings of the population differed when RE was calculated in terms of aluminum and in terms of biomass.

In order for the principle of allocation to be useful, there must be a limited supply of some necessary resource, and the allocation of this resource to different structures and activities must result in a cost to the plant such that an increase in the allocation to one results in a decrease to another (Harper 1977). Harper, therefore, argued that RE should be measured in terms of the particular resource that is limiting plant growth. Implicit in this approach is the assumption that if a resource is not limiting growth, a plant may not 'care' how it allocates this resource, and the amount allocated to a particular function may not be indicative of the value 'placed' on this function. In reality, it seems likely that at least among the macronutrients, some sort of approximate balance must be maintained between the various resources, and if a resource is present in the environment in excess, its uptake is regulated to maintain this balance. If this is the case, then it does not matter which resource is used as the currency in which to measure RE; the qualitative patterns should remain the same regardless of the chosen resource. The correlations between values of RE calculated in terms of various currencies (Abrahamson & Caswell 1982) support this argument to an extent. Nevertheless, the approach of calculating RE in terms of the limiting resource is an appealing one.

Unfortunately, this approach has practical difficulties. First, it may be difficult to determine what the limiting resource is in a particular environment. This is especially true in mesic environments or those in which micronutrients are limiting. Even in very extreme environments it is unlikely that there is ever only one limiting resource. Growth is probably limited by several different resources acting at different times in the life cycle or even within the same day. As a result, RE still has to be calculated in terms of several resources; if conflicting patterns are obtained, a choice among the various patterns has to be made. A second, more serious difficulty with this approach is that it is unlikely that the same resources will be limiting growth in different environments. Light may be the major limiting resource for a shallow-rooted herb on the forest floor where evapotranspiration is low and competition with deep-rooted trees for water and nutrients may be minimal. The same or similar species growing in an open field is more likely to be limited by water and nutrients because tree cover is less and it has to compete with a wide variety of other herbs for water and nutrients. In a situation such as this, it is not possible to compare the RE of plants from these two environments in terms of the same limiting resource.

F. Carbon as a Common Currency

A third approach to calculating RE is in terms of a resource that can integrate the allocation patterns of other resources. *Integration,* as it is defined here, simply means that shortages or sufficiencies of other resources will be reflected in the allocation pattern of the integrating resource. Money, for example, integrates the allocation patterns of goods and services in society. Money was developed as a means to facilitate the exchange of goods and services (that is, resources). The monetary value of a resource is set by supply and demand. If a resource is in short supply, its monetary value increases; conversely, if a resource is abundant, its monetary value decreases. A major advantage of a monetary system is that the values of diverse resources can be expressed in the same terms. This makes the evaluation of the cost of various activities and goals much simpler. In plants, it can be argued that carbon, at least to an extent, integrates the allocation patterns of other resources. Plants store the energy captured through photosynthesis in the form of reduced carbon compounds. Energy, in one form or another, is used to drive all biological processes, including the capture and subsequent utilization of all the various resources. It is, therefore, possible to express the cost of these activities in terms of carbon.

There are a number of examples of the way carbon is used by plants to 'purchase' resources in much the same way that money is used to purchase goods and services. Legumes export considerable quantities of carbon in the form of sugars to rhizobia, in return for which they receive nitrogen fixed by the bacteria (Schubert 1982). Similarly, the great majority of vascular plants export sugars to mycorrhizae and receive phosphorus and perhaps other mineral nutrients and water in return (Harley 1969; Trappe & Fogel 1977). On a more subtle level, plants can purchase water and mineral resources by allocating more carbon to root growth, thereby increasing the root–shoot ratio (Troughton 1977). Carbon is also used to 'purchase' nutrient ions at the molecular level. The cost of reducing a nitrate ion, for example, or the cost of taking up a cation from the soil solution can be expressed in terms of carbon. Although some of these mechanisms by which carbon can be used to purchase other resources such as the genetic changes that allow the formation of symbiotic relationships, can be developed only over evolutionary time, most of these mechanisms can also be manipulated over ecological time. Plants can probably vary carbon allocation at the molecular level within hours. Root–shoot ratios can vary over the life of a single individual. Even in plants that form symbiotic relationships with other organisms, individual plants may regulate the amount of carbon exported to the symbionts.

In order for carbon to be useful as a common currency in which to evaluate resource allocation patterns, not only must it be demonstrated that carbon is used to purchase other resources, but it should also be shown that the carbon cost per unit of resource increases as the resources become more limiting. This is equivalent to the increase in the monetary value of a good or service as this resource becomes more scarce (i.e. the law of supply and demand). If the carbon cost increases as the resource becomes more limiting, then the carbon allocation pattern will be biased toward the particular resources limiting growth. As the concept of resource allocation from the point of view of life history analysis is only meaningful if the resources are limiting, this is the preferred result.

There is good evidence that at the plant organ level, the carbon cost of resource uptake varies with the availability of the resources. It follows that a plant would have to allocate more and more carbon to root growth to extract the same amount of water and nutrients as these resources become more scarce. The cost of supporting symbionts also appears to vary with resource availability. Plants which allocate carbon to symbionts appear to regulate this flow to minimize it when the particular resource which the symbiont supplies is readily available. Legumes do not nodulate, or will do so only poorly when nitrogen is readily available in the soil. The degree of mycorrhizal infection decreases with the increasing availability of phosphorus. The carbon cost of supporting symbionts can be high (Mahon 1977; Schubert 1982), and where possible the plant avoids this cost by extracting the resources itself. Lambers (1979) presents experimental evidence that the respiratory costs associated with resource uptake and utilization costs actually vary and may be severalfold greater than the theoretical costs. Furthermore, the respiratory costs of maintenance processes are known to vary with the environment (Redmann & Reekie 1982). The relationship between maintenance costs and resource availability is not clear, however. In the case of water deficiency, for example, costs may either increase or decrease as water becomes less available.

Although we have only briefly summarized the evidence here, it appears that (a) the uptake of various resources does entail a carbon cost, and (b) this cost may increase as the availability of a resource decreases. This supports the idea that carbon can be used as a common currency in which to compare resource allocation patterns. One way of testing this would be to compare the allocation of carbon to that of various resources at different levels of these resources. The allocation of carbon would not necessarily reflect the allocation of any particular resource in environments in which a number of resources were limiting growth but would tend to represent the average allocation pattern. In environments in which one particular resource were limiting growth, however, it is predicted that the allocation pattern would tend to approach that of the limiting resource. This prediction is based on the assumption that the carbon cost of obtaining this particular resource would become so large as this resource became less available that it would overshadow the other costs.

III. EFFECT OF REPRODUCTION ON VEGETATIVE ACTIVITY

The amounts of carbon and other resources invested in reproductive structures constitute the direct costs of reproduction. The indirect costs are the consequences of this allocation, or in other words, the extent to which vegetative activity is reduced by reproduction, assuming the two activities are competing as discussed earlier.

There are a number of possible reasons why the correlation between direct and indirect costs of reproduction may be less close than has been assumed. In a plant that forgoes reproduction in favour of vegetative growth, any increase in vegetative growth would not simply be proportional to the resources diverted from reproduction; the new vegetative structures might in turn acquire more resources and grow further (i.e., the compound interest analogy). Therefore, the increase in vegetative growth would vary, depending upon the environment and the plant's innate ability to capture resources. A similar situation develops when looked at from the opposite perspective. In a plant that allocates resources to reproduction, the decrease in vegetative activity would depend not only upon the quantity of resources diverted to reproduction, but also upon the ability of the plant to compensate for this diversion through reproductive photosynthesis (Bazzaz, Carlson & Harper 1979). The correlation between direct and indirect costs might also break down if vegetative and reproductive growth were limited by different resources (Willson 1983). Watson (1984) has suggested that reproduction may not be resource-limited at all. If such is the case, then there may be no correlation between the direct and indirect costs of reproduction.

The Measurement of the Indirect Cost of Reproduction

The indirect costs of reproduction can be quantified demographically by comparing the growth and survivorship of individuals that do, or do not, reproduce (e.g. Law 1979; Piñero, Sarukhán & Alberdi 1982). This approach, although useful in field studies, has its limitations, the primary one being the correlative nature of these studies. Reproduction may be correlated with a number of other factors, making it difficult to attribute differences between reproductive and vegetative plants entirely to reproduction. It is possible, for example, that reproduction may be limited to only those plants occupying the most favourable sites. Comparisons between vegetative and reproductive plants in such a situation would be confounded by site differences. Antonovics (1980) has suggested the use of various experimental manipulations, such as removing reproductive buds or photoperiod control, to produce reproductive and vegetative plants under comparable growing conditions. By manipulating the photoperiod, Antonovics was able to determine the cost of reproduction in terms of lost leaf area in *Plantago lanceolata*. Individuals that produced many inflorescences had significantly lower numbers of leaves than did individuals that produced few inflorescences. The use of such experimental manipulations promises to be a valuable tool for comparing the consequences of reproduction in different plants and environments.

Although the manipulations suggested by Antonovics provide an experimental means of assessing the indirect costs of reproduction, these indirect measures are still subject to the same problems as the direct measures: (a) what structures and activities should be included as part of reproduction, and (b) what is the relevant currency in which to measure the cost of reproduction? Ideally, the indirect cost of reproduction should be measured in terms of lost vegetative activity, but in order to determine to what degree vegetative growth has been reduced, it is necessary to be able to distinguish between vegetative and reproductive structures. Furthermore, since respiration is part of vegetative activity, it could be argued that it is also necessary to separate vegetative and reproductive respiration. The indirect costs may also vary, depending upon the currency. Two plants may suffer the same costs of reproduction in terms of lost vegetative biomass, but the degree to which the nitrogen content of the biomass is reduced by allocation to reproductive structures may vary. If nitrogen is a limiting resource, the future growth and survival of a high-nitrogen plant is likely to be superior to that of a low-nitrogen plant. It is, therefore, important to take into consideration what the relevant currency is in calculating the indirect costs of reproduction.

The allometric relationships among leaves and other plant parts in vegetative plants can be used to estimate the vegetative biomass of reproductive plants. The respiration rates of the vegetative plants can be used in a similar manner to estimate vegetative respiration in the reproductive plants. Furthermore, if carbon allocation does reflect the allocation of other resources, carbon can be used as the currency in which to assess the indirect costs of reproduction.

The use of experimental manipulations for examining the effect of reproduction on vegetative processes has a further advantage over correlative techniques in that they provide a convenient reference point by which to compare costs between treatments. The loss in vegetative growth or activity (measured in the desired currency) can be expressed as a proportion of the total growth in the absence of reproduction. This ratio provides a measure of RE expressed in terms of vegetative growth (Table I). From the point of view of current life history theory, which is concerned with the tradeoff between vegetative and reproductive growth, this measure of RE is the most desirable.

IV. CONCLUSION

There are several questions concerning reproductive effort that emerge from this essay and that are being approached experimentally. They include:

1. a. In plants exhibiting both vegetative and reproductive growth, which structures should be considered part of RE?
   b. Is biomass allocation to flowers and fruits a good indicator of total biomass allocation to reproduction?
2. a. In plants exhibiting both vegetative and reproductive growth, what portion of the respiration should be considered part of RE?
   b. Is biomass allocation to reproduction a good indicator of total carbon allocation to reproduction?
3. a. In plants exhibiting both vegetative and reproductive growth, what portion of total photosynthesis can be attributed to reproduction?
   b. Is total carbon allocation to reproduction a good indicator of net allocation (i.e. total carbon minus reproductive photosynthesis)?

4. Is carbon allocation to reproduction (as measured by various means) a good indicator of the way that other resources are allocated to reproduction?
5. Do the direct costs of reproduction reflect the indirect costs?

Answers to these questions for several genotypes and species with contrasting features would take us a long way toward understanding plant life history strategies and their evolution.

REFERENCES

Abrahamson, W. G. (1975). Reproductive strategies in dewberries. *Ecology,* **56,** 721–726.

Abrahamson, W. G. & Caswell, H. (1982). On the comparative allocation of biomass, energy, and nutrients in plants. *Ecology,* **63,** 982–991.

Antonovics, J. (1980). Concepts of resource allocation and partitioning in plants. *Limits to Action: The Allocation of Individual Behavior* (Ed. by J. E. R. Staddon), pp. 1–25. Academic Press, New York.

Bazzaz, F. A. & Carlson, R. W. (1979). Photosynthetic contribution of flowers and seeds to reproductive effort of an annual colonizer. *New Phytologist,* **82,** 223–232.

Bazzaz, F. A., Carlson, R. W. & Harper, J. L. (1979). Contribution to the reproductive effort by photosynthesis of flowers and fruits. *Nature (London),* **279,** 554–555.

Bell, G. (1980). The costs of reproduction and their consequences. *American Naturalist,* **116,** 45–76.

Caswell, H. (1982a). Optimal life histories and the maximization of reproductive value: A general theorem for complex life cycles. *Ecology,* **63,** 1218–1222.

Caswell, H. (1982b). Life history theory and the equilibrium status of populations. *American Naturalist,* **120,** 317–339.

Chiariello, N. & Roughgarden, J. (1984). Storage allocation in seasonal races of an annual plant: Optimal versus actual allocation. *Ecology,* **65,** 1290–1301.

Gadgil, M. & Solbrig, O. T. (1972). The concept of r- and K-selection: Evidence from wild flowers and some theoretical considerations. *American Naturalist,* **106,** 14–31.

Hansen, P. (1970). ^{14}C-studies on apple trees. V. Translocation of labelled compounds from leaves to fruit and their conversion within the fruit. *Physiologia Plantarum,* **23,** 564–573.

Harley, J. L. (1969). *The Biology of Mycorrhiza.* Leonard Hill, London.

Harper, J. L. (1967). A Darwinian approach to plant ecology. *Journal of Ecology,* **55,** 247–270.

Harper, J. L. (1977). *Population Biology of Plants.* Academic Press, London.

Harper, J. L. & Ogden, J. (1970). The reproductive strategy of higher plants. I. The concept of strategy with special reference to *Senecio vulgaris* L. *Journal of Ecology,* **58,** 681–698.

Hickman, J. C. & Pitelka, L. F. (1975). Dry weight indicates energy allocation in ecological strategy analysis of plants. *Oecologia,* **21,** 117–121.

Hirshfield, M. F. & Tinkle, D. W. (1975). Natural selection and the evolution of reproductive effort. *Proceedings of the National Academy of Sciences of the U.S.A.,* **72,** 2227–2231.

Horn, H. S. (1978). Optimal tactics of reproduction and life history. *Behavioral Ecology* (Ed. by J. R. Krebs & N. B. Davies), pp. 411–429. Sinauer Associates, Sunderland, Massachusetts.

Jurik, T. W. (1983). Reproductive effort and CO_2 dynamics of wild strawberry populations. *Ecology,* **64,** 1329–1342.

King, D. & Roughgarden, J. (1982a). Graded allocation between vegetative and reproductive growth for annual plants in growing seasons of random length. *Theoretical Population Biology,* **21,** 1–16.

King, D. & Roughgarden, J. (1982b). Multiple switches between vegetative and reproductive growth in annual plants. *Theoretical Population Biology,* **21,** 194–104.

King, D. & Roughgarden, J. (1983). Energy allocation patterns of the California grassland annuals *Plantago erecta* and *Clarkia rubicunda*. *Ecology*, **64,** 16–24.

Lambers, H. (1979). Efficiency of root respiration in relation to growth rate, morphology and soil composition. *Physiologia Plantarum*, **46,** 194–202.

Law, R. (1979). The costs of reproduction in an annual meadow grass. *American Naturalist*, **113,** 3–16.

Lovett Doust, J. (1980). Experimental manipulation of patterns of resource allocation in the growth cycle reproduction of *Smyrnium olusatrum* L. *Biological Journal of the Linnaean Society*, **13,** 155–156.

Mahon, J. D. (1977). Respiration and the energy requirement for nitrogen fixation in nodulated pea roots. *Plant Physiology*, **60,** 817–821.

Michod, R. E. (1979). Evolution of life histories in response to age-specific mortality factors. *American Naturalist*, **113,** 531–550.

Penning de Vries, F. W. T. (1975). Use of assimilates in higher plants. *Photosynthesis and Productivity in Different Environments* (Ed. by J. P. Cooper), pp. 459–480. Cambridge University Press, Cambridge.

Piñero, D., Sarukhán, J. & Alberdi, P. (1982). The costs of reproduction in a tropical palm, *Astrocaryum mexicanum*. *Journal of Ecology*, **70,** 473–481.

Pitelka, L. F., Stanton, D. S. & Peckenham, M. O. (1980). Effects of light and density on resource allocation in a forest herb, *Aster acuminatus* (Compositae). *American Journal of Botany*, **67,** 942–948.

Redmann, R. E. & Reekie, E. G. (1982). Carbon balance in grasses. *Grasses and Grasslands: Systematics and Ecology* (Ed. by J. R. Estes, R. J. Tryl & J. N. Brunkens), pp. 195–231. University of Oklahoma Press, Norman.

Schaffer, W. H. & Gadgil, M. D. (1975). Selection for optimal life histories in plants. *Ecology and Evolution of Communities* (Ed. by M. L. Cody & J. M. Diamond), pp. 142–157. Belknap, Cambridge, Massachusetts.

Schaffer, W. M., Inouye, R. S. & Whittam, T. S. (1982). Energy allocation by an annual plant when the effects of seasonality on growth and reproduction are decoupled. *American Naturalist*, **120,** 787–815.

Schubert, K. R. (1982). *The Energetics of Biological Nitrogen Fixation*. Workshop Summaries I. American Society of Plant Physiologists, Rockville, Maryland.

Soule, J. D. & Werner, P. A. (1981). Patterns of resource allocation in plants with special reference to *Potentilla recta* L. *Bulletin of the Torrey Botanical Club*, **108,** 311–319.

Stearns, S. C. (1980). A new view of life-history evolution. *Oikos*, **35,** 266–281.

Thompson, K. & Stewart, A. J. A. (1981). The measurement and meaning of reproductive effort in plants. *American Naturalist*, **117,** 205–211.

Trappe, J. M. & Fogel, R. D. (1977). Ecosystematic functions of mycorrhizae. *The Belowground Ecosystem* (Ed. by J. K. Marshall), pp. 205–214. Colorado State University, Fort Collins.

Troughton, A. (1977). Relationship between the root and shoot systems of grasses. *The Belowground Ecosystem* (Ed. by J. K. Marshall), pp. 39–52. Colorado State University, Fort Collins.

Van Andel, J. & Vera, F. (1977). Reproductive allocation in *Senecio sylvaticus* and *Chamaenerion angustifolium* in relation to mineral nutrition. *Ecology*, **65,** 747–758.

Watson, M. A. (1984). Developmental constraints: Effect on population growth and patterns of resource allocation in clonal plants. *American Naturalist*, **123,** 411–426.

Werk, K. & Ehleringer, J. R. (1983). Photosynthesis by flowers in *Encelia farinosa* and *Encelia californica* (Asteraceae). *Oecologia*, **57,** 311–315.

Williams, G. C. (1975). *Sex and Evolution*. Princeton University Press, Princeton, New Jersey.

Willson, M. F. (1983). *Plant Reproductive Ecology*. Wiley, New York.

Index

D

E

F

G

R

S